Parasites and Pathogens of Insects

Volume 1: Parasites

Parasites and Pathogens of Insects

Volume 1: Parasites

Edited by

N. E. Beckage
S. N. Thompson
B. A. Federici
Department of Entomology
University of California
Riverside, California

Academic Press, Inc.
A Division of Harcourt Brace & Company
San Diego New York Boston London Sydney Tokyo Toronto

This book is printed on acid-free paper. ♾

Academic Press, Inc.
1250 Sixth Avenue, San Diego, California 92101-4311

United Kingdom Edition published by
Academic Press Limited
24–28 Oval Road, London NW1 7DX

Library of Congress Cataloging-in-Publication Data

Parasites and pathogens of insects / edited by N. E. Beckage, S. N. Thompson, B. A. Federici.
p. cm.
Includes bibliographical references and index.
ISBN 0-12-084441-9 (v. 1). --ISBN 0-12-084442-7 (v. 2)
1. Insects—Parasites. 2. Insects—Pathogens. 3. Host—parasites relationships. I. Beckage, N. E. (Nancy E.) II. Thompson, S. N. III. Federici, Brian A.
QL495. P265 1993
595.7'023–dc20 92-45034
CIP

PRINTED IN THE UNITED STATES OF AMERICA

93 94 95 96 97 98 EB 9 8 7 6 5 4 3 2 1

Contents

Chapter 3 Hormonal Interactions between Insect Endoparasites and Their Host Insects

Pauline O. Lawrence and Beatrice Lanzrein

Chapter 4 Reproductive Disturbances Induced by Parasites and Pathogens of Insects

Hilary Hurd

Chapter 5 Behavioral Effects of Parasites and Pathogens in Insect Hosts

David R. Horton and Janice Moore

Chapter 6 Redirection of Host Metabolism and Effects on Parasite Nutrition

S. N. Thompson

Chapter 7 Teratocytes: Developmental and Biochemical Characteristics

Douglas L. Dahlman and S. Bradleigh Vinson

Chapter 10 Venoms of Parasitic Hymenoptera as Investigatory Tools

Davy Jones and Thomas Coudron

Chapter 11 Biochemical and Molecular Basis of Mosquito Susceptibility to *Plasmodium* and Filarioid Nematodes

Bruce M. Christensen and David W. Severson

Chapter 12 Cellular Defense Responses of Insects: Unresolved Problems

Norman A. Ratcliffe

Chapter 13 The Response to Foreign Tissue Transplants in Insects

Richard D. Karp

Chapter 14 Regulation of Insect Hemolymph Phenoloxidases

Manickam Sugumaran and Michael R. Kanost

Contents of Volume 2

Contributors

Numbers in parentheses indicate the pages on which the authors' contributions begin.

Nancy E. Beckage (25), Department of Entomology, University of California, Riverside, California 92521

Bruce M. Christensen (245), Department of Animal Health and Biomedical Sciences, University of Wisconsin, Madison, Wisconsin 53706

Thomas Coudron (227), Biological Control of Insects Lab, USDA-ARS, Columbia, Missouri 65205

Douglas L. Dahlman (145), Department of Entomology, University of Kentucky, Lexington, Kentucky 40546

Jo-Ann G. W. Fleming (189), Department of Entomology, Texas A&M University, College Station, Texas 77843

David R. Horton (107), Yakima Agricultural Lab, USDA-ARS, Yakima, Washington 98902

Hilary Hurd (87), Department of Biological Sciences, University of Keele, Staffordshire, ST5 5BG United Kingdom

Davy Jones (227), Graduate Center for Toxicology, University of Kentucky, Lexington, Kentucky 40506

Michael R. Kanost (317), Department of Biochemistry, Kansas State University, Manhattan, Kansas 66506

Richard D. Karp (305), Department of Biological Sciences, University of Cincinnati, Cincinnati, Ohio 45221

Peter J. Krell (189), Department of Microbiology, University of Guelph, Guelph, Ontario N1T 2W1, Canada

Beatrice Lanzrein (59), Division of Developmental Biology, Zoological Institute, University of Berne, 3012 Berne, Switzerland

Pauline O. Lawrence (59), Department of Zoology, University of Florida, Gainesville, Florida 32611

M. Mackauer (1), Centre for Pest Management, Department of Biological Sciences, Simon Fraser University, Burnaby, British Columbia, Canada V5A 1S6

Janice Moore (107), Department of Biology, Colorado State University, Fort Collins, Colorado 80523

Norman A. Ratcliffe (267), Biomedical and Physiological Research Group, School of Biological Sciences, University College of Swansea, Singleton Park, Swansea, SA2 8PP Wales, United Kingdom

R. Sequeira (1), School of Biological Sciences, University of East Anglia, Norwich NR47TG, England

David W. Severson (245), Department of Animal Health and Biomedical Sciences, University of Wisconsin, Madison, Wisconsin 53706

Donald B. Stoltz (167), Department of Microbiology, Dalhousie University, Halifax, Nova Scotia, B3H 4H7 Canada

Manickam Sugumaran (317), Department of Biology, University of Massachusetts at Boston, Boston, Massachusetts 02125

S. N. Thompson (125), Department of Entomology, University of California, Riverside, California 92521

S. Bradleigh Vinson (145), Department of Entomology, Texas A&M University, College Station, Texas 77843

Preface

The focus of this two-volume set is the interface between insects and their associated parasites and pathogens, with particular emphasis placed on the basic biology, biochemistry, and molecular biology of these intimate and intriguing relationships. The initial inspiration for this book was the recognition that although the potential use as biological control agents of many parasites and pathogens of insects has been intensively investigated, few recent works have attempted to address the biochemical and molecular interactions occurring between insect hosts and their invaders. Yet these interactions clearly are critical to the ultimate outcome of the confrontation between any invader and its host. The parasite or pathogen must evade or suppress the host immune response and, while ultimately completing its development in the host, must not stress it prematurely such that the survival of both partners is compromised. In addition, the host environment must satisfy the parasites' nutritional and metabolic needs.

Our intent is to summarize developments and technological approaches currently being exploited to monitor the biochemical, immunological, metabolic, and behavioral alterations in insects infected with parasites and pathogens, with particular emphasis on interactions occurring at the molecular level. Additionally, our authors present several novel ideas for exploitation of this information in the manipulation of insect pests.

Pressure to minimize synthetic chemical insecticides in insect control programs has led to increased interest in the use and study of parasites and pathogens capable of limiting insect populations. This interest, in combination with rapid advances in the techniques now available to study biological systems, has resulted in an enormous increase in knowledge about the biochemistry and physiology of the parasites and pathogens that attack insects, as well as their interactions with the host. This knowledge is so diverse and extensive that most journal reviews deal only with highly specialized aspects of the overall field. While such reviews are of great value, our intent for these two volumes was to assemble a more extensive survey of this rapidly developing field by publishing reviews on selected topics dealing with interactions with parasites (Volume 1) and pathogens (Volume 2) of insects. It would be impossible, even in two volumes, to review all of the recent findings considered major advances. Therefore, in selecting our topics, we chose to focus on subject areas of long-standing interest (e.g., insect antibacterial proteins and parasite–host developmental interactions) and those dealing with recent breakthroughs that appear significant and likely to be of value in controlling insects (e.g., development of recombinant baculoviruses and formulation of "super" pathogens).

The intended audience for these volumes includes upper-level undergraduates with specialty interests in parasitology and entomology, graduate students, and post-graduate researchers who may use this information to devise new technologies for manipulation of insects of importance to agriculture and human health. We hope these volumes will find a niche on bookshelves in many personal and professional libraries focusing on parasitology, entomology, immunology, epidemiology, physiological ecology, evolutionary biology, and other areas dealing with various aspects of host–parasite relationships.

Insect parasitology and pathology have clearly become multidisciplinary fields. For example, the development of new technologies for gene transfer and arthropod transformation may benefit from the exploitation of naturally occurring transposable elements in insect virus genomes. Useful target genes for manipulation might include those that are critical to the normal functioning of the immune system as well as those parasite-associated factors invoking developmental disruption or sterility of insect pests (i.e., endocrine regulators). Genes associated with refractoriness in vector arthropods are also important and are under scrutiny as possible means for generating engineered vectors with reduced capacity to transmit parasites.

In contrast to the relatively limited information available on molecular host–parasite and host–pathogen interactions in insect hosts, a wealth of recently published material describes the relationships between parasites and pathogens of mammals with their respective hosts. In part, this difference reflects intense research efforts directed toward developing new therapeutic treatments for disease based on these interactions in the hope of identifying critical points of vulnerability that may be manipulated by drugs or other agents. A similar extensive literature describing the molecular interactions of plants with pathogens and parasites also now exists. Our knowledge of the molecular mechanisms of interactions between plants and plant pathogens is thus much more sophisticated than our understanding of the molecular mechanisms operating in insects as they confront invaders. While we have yet to identify virulence genes or avirulent mutants of those genes in species that attack insects, for example, such genes already have been isolated in species that are pathogenic to plants and mammals. Moreover, though mechanisms of virus resistance have been pinpointed in plants and mammals, our information about insect antiviral defenses remains rudimentary. Induction of resistance has been documented, but its mechanisms remain to be identified.

Nevertheless, we do have significant information about the complexity of some processes, such as the antibacterial defenses of insects, and how they parallel similar processes in mammalian hosts. Hormonal host–parasite relationships have been intensively scrutinized in several invertebrate species, including many insects, which appear to be particularly appropriate models for studying endocrine interactions. Moreover, the association of third-party elements with parasites is perhaps best illustrated by species that attack insects; for instance, both wasp and nematode parasites show intimate relationships with viruses and bacteria, respectively, during parasitism of insect hosts.

In the short term, these volumes are directed at filling a void in the literature by emphasizing basic interactions at the biochemical and molecular levels. In the long term, we expect that many of these interactions will provide avenues for exploitation to either enhance the rates of "beneficial" parasitism (in biological control, for example) or reduce the rate of disease transmission and the rate of infection of vertebrate hosts. Our hope is that the information assembled here will have significant impact on agriculture and human health and that these volumes will stimulate fresh approaches to the investigation of these fascinating and intricate interactions.

We express thanks to the authors for their timely contributions. In addition, the scope and depth of this work demanded that we enlist the assistance of many scientists to review the chapters. For their time and outstanding efforts in this endeavor, we sincerely thank the following individuals: Drs. Theodore Andreadis, Christopher Bayne, Gary Blissard, Drion Boucias, John Brown, Peter Bryant, John Burand, Thomas Coudron, Samuel Dales, Robin Denell, Douglas Dahlman, Gordon Gordh, Robert Granados, Patrick Greany, Leah Haimo, Alfred Handler, Kiyoshi Hiruma, Hilary Hurd, Davy Jones, Michael Kanost, James Kerwin, Karl Kramer, Leslie Lewis, Michael Locke, Robert Luck, Lois Miller, Louis Miller, Ed Platzer, John Postlethwait, Lynn Riddiford, Justin Schmidt, Don Stoltz, Michael Strand, Just Vlak, Bruce Webb, John Webster, Alan Wood, Timothy Yoshino, Rolf Ziegler, and Marlene Zuk.

In addition, we acknowledge the excellent secretarial staff of the Entomology Department of the University of California, Riverside. We express special thanks to Ms. Diana Hanson and Ms. Pam Hoatson for their efficient efforts in processing manuscripts and corresponding with our authors.

Lastly, but most importantly, the editors express deep appreciation to Dr. Phyllis B. Moses of Academic Press for her unwavering enthusiasm for this project. Her scientific input and editorial advice were invaluable to us first in developing the idea of assembling these volumes and subsequently in overseeing the project to fruition. We also acknowledge Academic Press for its sponsorship.

N. E. Beckage
S. N. Thompson
B. A. Federici

Chapter 1

Patterns of Development in Insect Parasites

M. Mackauer
Centre for Pest Management
Department of Biological Sciences
Simon Fraser University
Burnaby, British Columbia, Canada

R. Sequeira[1]
Centre for Pest Management
Department of Biological Sciences
Simon Fraser University
Burnaby, British Columbia, Canada

I. Introduction

Successful parasitism by insect parasitoids[2] requires the solution of three kinds of general problems: (1) which host to select; (2) how to integrate development and growth with those of the host; and (3) what life-history tactics to adopt. The first problem concerns behavioral adaptations relating to host selection and progeny allocation by the adult female (Charnov and Skinner, 1985; Waage and Godfray, 1985; van Alphen and Vet, 1986; Waage, 1986) and will not be considered here for this reason. Instead we will focus mainly on the third and, to a lesser extent, the second problem, both of which concern parasite growth and development.

Most early studies have emphasized mechanisms of parasite growth and development, with less attention paid to the adaptive significance of these mech-

[1]Current address: School of Biological Sciences, University of East Anglia, Norwich NR4 7TG, England.

[2]Henceforth we use the term *parasite* to mean *parasitoid*. These protelean parasites include a large group of mostly hymenopteran and dipteran species that, as immatures, are obligate parasites of insects and other arthropods and are free-living as adults. Hymenopterous parasites often are called parasitic wasps as opposed to the nonparasitic, aculeate wasps and bees.

Parasites and Pathogens of Insects
Volume 1: Parasites

anisms. In a causal approach, parasite development and growth may be characterized in terms either of the effects of parasitism on the host (e.g., Doutt, 1963; Salt, 1964; Smilowitz and Iwantsch, 1973; Cloutier and Mackauer, 1979, 1980; Lawrence, 1982; Thompson, 1982, 1983; Beckage and Templeton, 1985, 1986; Strand, 1986; Gunasena *et al.*, 1989; Vinson, 1990; Strand and Dover, 1991) or of the parasite's response to variations in the nutritional state and physiology of the host (e.g., Corbet, 1968; Weseloh, 1984; Hébert and Cloutier, 1990; Lawrence, 1990; Kouamé and Mackauer, 1991; Strand *et al.*, 1991; Sequeira and Mackauer, 1992a,b). In this regard, Lawrence (1986, 1990) considered host regulation (Vinson, 1975; Vinson and Iwantsch, 1980a) and flexibility of parasite development (Corbet, 1968; Weseloh, 1984) as alternate developmental strategies. This viewpoint stresses proximate mechanisms of parasite survival based on qualitative descriptions of the physiological and biochemical interactions between the parasites and their hosts.

However, functional constraints on the parasite's growth and development on or in different hosts can also be considered as variables within a broader evolutionary framework. This approach emphasizes questions about the fitness value of developmental characteristics and patterns of host utilization. Because protelean parasites depend exclusively on host-derived nutrients for their larval development and growth, natural selection would be expected to favor mechanisms that maximize the efficient utilization of these resources. Optimal resource allocation to adult body size, considered the perhaps most important component of Darwinian fitness in parasitic wasps (King, 1987), depends on the insect's growth rate and development time. The allocation of limited (host) resources to competing fitness functions may result in tradeoffs that determine an "optimal character set" and, in doing so, may shape the evolution of a species' life-history strategy (Sibly and Calow, 1986).

We use the term *strategy* in accordance with Dominey (1984) to refer to a set of general rules that specify which alternative pattern of responses will be adopted in a particular situation; these rules are typical for each species and determine its adaptedness to the environment, here the host. Tactics, by contrast, refer to several alternative options or mechanisms by which these evolutionary objectives are achieved; these option sets may vary among different individuals or phenotypes. Also, we distinguish between host suitability (Salt, 1938; Vinson and Iwantsch, 1980b) and host quality, two terms often used synonymously. A host species is suitable if it normally supports the successful development of parasite offspring; consequently, suitability is a characteristic of the host species and is genetically determined, or largely so. In comparison, we use the term quality to describe variations in the state or condition of the host that affect process dynamics, such as the rates of parasite growth and development. Such state variables include, or may be correlated with, host age, stage of development, size, sex, and nutritional status.

The chapter is organized as follows. First, we introduce the idiobiont–koinobiont dichotomy, which we use as a (macroevolutionary) organizing scheme. Second, we describe several developmental patterns that characterize broad differences between idiobiont and koinobiont parasites. Next, we discuss seasonal adaptations and the influence of host nutrition on parasite development, including superparasitism and starvation. Finally, we propose three models of parasite development in response to host constraints. We suggest that the essential components of any developmental strategy are the parasite's growth rate, development time, and adult biomass, which are constrained by host quality in an association-specific manner.

II. The Idiobiont–Koinobiont Dichotomy

Haeselbarth (1979) first drew attention to an important macroevolutionary division between parasites developing in hosts that continue to grow and metamorphose during the initial stages of parasitism (called koinophytes) and those that develop in nongrowing and paralyzed hosts (called idiophytes). Askew and Shaw (1986) introduced the terms koinobiont and idiobiont to describe these alternative host-exploitation strategies.

Because the host of an idiobiont does not feed, grow, or metamorphose during the course of the interaction, it contains a fixed amount of resources for the parasite larva, with large hosts being assumed of higher quality than small hosts. However, unless the host is killed by the female at oviposition, age-related variations in quality may result from developmental changes within a particular host stage, such as eggs and pupae (Strand, 1986; King, 1990a). By contrast, hosts parasitized by a koinobiont continue to feed, grow, and develop during much of the interaction. Consequently, host quality as a resource for the parasite larva is influenced by future host growth, which depends on the host's age and stage of development, rather than on its size, at the time of parasitization (Mackauer, 1986; King, 1989; Kouamé and Mackauer, 1991; Sequeira and Mackauer, 1992a).

Blackburn (1991) noted that koinobionts have longer pupal periods and preadult life spans than idiobionts, suggesting that these two groups have probably evolved under different constraints with regard to resource usage. Askew and Shaw (1986) compared idiobiont and koinobiont strategies as a correlate of host range. They proposed that koinobionts are more likely to show a narrow specialization because of their greater dependence on host physiology and development. Gauld and Bolton (1988) noted that idiobionts typically are synovigenic (i.e., females mature eggs continuously throughout life), produce relatively large, lecithal (or anhydropic) eggs containing sufficient resources for early embryonic development, and develop as ectoparasites on concealed hosts. Koinobionts, by contrast, are typically endoparasites, produce small nutrient-poor, alecithal (or hydropic) eggs, and attack free-moving hosts.

Idiobiont and koinobiont strategies show poor correlation with solitary and gregarious development (Askew and Shaw, 1986). Theoretical results (Godfray, 1987), and some experimental evidence (Cruz, 1981), suggest that solitary and gregarious development represent distinct reproductive and/or developmental strategies (Waage, 1986). However, many groups of parasites, including several genera such as *Cotesia*, contain both solitary and gregarious species. In most cases it is not clear whether such developmental variations are due to host-related conditions (le Masurier, 1987, 1991) or reflect phylogenetic constraints (Gauld, 1988; Gauld and Bolton, 1988), or both.

III. Patterns of Parasite Development and Growth

A species' life-history strategy represents a unique combination of responses (tactics) that are shaped by natural selection from an option set associated with each phenotypic character. Option sets may be empirically defined as the quantitative responses of their corresponding life-history characters. With regard to host–parasite interactions, it is convenient to distinguish between two response categories: the influence of the host's phenotype on the parasite's phenotype, and the fitness consequences of variations in the parasite's phenotypic attributes. The first category includes the influence of the species of host, of its stage of development, and of its nutritional status on the parasite's growth rate, development, and size at maturity.

In this and the following section, we use selected host–parasite associations to identify variables that are likely to represent major components of microevolutionary strategies.

A. Idiobionts

1. Egg Parasites (*Trichogramma*)

The generalist egg parasites of the genus *Trichogramma* are among the most thoroughly studied idiobionts. These species are easily reared in the laboratory on their habitual as well as on various factitious hosts (Bigler *et al.*, 1987; Schmidt and Smith, 1987). In his seminal studies, Salt (1940) demonstrated a close correlation between the size of host eggs and the adult size of the parasite. Eggs that were too small to satisfy all the nutritional requirements of the developing parasite larva produced "runts" with structural peculiarities characteristic of starvation. Also, development time varied with the egg's size and the thickness of its chorion.

Although species of *Trichogramma* exhibit distinct patterns of host-size and species usage (Hintz and Andow, 1990), accurate assessment of adaptive life-history variation and developmental strategies is made difficult by facultative gregarious development. Pak (1986) recognized six distinct patterns in the relationship between host age at parasitization and developmental success in *Trichogramma*. Five of these patterns were nonlinear, a fact indicating that host quality for parasite development is a nonlinear function of host age and, more important perhaps, that these quantitative relationships are association-specific. While parasite size and fecundity increased with host size (Reznik and Umarova, 1990), development time varied nonlinearly with host size, possibly reflecting differences in relative host quality rather than in absolute host size or age. Barrett and Schmidt (1991) suggested that, in small host eggs, the parasitoid's size is restricted by the nutrients and space available. However, in hosts above a threshold volume, the size of the resulting wasps depends on the number of eggs allocated by the female at oviposition.

Eggs may be considered as transitory host stages in that rapid embryogenesis can quickly deplete the amount and the availability of stored resources (Anderson, 1972). Thus, egg parasites are under selection to minimize development time and to maximize growth rate (Strand, 1986). Strand (1986) suggested that, in *Trichogramma* species, polyphagy may be a consequence of the physiological uniformity of insect eggs as a food resource, especially because eggs lack significant cellular defenses and are often unprotected. If this hypothesis is in fact correct, we would expect little variation in the growth rates of generalist (egg) parasites. However, more detailed studies of the growth, development, and bionomics of egg parasites may well reveal substantial host-related variation in parasite performance (Marston and Ertle, 1973; Bigler *et al.*, 1987; Reznik and Umarova, 1990; Barrett and Schmidt, 1991).

2. Pupal Parasites (*Pimpla, Coccygomimus*)

The growth and development of pupal endoparasites and of those that kill the host at or shortly after oviposition is less well studied. Arthur and Wylie (1959) showed that, in the ichneumonid *Pimpla turionellae*, body size increased with the pupal size of different host species, while development time was generally longer in large than in small pupae. Sandlan (1982) found that the adult size of *Coccygomimus turionellae* decreased with an increase in host age within a host species but was not correlated with size differences between host species; however, parasite development time increased with pupal age.

Host size-, age-, and species-usage patterns in pupal parasites, indexed by the percentage of parasitism, may vary within and between host species

(Hall and Fischer, 1988; King, 1990b). Furthermore, Legner (1969) reported that parasite development time was correlated, in an association-specific manner, with host size at parasitization in pupal endo- but not ectoparasites.

B. Koinobionts

Most of the literature on host–parasite developmental interactions is concerned with koinobionts, especially solitary endoparasites. In these species, the dynamics of parasitism is determined by the mechanisms that control the growth and development of the parasite immatures in response to host conditions, which in turn may be altered in response to parasitism. Such a "counterbalanced state of opposing conditions" was termed *enantiostasis* by Mangum and Towle (1977).

1. Aphidiid Wasps (*Aphidius, Ephedrus*)

The pea aphid, *Acyrthosiphon pisum*, supports a broad range of solitary endoparasites belonging to two hymenopteran families, Aphidiidae and Aphelinidae. Although many of these parasites can successfully develop in all the nymphal stages and in adult aphids, parasitism of different host stages and morphs has variable consequences for parasite larval ontogeny.

Sequeira and Mackauer (1992a) compared the growth trajectories of *Aphidius ervi* developing in different nymphal instars of apterous virginoparae. Parasite growth rates varied nonlinearly with the host instar, with the growth of early larvae being highest in second instars and lowest in third instars of the pea aphid. Using parasite growth as a measure of host quality, Sequeira and Mackauer (1992a) ranked the four nymphal instars in the order $L_2 > L_4 > L_1 > L_3$. This ranking order was consistent with differences in reproductive achievement (as indexed by the intrinsic rate of increase, r_m) between *A. ervi* females reared on these instars (Sequeira and Mackauer, 1992b). Changes in parasite fitness attributes in response to variation in host resources are indicative of the parasite's life-history strategy. First- and fourth-instar pea aphids differ in biomass by one order of magnitude or more. In *A. ervi,* adult dry mass was correlated with maximum larval mass achieved just prior to pupation (Sequeira and Mackauer, 1992a). Parasite mass and development time were correlated with host size (Sequeira and Mackauer, 1992b). Dry mass and development time were positively correlated in L_1–L_3 aphids but not correlated in large L_4 aphids. The period from oviposition to pupation was independent of host size at parasitization. These results clearly show that the parasite responds to a dynamic nutritional and physiological host environment through variations in the larval growth trajectories. Thus, in *A. ervi* developing in hosts below a certain size threshold (L_1–L_3), the optimization of body size took precedence over the minimization of development time. However, parasites developing in hosts above a threshold size utilized the additional resources both to increase body mass and to decrease development time.

Similarly, in *Ephedrus californicus,* adult body size increased with pea aphid size up to a threshold value. Once the larva had achieved this threshold in dry mass, additional host resources were allocated toward the simultaneous minimization of development time and maximization of body mass. In contrast to *A. ervi,* the time-to-pupation was longer in L_2 than in L_1 and L_4 hosts, a difference indicating that *E. californicus* did not adjust the larval growth trajectory to host conditions (Sequeira and Mackauer, 1993).

Reproductive success in aphid parasites is frequently assumed to be correlated with adult size (Cloutier *et al.,* 1981; Liu, 1985; Irvine, 1991). Although body size can be a useful index of a wasp's potential fecundity, such relationships may be nonlinear and hence must be interpreted with caution, possibly being an artifact of the step-size selected for the variables (Leather, 1988). For example, in *A. ervi,* the intrinsic rate of increase for cohorts of females reared on different aphid instars was a quadratic function of body dry mass (R. Sequeira and M. Mackauer, unpublished). It is worth noting that adult size is highly plastic in aphidiid wasps. In *Aphidius smithi,* very small but viable and fertile offspring developed from eggs deposited in aphid embryos while still inside their mother (Mackauer and Kambhampati, 1988). This finding suggests that the larval development of this parasite may not be constrained by the need to achieve a critical size for molting or pupation.

Differential development rates may have a profound effect on competitive interactions among different species of parasites that exploit the same host resource, independent of environmental conditions. For example, developing in third nymphal instars of the pea aphid, *Praon pequodorum* required more time to complete the embryonic stage, and showed greater developmental variation, than *A. smithi;* however, both parasites molted to the second larval instar at approximately the same time (Chow and Mackauer, 1984). Because *P. pequodorum* physically attacks and kills potential competitors during the early first larval stage, a slow embryonic development increased the probability of its survival in hosts already parasitized by *A. smithi* over a broad range of age differences.

2. *Hyposoter exiguae*

The ichneumonid *H. exiguae* develops as a solitary endoparasite in larvae of various lepidopteran species. The ovipositional and developmental success of the parasite appears to vary with the species and the stage of host selected. For example, Puttler (1961) reported that wasps deposited their eggs in the first three instars of *Peridroma saucia,* although larval development was often inhibited in third-instar hosts. By comparison, *Heliothis zea* was successfully parasitized in instars 1–4, but the parasitoid was generally unable to oviposit in fifth instars because of their aggressiveness (Campbell and Duffey, 1979).

The interactions between *H. exiguae* and its various host species are characterized by host developmental arrestment, which may vary with the species of host (Thompson, 1982, 1983, 1985a,b; Beckage and Templeton, 1985). In *Trichoplusia ni,* the parasite larva emerged from the fourth or the fifth host instar, regardless of the host stage at parasitization. In contrast, parasite emergence from *Manduca sexta* larvae varied with host age at parasitization (Beckage and Templeton, 1985). These differences in host exploitation appear to be related to host species-specific nutritional and physiological interactions (Beckage and Templeton, 1985), including effects mediated by polydnavirus and other parasite-derived products injected at oviposition (Vinson, 1990; Fleming, 1992).

In comparison with physiological and biochemical interactions, the relationship between adult life-history attributes and the larval ontogeny of *H. exiguae,* especially as influenced by variation in host size (or quality), has received less attention. Total development time from egg to adult eclosion appears to decrease (and body size appears to increase) with host size or age at parasitization (Puttler, 1961; Smilowitz and Iwantsch, 1973; Campbell and Duffey, 1979). What evidence is available suggests that *H. exiguae* requires a critical host biomass to mature or molt to the next larval stage (Beckage and Templeton, 1985; Bloem and Duffey, 1990), although larval ontogeny appears not to be directly influenced by the host's endocrine system.

3. *Microplitis* Species

The interactions between solitary braconid wasps and their lepidopteran hosts are well exemplified by species of *Microplitis* (Jones and Lewis, 1971; Webb and Dahlman, 1985, 1986; Strand and Dover, 1991). Strand *et al.* (1988) compared the responses of *Heliothis virescens* larvae to parasitism by *M. croceipes* and *M. demolitor.* Unlike *H. exiguae,* these parasites feed exclusively on host hemolymph throughout their larval development. The growth trajectory of unparasitized *H. virescens* is sigmoid; however, in larvae parasitized by *M. demolitor* and *M. croceipes,* the growth rate is linear and severely depressed. In comparison, parasitism of *H. zea* larvae by *M. croceipes* results in a growth trajectory similar in shape, but reduced in magnitude, to that of unparasitized larvae (Jones and Lewis, 1971). Strand *et al.* (1988) concluded that the parasite's growth rate is determined by the host instar at parasitization.

Different *Microplitis* species exhibit distinct patterns of host-size preference (Rajapakse and Ashley, 1985; Tillman and Powell, 1989). For example, *M. croceipes* preferred third-instar larvae of *Heliothis* species, with the acceptance rate of other instars declining symmetrically about the third instar (Lewis, 1970; Hopper and King, 1984). Development time was correlated with instar preference; parasites developing in third-instar hosts developed fastest (Hopper and King, 1984) and achieved the largest body size (Hopper, 1986).

IV. Parasite Development and Host Ecology

A. Seasonal Adaptations

Many insects enter a state of arrested development and reproduction to survive during periods of unfavorable conditions, such as low and high temperatures. Tauber *et al.* (1984) distinguished between two kinds of adaptations to hazardous seasonal conditions: quiescence, a period during which growth and development are only delayed until conditions are favorable again, and dormancy, during which growth and reproduction are suppressed by diapause.

1. Temperature Requirements for Development

Obligate parasites cannot survive in the absence of their hosts and, thus, must synchronize their development and reproduction with host abundance. Synchrony requires that two conditions be satisfied: (1) the parasite must initiate development in response to the same, or to a closely correlated, set of conditions as the host; and (2) the parasite must cease development when host abundance declines or, alternatively, shift to a different host species. In variable climates, early-season synchrony is particularly important. Campbell *et al.* (1974) compared the temperature requirements of several species of aphid parasites with those of their hosts. Parasites had a higher temperature threshold than the aphids, a difference that ensures their emergence from diapause after hosts become available in spring. Interestingly, the primary parasites had a considerably lower threshold than the hyperparasite *Asaphes lucens,* consistent with this species' relative late appearance in the season. Variations in the temperature requirements between different populations indicate that populations are locally adapted (Campbell *et al.,* 1974; Kambhampati and Mackauer, 1989). Apparently host size may influence a parasite's temperature threshold for development. Allen and Keller (1991) found that *Cotesia urabae,* which is a solitary koinobiont endoparasite, had a higher threshold for egg-larval development when developing in small rather than in mid-size larvae of its host, the noctuid *Uraba lugens.*

2. Parasite Diapause

In addition to external abiotic cues, which may be mediated by changes in the host's endocrine system, the initiation and maintenance of diapause in parasites often is influenced by host cues. Several host conditions have been implicated in diapause induction, especially in endoparasites, for example, host age (in *Cotesia glomerata;* Schoonhoven, 1962), host size (in *Aphidius nigripes;* Brodeur and McNeil, 1989), and host species (in *Nasonia vitripennis;*

Saunders *et al.*, 1970). Brown *et al.* (1990) showed that, in larvae of the codling moth, *Cydia pomonella*, exposure to short day length resulted in dormancy caused by failure of the endocrine system. The changed endocrine milieu, in turn, induced diapause in first-instar larvae of the braconid endoparasite, *Ascogaster quadridentata*. A direct effect of the host on diapause initiation was demonstrated recently in *Aphidius matricariae* and *Praon volucre*, two solitary endoparasites of the green peach aphid, *Myzus persicae* (Polgár *et al.*, 1991). Under identical long-day conditions, parasite larvae developing in nymphs of oviparae entered a prepupal diapause, whereas those developing in parthenogenetically reproducing morphs (fundatrices, fundatrigeniae, and virginoparae) did not. The authors suggested that, in the fall, the production of oviparae in holocyclic aphids, which is usually associated with short-day conditions, can serve as an unambiguous signal of declining host numbers to these parasites.

3. Developmental Arrestment

Hagen (1964) discussed various examples of delayed development in insect parasites. Several species of Trigonalidae (Hymenoptera) and Tachinidae (Diptera) lay so-called microtype eggs directly on foliage. For the egg to hatch, it must be ingested by a potential host; the larva then penetrates the gut wall and develops endoparasitically (Wishart, 1956; Coppel, 1958). Delayed larval development has also been found in some aphid hyperparasites, for example, *Alloxysta vitrix* and *A. megourae*. Females deposit their eggs in the larva, or occasionally the embryonated egg, of a primary aphid parasite such as *Aphidius*. However, the egg does not hatch until the primary parasite has completely consumed and mummified the host aphid; the hyperparasite larva then feeds on and eventually kills the primary larva (Matejko and Sullivan, 1979; Sullivan, 1987).

Developmental arrestment may also occur at a later stage in parasite development, for example, during the first instar. The parasite larva continues to develop and grow only after the host has reached a specific size or instar (Hagen, 1964; Allen and Keller, 1991). It is not always clear whether such examples represent normal variations in the parasite's growth rate, perhaps correlated with differences in host quality, or an arrestment of growth in the absence of a specific (hormonal) stimulus (Lawrence, 1990; Strand *et al.*, 1991).

B. Influence of Host Nutrition

Any functional constraints on option sets and trade-offs between life-history characters may be revealed when the host–parasite system is subjected to additional stress. Two approaches would seem particularly promising for such studies, especially in endoparasites: the use of superparasitism and host starvation. In superparasitism, not all parasite offspring can complete development because either the host resources are insufficient to support all larvae (typically gregarious species) or the larvae compete with each other with only one surviving (typically solitary species) (Salt, 1961; Mackauer, 1990). In contrast, starvation causes a general reduction in insect growth and development (Slansky and Scriber, 1985), a change that presumably lowers host quality.

1. Superparasitism

Parasitism by many endoparasites results in diminished host growth and consumption rates (Thompson, 1985a). Parasitism of the pea aphid by *A. smithi* thus may be atypical in that parasitized aphids ingested more food (but assimilated it less efficiently) than their unparasitized counterparts; all aphids were maintained on an artificial diet (Cloutier and Mackauer, 1979). Because superparasitized aphids incorporated food more efficiently and grew faster than singly parasitized aphids (Cloutier and Mackauer, 1980), the growth potential of the surviving parasite larva may be higher. This supposition was confirmed recently. Males of *A. ervi* that, as larvae, developed in superparasitized pea aphids had greater adult mass than those from singly parasitized hosts; however, development time did not differ between the two groups (Bai and Mackauer, 1992). These results suggest that, in some host–parasite associations at least, the host may respond to the added demands of the developing parasite larva(e) by increased food uptake.

In contrast, in gregarious species, the surviving wasps from superparasitized hosts generally are smaller than their counterparts from nonsuperparasitized hosts; however, development time is less commonly affected. Wylie (1965) showed that, in the ectoparasite *N. vitripennis*, the mean percentage of larvae that survived in same-sized house fly pupae declined with the parasite load. Supernumerary larvae were eliminated by starvation, usually in the last instar. In cases of heavy superparasitism (≥50 larvae per pupa initially), the host was completely consumed before the parasite larvae reached the last instar, and hence all parasites died. The starved larvae of gregarious species are usually not eaten by the survivors, a behavior that can result in the wastage of host resources (Salt, 1961). Because mortality was higher among female than male larvae in *Eupteromalus dubius,* Wylie (1976) suggested that superparasitism may bias the offspring sex ratio. However, Taylor (1988) found no evidence of differential offspring mortality in *Bracon hebetor,* concluding that any bias reported for other species (Waage, 1986; King, 1987) may have been due to maternal sex ratio manipulation.

Beckage and Riddiford (1978, 1983) examined the growth and development of the gregarious endoparasite *C. congregata* in relation to the parasite load and the nutritional status of its host, *M. sexta* larvae. In first-instar hosts, the number of parasite larvae determined the timing of their emergence from the host. Emergence usually occurred during the fourth or fifth host instar; however, parasites emerged in the third or a supernumerary sixth instar in cases of an extremely light or heavy parasite load, respectively. The first larval molt of the parasite occurred after host ecdysis to the terminal stage, regardless of whether that stage was the fourth, fifth, or sixth instar. Whereas the total wet mass of the host–parasite complex was positively correlated with the parasite load, the mass of individual larvae or of adult wasps was less in heavily parasitized hosts. In contrast, development time appeared to be independent of the parasite load or the host instar parasitized in *C. glomerata,* an endoparasite of *Pieris rapae;* the gregarious larvae emerged always from fifth-instar caterpillars (Führer and Keja, 1976).

2. Host Starvation

The parasite *C. congregata* failed to emerge if newly ecdysed, terminal-stage host larvae, *M. sexta,* were starved. By postponing the starvation treatment to a later time, Beckage and Riddiford (1983) were able to increase the percentage of parasites reaching the adult stage; they suggested that host larvae must feed during a specific period to satisfy the developmental requirements of the parasite. Similarly, the development time of the tachinid *Compsilura concinnata* was unaffected by starvation if the parasite developed in a large gypsy moth larva; however, parasites emerged earlier from severely stressed hosts (Weseloh, 1984). In the aphid parasite *E. californicus,* wasps emerged earlier from starved aphids, and they were smaller, than their counterparts from nonstarved hosts (Kouamé and Mackauer, 1991). By contrast, the development time of the solitary ichneumonid *Campoletis sonorensis* increased if fourth-instar host larvae, *Heliothis virescens,* were starved (Dover and Vinson, 1990).

Crowding may have a similar effect as host starvation in that both act as general stress factors. For example, in *Leptopilina boulardi,* a cynipid endoparasite of *Drosophila* larvae, adult size and development time were reduced in hosts maintained under crowded conditions (Wajnberg *et al.,* 1990).

V. Developmental Strategies

Diverse host exploitation and developmental strategies have evolved in insect parasites that adapt the immatures' metabolic requirements to the available biomass, growth potential, and survival of the parasitized host. A high degree of nutritional and physiological integration between the host and the parasite system is characteristic of such associations (Thompson, 1985b, 1990; Barrett and Schmidt, 1991). We have shown in Sections III and IV that the parasite's growth rate, development time, and adult biomass are the essential components of any developmental strategy; they are constrained by host quality in an association-specific manner.

Metabolic processes that govern growth, development, and ultimately fitness are usually size-dependent (Peters, 1983; Schmidt-Nielsen, 1984). In insect parasites, body size is generally correlated with life-history and demographic characters (King, 1987), a fact suggesting that body size is the main target of selection. However, adult size and development rate may have independent, and even opposite, effects on fitness (Roff, 1981; Sibly and Calow, 1986). Consequently, the particular values adopted by a parasite species for these attributes are condition-specific and determined, for example, by any fitness trade-off between size and development time. Sibly and Calow (1986) noted that selection either may push phenotypes and their underlying physiologies to their limit imposed by size-related constraints, or may optimize phenotypes below size constraints in response to fitness trade-offs between individual characters. Thus, a parasite's "decision" to grow bigger, or to mature earlier, or to compromise both, will depend on adaptive variability in the growth rate. When the growth rate is at its physiological maximum, an increase in body size can only be achieved at the cost of an increase in development time. Alternatively, a larger body size combined with a decrease in development time requires an increased growth rate.

Empirical data (see Sections III and IV) suggest that three broadly different developmental patterns have evolved among insect parasites in response to host constraints. The first model (Fig. 1) assumes that the host is a closed resource environment and that parasite growth and development are constrained solely by host quality. The host contains a finite amount of resources that is determined at the time of parasitization. Host quality does not vary, or varies very little, during the course of the interaction; however, quality may vary between different species and ages of hosts. The parasite's strategy is to maximize adult biomass (assumed to be the chief determinant of fitness) per unit of host resources. The growth trajectory is relatively invariant and maximized for a given set of host and environmental conditions. Therefore, parasite adult size is positively correlated with host quality; large hosts give rise to large (and presumably more fecund) parasites that, however, may require more time to complete development than small individuals.

The second model (Fig. 2) assumes that the host represents an open resource environment and, as for Model 1, that parasite growth and development are constrained solely by host quality. Host quality and resource availability vary over time and with host age at parasitization. Host feeding and growth are expected to continue during the initial phases of parasitism, albeit at modified rates. In low-quality hosts, parasite development may include a

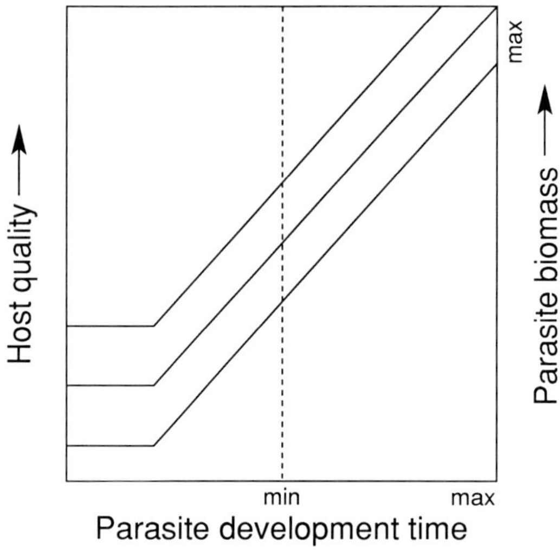

Figure 1 Parasite growth and development in a closed resource environment: biomass per unit of host resources is maximized (Model 1). The parasite grows at a constant rate that is independent of host quality. Parasite biomass is directly proportional to host quality (and development time).

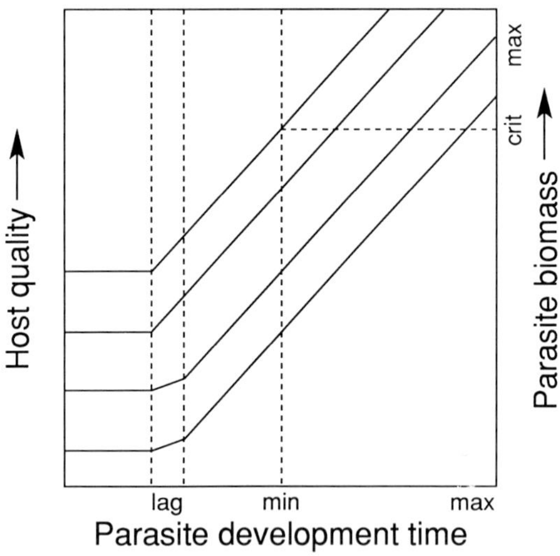

Figure 2 Parasite growth and development in an open resource environment: biomass per unit of host resources is maximized (Model 2). The parasite must reach a critical mass within a variable development time. By delaying the beginning of exponential growth (=lag phase) until the host has reached a larger size, the parasite may increase its growth potential in low-quality hosts but cannot offset any initial delay.

lag phase during which time the host continues to grow. As in Model 1, the parasite's strategy is to maximize biomass per unit of host resources. Apparently, parasites grow at their physiological limit. Whereas fitness gains from increased biomass must be balanced against fitness losses from increased development time, the former are assumed to be of relatively greater importance for fitness. This strategy favors the evolution of a minimum or critical body size for successful development to the adult stage.

The third model (Fig. 3) assumes that the host represents an open resource environment and that parasite growth and development are resource- and time-limited. The dynamics of host quality are the same as in the second model; however, this model assumes a developmental strategy that optimizes a set of characters that jointly determine fitness, for example, body size and development time. Time constraints can take several forms, such as a fixed period of host survival after parasitization or a sharp decline in adult parasite survival with increased age. In low-quality or small hosts, parasite growth may be resource-limited during either early development (lag phase) or late development (precocious death). The parasite's growth trajectory is variable, reflecting host age- or stage-specific constraints on nutritional resources.

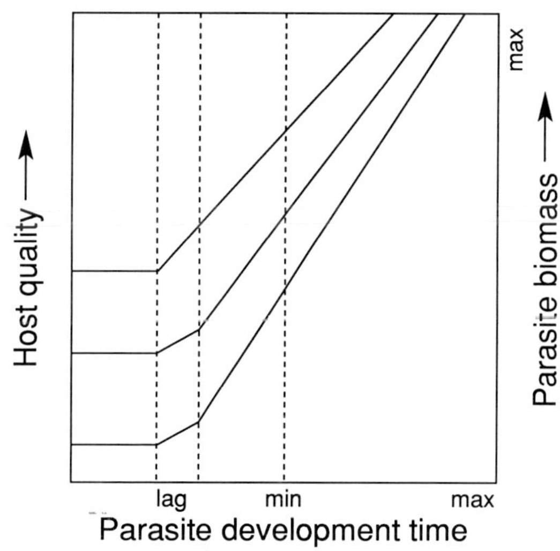

Figure 3 Parasite growth and development in an open resource environment: the optimal phenotype is determined by a trade-off between adult biomass and development time (Model 3). As in Model 2, a parasite developing in a low-quality host may delay the beginning of exponential growth (=lag phase) until the host has reached a larger size. This initial delay is offset by compensatory feeding (and increased growth) during late larval development.

Thus, an initial delay in the onset of exponential parasite growth makes a compensatory increase in the growth rate during later stages necessary in order to balance any fitness gains (from increased biomass) and losses (from increased development time).

Whereas Model 1 conforms to the developmental pattern of most idiobionts, Models 2 and 3 are representative of koinobiont life histories. Host biomass at the time of parasitization represents the main constraint on idiobiont adult biomass and development time. Egg parasites typically cause the arrestment of host embryogenesis (Strand, 1986), thus ensuring that host quality remains largely unchanged during the course of interaction. The strategy adopted by the parasite is to grow until host resources are expended (Fig. 1). In some species, development may be compromised by the parasite's inability to utilize all the resources contained in large hosts (Salt, 1964; Wylie, 1965). Several koinobionts developing in lepidopteran hosts require a critical host size/biomass for completion of larval development (Beckage and Templeton, 1985; Bloem and Duffey, 1990) or molting (Slansky, 1986). However, some changes in parasite growth and development may be mediated by the host's endocrine system (Corbet, 1968; Beckage, 1985; Lawrence, 1990; Strand *et al.*, 1991). The developmental characteristics of *H. exiguae* and *Microplitis* species appear to be consistent with the pattern shown in Fig. 2.

In comparison, the developmental strategy of *A. ervi* appears to be based on trade-offs that result in the optimization of both adult biomass and development time (Fig. 3). The survivorship of parasitized aphids is independent of age at parasitization (Campbell and Mackauer, 1975; Liu and Hughes, 1984), a condition limiting resource availability to a fixed period before the host dies. In low-quality hosts, parasite larvae experienced slow initial growth but completed development at a compensatory rate. Under different option sets, parasite body size was optimized before development time (Sequeira and Mackauer, 1992a,b).

VI. Conclusions

From the parasite's perspective, a host is obviously more than its nutritional quality, including its availability, behavior, and associated competitors and natural enemies. However, to understand the evolution of different life-history strategies in insect parasites, we need to know the functional constraints on their development and reproduction in various host species. In this regard, adult biomass may indeed serve as a useful index of parasite fitness. As we have shown, the best strategy for the maximization of biomass will be determined by the balance between time and resource limitations on parasite growth. Optimality theory (Maynard Smith, 1982; Parker and Maynard Smith, 1990) assumes that, for different species, only certain constrained combinations of fitness-related characters can be realized in practice. These option sets differ between idiobiont and koinobiont and, possibly as well, between solitary and gregarious species of insect parasites. For example, a strategy based on a flexible growth trajectory may enable a koinobiont to respond to variable host quality but would have little utility in idiobionts.

A distinction between "conformers" and "regulators," as proposed by Lawrence (1986, 1990) and Vinson (1990), focuses on the proximate mechanisms by which parasites exploit host resources. By definition, all forms of parasitism are expected to alter host physiology and ontogeny, albeit in ways that may be obvious and pathological in some associations ("regulators") and subtle in others ("conformers"). Similarly, the involvement in host alteration of parasite-derived products, such as polydnavirus and venoms (Stoltz, 1986; Strand and Dover, 1991; Fleming, 1992), does not in principle change the nature of the interactions, although these mechanisms enlarge the option set for host exploitation.

Detailed studies of the dynamics of parasite development under different constraints have been made on only few species, notably solitary koinobionts. Thus, the three models presented here are based on fragmentary information and should be seen as preliminary only. In particular, the models and their underlying assumptions must be tested empirically. Of interest would be a comparison of the growth trajectories of individual larvae between solitary and gregarious species, especially as related to superparasitism and nutritional stress resulting from, for example, starvation and crowding. Another area that is largely unexplored and needs attention is the timing of the end of larval feeding and of host death. Such comparative studies of association-specific functional constraints on parasite ontogeny may help explain the survival value of different life-history strategies in insect parasites.

Acknowledgment

We thank the Natural Sciences and Engineering Research Council of Canada for financial support.

References

Allen, G. R., and Keller, M. A. (1991). *Uraba lugens* (Lepidoptera: Noctuidae) and its parasitoids (Hymenoptera: Braconidae): Temperature, host size, and development. *Environ. Entomol.* **20:**458–469.

Anderson, D. T. (1972). The development of holometabolous insects. *In* "Developmental Systems: Insects" (S. J. Counce and C.H.W. Waddington, eds.), Vol. 1, pp. 165–242. Academic Press, London.

Arthur, A. P., and Wylie, H. G. (1959). Effects of host size on sex ratio, developmental time and size of *Pimpla turionellae* L. *Entomophaga* **4:**297–301.

Askew, R. R., and Shaw, M. R. (1986). Parasitoid communities: Their size, structure and development. *In* "Insect Parasitoids" (J. Waage and D. Greathead, eds.), pp. 225–264. Academic Press, London.

Bai, B., and Mackauer, M. (1992). Influence of superparasitism on development rate and adult size in a solitary parasitoid wasp, *Aphidius ervi*. *Funct. Ecol.* **6:**302–307.

Barrett, M., and Schmidt, J. M. (1991). A comparison between the amino acid composition of an egg parasitoid wasp and some of its hosts. *Entomol. Exp. Appl.* **59:**29–41.

Beckage, N. E. (1985). Endocrine interactions between endoparasitic insects and their hosts. *Annu. Rev. Entomol.* **30:**371–413.

Beckage, N. E., and Riddiford, L. M. (1978). Developmental interactions between the tobacco hornworm *Manduca sexta* and its braconid parasite *Apanteles congregatus*. *Entomol. Exp. Appl.* **23:**139–151.

Beckage, N. E., and Riddiford, L. M. (1983). Growth and development in the endoparasitic wasp *Apanteles congregatus:* Dependence on host nutritional status and parasite load. *Physiol. Entomol.* **8:**231–241.

Beckage, N. E., and Templeton, T. J. (1985). Temporal synchronization of emergence of *Hyposoter exiguae* and *H. fugitivus* (Hymenoptera: Ichneumonidae) with apolysis preceding larval moulting in *Manduca sexta* (Lepidoptera: Sphingidae). *Ann. Entomol. Soc. Am.* **78:**775–782.

Beckage, N. E., and Templeton, T. J. (1986). Physiological effects of parasitism by *Apanteles congregatus* in terminal-stage tobacco hornworm larvae. *J. Insect Physiol.* **32:**299–314.

Bigler, F., Meyer, A., and Bosshart, S. (1987). Quality assessment in *Trichogramma maidis* Pintureau et Voegele reared from eggs of the factitious hosts *Ephestia kuehniella* Zell. and *Sitotroga cerealella* (Olivier). *J. Appl. Entomol.* **104:**340–353.

Blackburn, T. M. (1991). A comparative examination of life-span and fecundity in parasitoid Hymenoptera. *J. Anim. Ecol.* **60:**151–164.

Bloem, K. A., and Duffey, S. S. (1990). Effect of protein type and quantity on growth and development of larval *Heliothis zea* and *Spodoptera exigua* and the endoparasitoid *Hyposoter exiguae*. *Entomol. Exp. Appl.* **54:**141–148.

Brodeur, J., and McNeil, J. N. (1989). Biotic and abiotic factors involved in diapause induction of the parasitoid, *Aphidius nigripes* (Hymenoptera: Aphidiidae). *J. Insect Physiol.* **35:**969–974.

Brown, J. J., Reed-Larsen, D., and Ahl, J. (1990). Physiological relationship between a diapausing endoparasitoid (*Ascogaster quadridentata*) and its dormant host (*Cydia pomonella*). *Arch. Insect Biochem. Physiol.* **13:**229–238.

Campbell, A., and Mackauer, M. (1975). The effects of parasitism by *Aphidius smithi* (Hymenoptera: Aphidiidae) on reproduction and population growth of the pea aphid (Homoptera: Aphididae). *Can. Entomol.* **107:**919–926.

Campbell, A., Frazer, B. D., Gilbert, N., Gutierrez, A. P., and Mackauer, M. (1974). Temperature requirements of some aphids and their parasites. *J. Appl. Ecol.* **11:**431–438.

Campbell, B. C., and Duffey, S. S. (1979). Effect of density and instar of *Heliothis zea* on parasitization by *Hyposoter exiguae*. *Environ. Entomol.* **8:**127–130.

Charnov, E. L., and Skinner, S. W. (1985). Complementary approaches to the understanding of parasitoid oviposition decisions. *Environ. Entomol.* **14:**383–391.

Chow, F. J., and Mackauer, M. (1984). Inter- and intraspecific competition in *Aphidius smithi* and *Praon pequodorum* (Hymenoptera: Aphidiidae). *Can. Entomol.* **116:**1097–1107.

Cloutier, C., and Mackauer, M. (1979). The effect of parasitism by *Aphidius smithi* (Hymenoptera: Aphidiidae) on the food budget of the pea aphid, *Acyrthosiphon pisum* (Homoptera: Aphididae). *Can. J. Zool.* **57:**1605–1611.

Cloutier, C., and Mackauer, M. (1980). The effect of superparasitism by *Aphidius smithi* (Hymenoptera: Aphidiidae) on the food budget of the pea aphid, *Acyrthosiphon pisum* (Homoptera: Aphididae). *Can. J. Zool.* **58:**241–244.

Cloutier, C., McNeil, J. N., and Regnière, J. (1981). Fecundity, longevity, and sex ratio of *Aphidius nigripes* (Hymenoptera: Aphidiidae) parasitizing different stages of its host, *Macrosiphum euphorbiae* (Homoptera: Aphididae). *Can. Entomol.* **113:**193–198.

Coppel, H. C. (1958). Studies on dipterous parasites of the spruce budworm, *Choristoneura fumiferanae* (Clem.) (Lepidoptera: Tortricidae). VI. *Phorocera incrassata* Smith (Diptera: Tachinidae). *Can. J. Zool.* **36:**453–462.

Corbet, S. A. (1968). The influence of *Ephestia kuehniella* on the development of its parasite *Nemeritis canescens*. *J. Exp. Biol.* **48:**291–304.

Cruz, Y. P. (1981). A sterile defender morph in a polyembryonic hymenopterous parasite. *Nature (London)* **294:**446–447.

Dominey, W. J. (1984). Alternative mating tactics and evolutionary stable strategies. *Am. Zool.* **24:**385–396.

Doutt, R. L. (1963). Pathologies caused by insect parasites. *In* "Insect Pathology" (E. A. Steinhaus, ed.), Vol. 2, pp. 393–422. Academic Press, New York.

Dover, B. A., and Vinson, S. B. (1990). Effect of host ligation and starvation on the development and emergence of the parasitoid *Campoletis sonorensis*. *Entomol. Exp. Appl.* **57:**209–213.

Fleming, J.-A. G. W. (1992). Polydnaviruses: Mutualists and pathogens. *Annu. Rev. Entomol.* **37:**401–425.

Führer, E., and Keja, T. D. (1976). Physiologische Wechselbeziehungen zwischen *Pieris brassicae* und dem Endoparasiten *Apanteles glomeratus*. Der Einfluss der Parasitierung auf Wachstum und Körpergewicht des Wirtes. *Entomol. Exp. Appl.* **19:**287–300.

Gauld, I. D. (1988). Evolutionary patterns of host utilization by ichneumonoid parasitoids (Hymenoptera: Ichneumonidae and Braconidae). *Biol. J. Linn. Soc.* **35:**351–377.

Gauld, I., and Bolton, B., eds. (1988). "The Hymenoptera." Oxford Univ. Press, New York.

Godfray, H.C.J. (1987). The evolution of clutch size in parasitic wasps. *Am. Nat.* **129:**221–233.

Gunasena, G. H., Vinson, S. B., and Williams, H. J. (1989). Interrelationships between growth of *Heliothis virescens* (Lepidoptera: Noctuidae) and that of its parasitoid, *Campoletis sonorensis* (Hymenoptera: Ichneumonidae). *Ann. Entomol. Soc. Am.* **82:**187–191.

Haeselbarth, E. (1979). Zur Parasitierung der Puppen von Forleule (*Panolis flammea* [Schiff.]), Kiefernspanner (*Bupalus piniarius* [L.]) and Heidelbeerspanner (*Boarmia bistortana* [Goeze]) in bayerischen Kiefernwäldern. *Z. Angew. Entomol.* **87:**186–202, 311–322.

Hagen, K. S. (1964). Developmental stages of parasites. *In* "Biological Control of Insect Pests and Weeds" (P. DeBach and E. I. Schlinger, eds.), pp. 168–246. Reinhold, New York.

Hall, R. D., and Fischer, F. J. (1988). Laboratory studies on the biology of *Spalangia nigra* (Hymenoptera: Pteromalidae). *Entomophaga* **33:**495–504.

Hébert, C., and Cloutier, C. (1990). Host instar as a determinant of preference and suitability for two parasitoids attacking late instars of the spruce budworm (Lepidoptera: Tortricidae). *Ann. Entomol. Soc. Am.* **83:**734–741.

Hintz, J. L., and Andow, D. A. (1990). Host age and host selection by *Trichogramma nubilale*. *Entomophaga* **35:**141–150.

Hopper, K. R. (1986). Preference, acceptance, and fitness components of *Microplitis croceipes* (Hymenoptera: Braconidae) attacking various instars of *Heliothis virescens* (Lepidoptera: Noctuidae). *Environ. Entomol.* **15:**274–280.

Hopper, K. R., and King, E. G. (1984). Preference of *Microplitis croceipes* (Hymenoptera: Braconidae) for instars and species of *Heliothis* (Lepidoptera: Noctuidae). *Environ. Entomol.* **13:**1145–1150.

Irvine, M. T. (1991). Preference of *Ephedrus californicus* for different instars of the pea aphid and its relationship to parasitoid fitness. Master of Pest Management Thesis, Simon Fraser University, Burnaby, B.C.

Jones, R. L., and Lewis, W. J. (1971). Physiology of the host–parasite relationship between *Heliothis zea* and *Microplitis croceipes*. *J. Insect Physiol.* **17:**921–927.

Kambhampati, S., and Mackauer, M. (1989). Multivariate assessment of inter- and intraspecific variation in performance criteria of several pea aphid parasites (Hymenoptera: Aphidiidae). *Ann. Entomol. Soc. Am.* **82:**314–324.

King, B. H. (1987). Offspring sex ratios in parasitoid wasps. *Q. Rev. Biol.* **62:**367–396.

King, B. H. (1989). Host-size-dependent sex ratios among parasitoid wasps: Does host growth matter? *Oecologia* **78:**420–426.

King, B. H. (1990a). Sex ratio manipulation by the parasitoid wasp *Spalangia cameroni* in response to host age: A test of the host-size model. *Evol. Ecol.* **4:**149–156.

King, B. H. (1990b). Interspecific differences in host (Diptera: Muscidae) size and species usage among parasitoid wasps (Hymenoptera: Pteromalidae) in a poultry house. *Environ. Entomol.* **19:**1519–1522.

Kouamé, K. L., and Mackauer, M. (1991). Influence of aphid size, age and behaviour on host choice by the parasitoid wasp *Ephedrus californicus:* A test of host-size models. *Oecologia* **88:**197–203.

Lawrence, P. O. (1982). *Biosteres longicaudatus:* Developmental dependence on host (*Anastrepha suspensa*) physiology. *Exp. Parasitol.* **53:**396–405.

Lawrence, P. O. (1986). Host–parasitoid hormonal interactions: An overview. *J. Insect Physiol.* **32:**295–298.

Lawrence, P. O. (1990). The biochemical and physiological effects of insect hosts on the development and ecology of their parasites: An overview. *Arch. Insect Biochem. Physiol.* **13:**217–228.

Leather, S. R. (1988). Size, reproductive potential and fecundity in insects: Things aren't as simple as they seem. *Oikos* **51:**386–389.

Legner, E. F. (1969). Adult emergence interval and reproduction in parasitic Hymenoptera influenced by host size and density. *Ann. Entomol. Soc. Am.* **62:**220–226.

le Masurier, A. D. (1987). A comparative study of the relationship between host size and brood size in *Apanteles* spp. (Hymenoptera: Braconidae). *Ecol. Entomol.* **12:**383–393.

le Masurier, A. D. (1991). Effect of host size on clutch size in *Cotesia glomerata*. *J. Anim. Ecol.* **60:**107–118.

Lewis, W. J. (1970). Study of species and instars of larval *Heliothis* parasitized by *Microplitis croceipes*. *J. Econ. Entomol.* **63:**363–365.

Liu, S. S. (1985). Development, adult size and fecundity of *Aphidius sonchi* reared in two instars of its aphid host, *Hyperomyzus lactucae*. *Entomol. Exp. Appl.* **37:**41–48.

Liu, S. S., and Hughes, R. D. (1984). Effect of host age at parasitization by *Aphidius sonchi* on the development, survival, and reproduction of the sowthistle aphid, *Hyperomyzus lactucae*. *Entomol. Exp. Appl.* **36:**239–246.

Mackauer, M. (1986). Growth and developmental interactions in some aphids and their hymenopterous parasites. *J. Insect Physiol.* **32:**275–280.

Mackauer, M. (1990). Host discrimination and larval competition in solitary endoparasitoids. *In* "Critical Issues in Biological Control" (M. Mackauer, L. E. Ehler, and J. Roland, eds.), pp. 41–62. Intercept, Andover, Hants.

Mackauer, M., and Kambhampati, S. (1988). Parasitism of aphid embryos by *Aphidius smithi:* Some effects of extremely small host size. *Entomol. Exp. Appl.* **49:**167–173.

Mangum, C., and Towle, D. (1977). Physiological adaptation to unstable environments. *Am. Sci.* **65:**67–75.

Marston, N., and Ertle, L. R. (1973). Host influence on the bionomics of *Trichogramma minutum*. *Ann. Entomol. Soc. Am.* **66:**1155–1162.

Matejko, I., and Sullivan, D. J. (1979). Bionomics and behavior of *Alloxysta megourae*, an aphid hyperparasitoid (Hymenoptera: Cynipidae). *J. N.Y. Entomol. Soc.* **87:**275–282.

Maynard Smith, J. (1982). "Evolution and the Theory of Games." Cambridge Univ. Press, Cambridge.

Pak, G. A. (1986). Behavioural variations among strains of *Trichogramma* spp. A review of the literature on host age selection. *J. Appl. Entomol.* **101:**55–64.

Parker, G. A., and Maynard Smith, J. (1990). Optimality theory in evolutionary biology. *Nature (London)* **348:**27–33.

Peters, R. H. (1983). "The Ecological Implications of Body Size." Cambridge Univ. Press, Cambridge.

Polgár, L., Mackauer, M., and Völkl, W. (1991). Diapause induction in two species of aphid parasitoids: The influence of aphid morph. *J. Insect Physiol.* **37:**699–702.

Puttler, B. (1961). Biology of *Hyposoter exiguae* (Hymenoptera: Ichneumonidae), a parasite of lepidopterous larvae. *Ann. Entomol. Soc. Am.* **54:**25–30.

Rajapakse, R.H.S., and Ashley, T. R. (1985). Biology and host acceptance of *Microplitis manilae* (Hym: Braconidae) raised on fall armyworm larvae *Spodoptera frugiperda* (Lepidoptera: Noctuidae). *Fla. Entomol.* **68:**653–657.

Reznik, S. Y., and Umarova, T. Y. (1990). The influence of host's age on the selectivity of parasitism and fecundity of *Trichogramma. Entomophaga* **35:**31–37.

Roff, D. A. (1981). On being the right size. *Am. Nat.* **118:**405–422.

Salt, G. (1938). Experimental studies in insect parasitism. VI. Host suitability. *Bull. Entomol. Res.* **29:**223–246.

Salt, G. (1940). Experimental studies in insect parasitism. VII. The effects of different hosts on the parasite *Trichogramma evanescens* Westwood. *Proc. R. Entomol. Soc. London, Ser. A* **15:**81–95.

Salt, G. (1961). Competition among insect parasitoids. *Symp. Soc. Exp. Biol.* **15:**96–119.

Salt, G. (1964). The ichneumonid parasite *Nemeritis canescens* (Gravenhorst) in relation to the wax moth *Galleria mellonella* (L.). *Trans. R. Entomol. Soc. London* **116:**1–14.

Sandlan, K. P. (1982). Host suitability and its effects on parasitoid biology in *Coccygomimus turionellae* (Hymenoptera: Ichneumonidae). *Ann. Entomol. Soc. Am.* **75:**217–227.

Saunders, D. S., Sutton, D., and Jarvis, R. A. (1970). The effect of host species on diapause induction in *Nasonia vitripennis. J. Insect Physiol.* **16:**405–416.

Schmidt, J. M., and Smith, J.B.B. (1987). The measurement of exposed host volume by the parasitoid wasp *Trichogramma minutum* and the effects of wasp size. *Can. J. Zool.* **65:**2837–2845.

Schmidt-Nielsen, K. (1984). "Scaling: Why is Animal Size So Important?" Cambridge Univ. Press, Cambridge.

Schoonhoven, L. M. (1962). Synchronization of a parasite/host system, with special reference to diapause. *Ann. Appl. Biol.* **50:**617–621.

Sequeira, R., and Mackauer, M. (1992a). Nutritional ecology of an insect host–parasitoid association: The pea aphid—*Aphidius ervi* system. *Ecology* **73:**183–189.

Sequeira, R., and Mackauer, M. (1992b). Covariance of adult size and development time in the parasitoid wasp *Aphidius ervi* in relation to the size of its host, *Acyrthosiphon pisum. Evol. Ecol.* **6:**34–44.

Sequeira, R., and Mackauer, M. (1993). The nutritional ecology of a parasitoid wasp: *Ephedrus californicus* Baker (Hymenoptera: Aphididae). *Can. Entomol.* **125** (in press).

Sibly, R. M., and Calow, P. (1986). "Physiological Ecology of Animals. An Evolutionary Approach." Blackwell, Oxford.

Slansky, F., Jr. (1986). Nutritional ecology of endoparasitic insects and their hosts: An overview. *J. Insect Physiol.* **32:**255–261.

Slansky, F., Jr., and Scriber, J. M. (1985). Food consumption and utilization. *In* "Comprehensive Insect Physiology, Biochemistry and Pharmacology" (G. A. Kerkut and L. I. Gilbert, eds.), Vol. 4, pp. 87–163. Pergamon, Oxford.

Smilowitz, Z., and Iwantsch, G. F. (1973). Relationships between the parasitoid *Hyposoter exiguae* and the cabbage looper, *Trichoplusia ni:* Effects of host age on developmental rate of the parasitoid. *Environ. Entomol.* **2:**759–763.

Stoltz, D. B. (1986). Interactions between parasitoid-derived products and host insects: An overview. *J. Insect Physiol.* **32:**347–350.

Strand, M. R. (1986). The physiological interactions of parasitoids with their hosts and their influence on reproductive strategies. *In* "Insect Parasitoids" (J. Waage and D. Greathead, eds.), pp. 97–136. Academic Press, London.

Strand, M. R., and Dover, B. A. (1991). Developmental disruption of *Pseudoplusia includens* and *Heliothis virescens* larvae by the calyx fluid and venom of *Microplitis demolitor. Arch. Insect Biochem. Physiol.* **18:**131–145.

Strand, M. R., Johnson, J. A., and Culin, J. D. (1988). Developmental interactions between the parasitoid *Microplitis demolitor* (Hymenoptera: Braconidae) and its host *Heliothis virescens* (Lepidoptera: Noctuidae). *Ann. Entomol. Soc. Am.* **81:**822–830.

Strand, M. R., Baerecke, E. H., and Wong, E. A. (1991). The role of host endocrine factors in the development of polyembryonic parasitoids. *Biol. Control* **1:**144–152.

Sullivan, D. J. (1987). Insect hyperparasitism. *Annu. Rev. Entomol.* **32:**49–70.

Tauber, M. J., Tauber, C. A., and Masaki, S. (1984). Adaptations to hazardous seasonal conditions: Dormancy, migration, and polyphenism. *In* "Ecological Entomology" (C. B. Huffaker and R. L. Rabb, eds.), pp. 149–183. Wiley (Interscience), New York.

Taylor, A. D. (1988). Host effects on larval competition in the gregarious parasitoid *Bracon hebetor. J. Anim. Ecol.* **57:**163–172.

Thompson, S. N. (1982). Effects of parasitization by the insect parasite *Hyposoter exiguae* on the growth, development and physiology of its host *Trichoplusia ni. Parasitology* **84:**491–510.

Thompson, S. N. (1983). The nutritional physiology of *Trichoplusia ni* parasitized by the insect parasite, *Hyposoter exiguae,* and the effects of parallel-feeding. *Parasitology* **87:**15–28.

Thompson, S. N. (1985a). Biochemical and physiological effects of metazoan endoparasites on their host species. *Comp. Biochem. Physiol. B* **74B:**183–211.

Thompson, S. N. (1985b). Metabolic integration during the host associations of multicellular animal endoparasites. *Comp. Biochem. Physiol. B* **81B:**21–42.

Thompson, S. N. (1990). Physiological alterations during parasitism and their effects on host behaviour. *In* "Parasitism and Host Behaviour" (C. J. Barnard and J. M. Behnke, eds.), pp. 64–94. Taylor & Francis, London.

Tillman, P. G., and Powell, J. E. (1989). Comparison of acceptance of larval instars of the tobacco budworm (Lepidoptera: Noctuidae) by *Microplitis croceipes, Microplitis demolitor, Cotesia kasak* (Hymenoptera: Braconidae) and *Hyposoter didymator* (Hymenoptera: Ichneumonidae). *J. Agric. Entomol.* **6:**201–209.

van Alphen, J.J.M., and Vet, L.E.M. (1986). An evolutionary approach to host finding and selection. *In* "Insect Parasitoids" (J. Waage and D. Greathead, eds.), pp. 23–61. Academic Press, London.

Vinson, S. B. (1975). Biochemical coevolution between parasitoids and their hosts. *In* "Evolutionary Strategies of Parasitic Insects and Mites" (P. W. Price, ed.), pp. 14–48. Plenum, New York.

Vinson, S. B. (1990). Physiological interactions between the host genus *Heliothis* and its guild of parasites. *Arch. Insect Biochem. Physiol.* **13:**63–81.

Vinson, S. B., and Iwantsch, G. F. (1980a). Host regulation by insect parasitoids. *Q. Rev. Biol.* **55:**145–165.

Vinson, S. B., and Iwantsch, G. F. (1980b). Host suitability for insect parasitoids. *Annu. Rev. Entomol.* **25:**397–419.

Waage, J. K. (1986). Family planning in parasitoids: Adaptive patterns of progeny and sex allocation. *In* "Insect Parasitoids" (J. Waage and D. Greathead, eds.), pp. 63–95. Academic Press, London.

Waage, J. K., and Godfray, H.C.J. (1985). Reproductive strategies and population ecology of insect parasitoids. *In* "Behavioural Ecology: Ecological Consequences of Adaptive Behaviour" (R. M. Sibly and R. H. Smith, eds.), pp. 449–470. Blackwell, Oxford.

Wajnberg, E., Bouletreau, M., Prévost, G., and Fouillet, P. (1990). Developmental relationships between *Drosophila* larvae and their endoparasitoid *Leptopilina* (Hymenoptera: Cynipidae) as affected by crowding. *Arch. Insect Biochem. Physiol.* **13:**239–245.

Webb, B. A., and Dahlman, D. L. (1985). Developmental pathology of *Heliothis virescens* larvae parasitized by *Microplitis croceipes:* Parasite mediated host developmental arrest. *Arch. Insect Biochem. Physiol.* **2:**131–143.

Webb, B. A., and Dahlman, D. L. (1986). Ecdysteroid influence on the development of the host *Heliothis virescens* and its endoparasite *Microplitis croceipes. J. Insect Physiol.* **32:**339–345.

Weseloh, R. M. (1984). Effect of size, stress, and ligation of gypsy moth (Lepidoptera: Lymantriidae) larvae on development of the tachinid parasite *Compsilura concinnata* Meigen (Diptera: Tachinidae). *Ann. Entomol. Soc. Am.* **77:**423–428.

Wishart, G. (1956). Effects of hydrogen ion concentration on hatching of eggs of *Aplomya caesar* (Ald.) (Diptera: Tachinidae). *Can. Entomol.* **88:**655–656.

Wylie, H. G. (1965). Effects of superparasitism on *Nasonia vitripennis* (Wlk.) (Hymenoptera: Pteromalidae). *Can. Entomol.* **97:**326–331.

Wylie, H. G. (1976). Observations on life history and sex ratio variability of *Eupteromalus dubius* (Hymenoptera: Pteromalidae), a parasite of cyclorrhaphous Diptera. *Can. Entomol.* **108:**1267–1274.

Chapter 2

Games Parasites Play: The Dynamic Roles of Proteins and Peptides in the Relationship between Parasite and Host

Nancy E. Beckage
Department of Entomology
University of California
Riverside, California

I. Parasitism-Induced Changes: Host Responses or Parasite-Directed Host Manipulation?

Parasitism inflicts multiple changes in the physiological and biochemical functions of host insects aside from altering their behavior, hormone titers, and development as emphasized elsewhere in this volume. Given the sophisticated biochemical and molecular techniques now available, monitoring the effects of parasitism on host protein synthesis, and confirmation that parasitism sometimes induces synthesis of novel proteins in the host, now constitute some of the most readily documentable changes occurring during insect endoparasitism. The goals of this review are to summarize the major lines of evidence demonstrating that parasitized insects exhibit major changes involving proteins and peptides, and pose questions about the nature of their potential biological significance and impact upon the host–parasite relationship.

Importantly, a broad taxonomic survey reveals that parasitism induces alterations in the proteins and peptides present in a wide variety of organisms aside from insects, indicating they likely represent major players in the relationship between parasite and host in many systems. Many mammals, in addition to several classes of invertebrates, including molluscs and crustaceans, exhibit major changes in their blood components or hemolymph during parasitism, arguing that similar biochemical changes occur in widely ranging taxa as a result of parasitism. In some systems, the parasites themselves directly intervene in causing these changes, whereas in others their influence appears to be more indirect. Several parasites developing in mammalian hosts (or even plants) have been shown to secrete proteins and enzymes that facilitate, or are absolutely required for, invasion of host tissues or maintenance of an appropriate biochemical environment that markedly enhances parasite success.

One of the most critical features of parasitic invasion obviously is that the parasites successfully escape or suppress the host immune response, and some of the molecules produced or otherwise associated with parasites are thought to serve important roles in actively disengaging the host immune response, thereby preventing immunodetection. Yet other species of parasites are thought to employ host "mimicry" to circumvent host defenses by incorporating hostlike antigens (either parasite- or host-derived molecules) onto their surface, thereby allowing the parasites to escape recognition as foreign during development in host tissues. Proteins thus play important roles in "masking" the surface of parasites or otherwise providing protection.

Unfortunately, insects have been considerably less intensively studied compared to mammals with respect to the precise mechanisms whereby their parasites cause host immunosuppression or escape immunorecognition and rejection. Consequently, our biochemical insights into the nature of the critical molecules and mechanisms functioning at the host–parasite immunological interface in insect systems remain very limited. In one instance, it appears that the viruslike particles injected by a wasp parasitoid (*Venturia canescens*) express "hostlike" epitopes on the surface of the virions. Following oviposition, large numbers of virions adhere to the surface of the developing parasite eggs, thus protecting the accompanying parasites by a process akin to mimicry of host antigens (Schmidt and Schuchmann-Feddersen, 1989). However, these virus particles are unique among the polydnaviruslike elements carried by parasitoids in that they appear to lack any nucleic acids yet have evolved potent protective strategies based upon host protein recognition processes (Schmidt *et al.*, 1990).

The mechanisms operating in other species of parasitized insects are less clear. For example, although it is known that the hemocytes of parasitized insects show extensive behavioral transformations that preclude their involvement in encapsulation (Stoltz, Volume 1, Chapter 8), the transducing pathways whereby the parasitoid-associated polydnaviruses cause these cellular changes remain wholly speculative. The molecular mechanisms operating to suppress host humoral and cellular defenses during parasitism thus remain to be defined. This situation hopefully soon will be dramatically changed as we exploit the recent rapid proliferation of probes and techniques available for us to dissect the dynamic nature and various levels of molecular interactions operating between parasites and their respective hosts. On the basis of our knowledge of mammalian systems, we might suspect that proteins, peptides, glycoproteins, lectins, lipophosphoglycans, and other cell-surface-associated molecules might play major roles.

In mammals, recent advances in developmental biology have proven insightful in that many of the same molecules heretofore recognized as major players in cell–cell recognition processes in immunity also appear to play critical roles in cell proliferation, tissue organization, and other developmental processes. Cytokines and several growth factors plus their receptors ultimately may be shown to have such dual roles in mammals. For example, members of transforming growth factor β-family play a role in defense to pathogens, for example, in determining macrophage susceptibility to leishmania parasites (Barral-Netto *et al.,* 1992), while also functioning during embryonic development (Ferguson and Anderson, 1992; Jessell and Melton, 1992). In insects, a similar dual functionality of some molecules appears likely (Natori, 1990). In *Drosophila,* a neuroglian belonging to the immunoglobulin superfamily plays an important role during glial and neural cell adhesion during development of the nervous system (Bieber *et al.,* 1989) and another insect immunoglobulinlike molecule called hemolin appears to bind to the surface of bacteria and target them for destruction (Sun *et al.,* 1990). Hence, some of the mechanisms involved in distinguishing "self" versus "non-self" during animal development versus infection may involve some of the same molecular signals and recognition processes directed at identifying different degrees of "foreignness" in the animal.

Several authors studying insect systems have identified "parasitism-specific" or "parasitism-induced" proteins in insect hosts, which become detectable in the host at varying times postparasitization (see Section II). Their *de novo* synthesis may begin as early as a few hours after the host is parasitized, or they may not appear in the hemolymph until the parasites have nearly matured, with the precise temporal pattern of their appearance varying according to the species of host and the parasites involved. Several species of lepidopterans (Cook *et al.,* 1984; Beckage *et al.,* 1987, 1989; Soldevila and Jones, 1991, 1993) and dipterans (Lawrence, 1990) parasitized by endoparasitic wasps have been shown to synthesize parasitism-specific proteins. In some species (e.g., *Manduca sexta,* see Section II) the host hemolymph shows evidence of the presence of different sets of "early" and "late" parasitism-specific proteins, with the temporal pattern of their expression being precisely regulated. Importantly, the proteins appear unique to parasitized larvae, being noninducible by other forms of "physiological stress" such as heat or cold shock (Fittinghoff and Riddiford, 1990; Joplin *et al.*, 1990), pathogenic infection (Beckage *et al.*, 1989; see the following discussion), or physical trauma and injury (i.e., sham injections) (Ferkovich *et al.*, 1983). While new proteins are produced in response to the latter challenges, careful analysis reveals that they differ in molecular weight and other characteristics from those induced by parasitization. For example, although wounding occurs following puncture of the hemocoel during oviposition, the molecules that have thus far been characterized as being synthesized in response to physical injury (e.g., sarcotoxins—see Kanai and Natori, 1989) appear notably different from those inducible by parasitization. Thus, they do not represent generalized stress proteins but instead constitute a specific response.

A survey of recent investigations suggests that the absence of any quantitative or qualitative effect of parasitism on host hemolymph enzymes or constitutive proteins likely represents the exception rather than the rule. The induction of hormone-related changes also is now recognized as being commonly associated with insect endoparasitism. Hypothetically, the antibacterial proteins induced by bacterial challenge (see references in Faye and Hultmark, Volume 2, Chapter 2; Kanost *et al.*, 1990) might have close parallels in insects attacked by multicellular organisms, such that hosts respond by producing novel proteins or peptides, which are specifically (and uniquely) associated with the parasitized state. However, the situation is more complicated with respect to parasitism in that in some cases the parasites have "symbiotic" or third-party elements such as viruses or bacteria associated with them, thus confounding interpretation of experiments designed to pinpoint the origin(s) of the cognizant molecules (i.e., genes of the host or its parasites or the element involved). Moreover, these proteins often appear correlated with successful parasitism, rather than a lack thereof. During normal development of the parasites, the proteins are produced, and their synthesis hence is correlated with parasite protection and absence of encapsulation. Thus, the pattern of their occurrence differs from that of the antibacterial proteins, which constitute an efficacious "protective" host response that checks pathogen invasion.

An analysis of this complex scenario would not be complete without reference to parasitism-induced modifications in the host's endogenous proteins. Titers of many of those molecules are significantly altered during parasitism. Factors associated with parasitism may affect the rates of transcription or translation of host genes and be responsible for inducing these changes. In the least complicated situation, the simple presence of the parasites within the body cavity of the host may be sufficient to affect the rates of transcription or translation, thereby having an impact on host protein synthesis. Time course studies suggest changes occur early, before the wasps hatch. Later, depletion of amino acids and other nutrients due to parasite consumption of host hemolymph constituents conceivably could modify the rate of host protein synthesis, since the pools of precursors required for optimal transcription and translation of host mRNAs might consequently be present in suboptimal concentrations or ratios due to diversion to "competing" tissues, that is, the parasites. Thus, some metabolic effects (Thompson, Volume 1, Chapter 6) likely are tightly interwoven with parasitism-induced modifications in host protein production and transport.

Additionally, factors secreted or otherwise associated with the parasites, such as teratocytes (Dahlman and Vinson, Volume 1, Chapter 7) and polydnaviruses (Stoltz, Volume 1, Chapter 8; Fleming and Krell, Volume 1, Chapter 9), may also have a regulatory influence. Examples of the endogenous proteins now known to be affected by insect endoparasitism include those intimately associated with larval development and metamorphosis (e.g., arylphorins), reproduction (e.g., vitellogenin), and the host immune response (e.g., phenoloxidase). As some of these changes are described in detail in other chapters, only a brief summary is presented here to illustrate the nature and extent to which the host's endogenous proteins are modified during endoparasitism.

An unresolved, but nonetheless intriguing, question concerns the degree to which these changes reflect activity of parasite (or parasite-associated element) "designer genes" whose expression directly manipulates the physiology of the host for the parasites' benefit. The concept of "host regulation" is frequently invoked by physiologists and parasitologists to explain the adaptive significance of many of the immunological, metabolic, and behavioral changes observed in host organisms during parasitism; yet the supporting evidence for such regulatory phenomena is disappointingly sparse, especially with respect to insect systems. Unfortunately, the truly intimate coevolutionary nature of the host–parasite relationship renders interpretation of many of the biochemical changes seen in host insects exceedingly problematic, due to the complexities involved in unraveling the web of interactions. Simultaneously one must consider the intriguing interplay of the biochemical "attack" strategies evolved by parasites and the beguiling counterstrategies evolved by host insects for their defense, not to mention the roles of parasite-associated elements.

Among plant pathologists, the idea of gene-for-gene coevolution between parasites and hosts has now been widely accepted since this concept was first proposed by Flor (1942). Since its inception, this working hypothesis has generated much debate and experimental research, and fortuitously has led to the identification of several "virulence" and "avirulence" genes in plant pathogens, hence verifying its usefulness as an experimental paradigm (for review, see Thompson and Burdon, 1992). Virulent genotypes recently have also been identified in other parasites, for example, protozoan parasites of mammals (Sibley and Boothroyd, 1992). This conceptual framework seemingly has been less warmly embraced by insect parasitologists and parasitoid specialists, who have interpreted the panorama of host changes as symptomatic of host "stress," "redirection," or "regulation" without any direct reference to their importance as indicators of evolution of parasite virulence characteristics required for host invasion or long-term survival in what might otherwise be a noncompatible host environment.

However, as recent research on malaria has emphasized, a delicate interplay of host genes encoding factors related to susceptibility and resistance characteristics, interacting with the products of parasite virulence and avirulence genes, likely determines the outcome of many host–parasite encounters (Sayles and Wassom, 1988). The delicate counterbalance of parasite offense versus host defense presumably requires a similar strategy of gene-for-gene coevolution that ultimately renders the relationship a "successful" one when the parasites become capable of development to maturity within a previously "hostile" or nonhabitual host species (or strain thereof). The parasite thus may be considered the partner with the "driver" genes directing the host toward evolution of resistance-related characters.

The identification of parasite-associated "virulence" or "killer" genes, and their corresponding avirulant mutant alleles, unfortunately still awaits us with respect to systems involving parasites of insect hosts. The successful expression of gene products by their associated "third-party" elements (e.g., polydnaviruses associated with wasps, or bacteria associated with nematodes) also likely is critical to determining the outcome of parasitic interactions involving a variety of insect hosts. Thus, in these complex systems involving several "partners" it appears likely that both the element (such as a polydnavirus) and the parasite must be capable of completing its life cycle to ensure successful propagation in the host system, which culminates in maturation of the parasites followed by dissemination and infection of new hosts.

II. Parasitism- and Polydnavirus-Induced Peptides and Proteins in Insect Hosts

Many studies have shown that insect hemolymph proteins are altered during parasitism (see references in following discussion, plus earlier studies by Fisher and Ganesalingam, 1970; King and Rafai, 1970; Vinson and Barras, 1970; Barras *et al.*, 1972; Brewer *et al.*, 1973; Smilowitz, 1973; Smilowitz and Smith, 1977; Vinson and Iwantsch, 1980). In a few cases of parasitism of lepidopteran larvae by braconid and ichneumonid wasps, novel "parasitism-specific" hemolymph proteins have been shown to be synthesized *de novo* following parasitization, as indicated by *in vivo* amino acid radiolabeling of the protein(s) produced in newly parasitized hosts (Cook *et al.*, 1984; Beckage *et al.*, 1987). This line of evidence suggests that they are not molecules shuttled from the adult parasitoid to the host during parasitization, but instead represent newly synthesized molecules. Most work has focused on assessing changes in hemolymph-borne molecules, which are therefore emphasized in this chapter; a variety of intracellular or tissue-associated changes likely also occur but have been considerably less well studied.

The polydnaviruses injected by the wasp parasitoid *Campoletis sonorensis* together with the eggs initiate viral gene transcription following transfer of the virus from the wasp "vector" to the lepidopteran "host." Different subsets of viral genes are expressed in each species; a third set of "constitutive" genes appears to be expressed initially in the wasp ovary and expression continues following transfer to the lepidopteran host (Theilmann and Summers, 1988; Fleming and Krell, Volume 1, Chapter 9). In this system, mRNA transcripts have been isolated by hybridization of extracted RNAs to viral probes, although the corresponding proteins await future characterization (Blissard *et al.*, 1986; Theilmann and Summers, 1988). These may represent the novel proteins described in earlier studies of parasitized *Heliothis* larvae (Vinson and Barras, 1970; Barras *et al.*, 1972; Brewer *et al.*, 1973). Interestingly, some particularly abundant "parasitism-specific" proteins detectable in host insects are indeed polydnavirus-induced transcripts, which are not produced when the virus is inactivated (Cook *et al.*, 1984; Beckage *et al.*, 1987; Harwood, 1993; Harwood *et al.*, 1993a). Other molecules apparently are synthesized by the developing parasite larvae or their associated teratocyes, and obviously are generated in sufficient amounts to be detectable in the host's blood (Lawrence, 1990; Soldevila and Jones, 1993). Thus, the origin and site(s) of production of these proteins are variable and likely species-specific, with some proteins probably having multiple expression loci within the host (wasp, polydnavirus, host).

The molecular weights of the parasitization-induced molecules described thus far vary from <10,000 daltons to >150,000 daltons, and within a single host, multiple proteins or peptides may be produced according to a tightly regulated temporal pattern varying according to the stage of parasitism. For example, different sets of parasitism-specific proteins are produced in "early" versus "late" stage host larvae of the tobacco hornworm, *Manduca sexta*, following parasitization by the braconid wasp *Cotesia congregata* (Dahlman and Greene, 1981; Beckage *et al.*, 1987, 1989; Beckage and Templeton, 1986; Beckage and Kanost, 1993; Harwood, 1993; Harwood and Beckage, 1993). Following the disappearance of the "early" polypeptides several days post-parasitization, other "late" appearing proteins subsequently appear in last-instar hosts, which continue to circulate in the host's hemolymph until it eventually dies several days following the wasps' emergence (Beckage and Templeton, 1986) due to neurendocrine disruption (Zitnan *et al.*, 1993).

The temporal pattern of appearance of the "late" proteins mirrors that seen in other systems. In cabbage looper larvae parasitized by *Chelonus* near *curvimaculatus*, a novel high-molecular-weight protein (ca. 160 kDa on SDS–PAGE gels) appears in the hemolymph of newly molted mature host larvae, which then, as result of parasitism, are induced to undergo precocious metamorphosis and wander following what would normally be the penultimate instar (Jones *et al.*, 1985, 1986; Jones, 1989; Soldevila and Jones, 1991). This protein is of low abundance relative to other hemolymph proteins, and begins to appear at head capsule slippage preceding the last larval ecdysis; it is undetectable in nonparasitized larvae of the same species, regardless of their stage of development. Moreover, the protein persists in the hemolymph of the host until the wasp emerges (Soldevila and Jones, 1991, 1993), similar to the "late" proteins characteristic of parasitized *M. sexta*.

The protein fails to appear in pseudoparasitized *T. ni* larvae, defined as larvae in which no parasite is present but into which venom and polydnavirus have been injected, the parasite having been killed by chilling or other experimental manipulations (Jones *et al.*, 1986), arguing that the presence of a developing parasite is requisite for its synthesis (Soldevila and Jones, 1991, 1993). Recent data show the protein actually originates from the parasites and the wasps show a positive reaction with antibodies directed against the protein. Evidence also exists for its *de novo* biosynthesis by the wasps, which could also acquire the protein by ingestion of their host's hemolymph. Hence, Soldevila and Jones (1991, 1993) concluded that the most likely scenario is that the protein is parasite-derived. Earlier, it was thought that the presence of the mature wasps (or their associated factors) somehow induces the host's tissues to synthesize the protein beginning when the host initiates its final larval molt.

The predominant "early" expressed parasitism-specific polypeptides seen in *M. sexta* are in the size range of 33–38 kDa and are very abundant proteins, representing up to 8–10% of the total circulating hemolymph protein analyzed at 24 hr postparasitization. One of the three major bands detectable in this size range has been confirmed to be a polydnaviral gene product (Harwood, 1993; Harwood *et al.*, 1993a). Several lower-molecular-weight polypeptides in the 17- to 20-kDa range also are induced during the same temporal window, and likewise disappear synchronously with the higher-molecular-weight proteins several days following oviposition (Beckage *et al.*, 1989). The simultaneous up- and down-regulation of several of this array of "early" polypeptides is striking and suggests they may be coordinately regulated (Beckage *et al.*, 1987, 1989). Moreover, since all of them appear inducible by injection of purified polydnavirus into nonparasitized *M. sexta* larvae, our observations suggest they might have a common origin, that is, coordinately regulated viral genes.

In parasitized tobacco hornworms a very large polypeptide (>120 kDa) appears as a major dense band on SDS–PAGE gels beginning a few days following the host's ecdysis to its final instar, just a few days prior to the wasps' emergence (Beckage and Templeton, 1986). Interestingly, this proteins appears synchronously with a much less abundant ca. 100-kDa polypeptide a few days prior to wasp emergence, just at the time the wasps undergo their first larval ecdysis to the second instar (Beckage and Templeton, 1986). The onset of synthesis of these proteins seems remarkably well-synchronized with this L_1–L_2 transition in the wasps, suggesting they may be of parasite origin or synthesized by host tissues in response to the presence of this specific parasite stage. Moreover, their appearance is suppressed in hosts treated with the insect growth regulator azadirachtin, which acts as an ecdysis inhibitor so the parasites produce a second-instar cuticle but die as pharate L_2 larvae trapped within the host (Beckage *et al.*, 1988). Failure of the parasites to ecdyse prevents appearance of the "late" proteins, further corroborating their correlation with onset of the parasites' L_2 stage. Yet another parasitism-specific polypeptide of 56 kDa molecular weight is present in host *M. sexta* hemolymph beginning at its ecdysis to the last instar (Beckage and Templeton, 1986). Its timing of appearance is reminiscent of the protein detected in parasitized cabbage loopers by Soldevila and Jones (1991, 1993); however, the relative abundance and molecular weights of the proteins isolated in the two species are significantly different, suggesting they may be analogous but not identical parasitism-associated proteins.

In *M. sexta* we hypothesized that the parasites might be the source of these "late" proteins; however, culturing the parasites *in vitro* with [^{35}S]methionine followed by fluorogram analysis of proteins secreted into the medium over a 24-hr period produced no evidence of their *de novo* production by the wasps, although several proteins were released by the wasps (Fig. 1). The teratocytes developing in the host hemocoel also do not appear to be the synthetic source (de Buron *et al.*, 1993). Our current working hypothesis is that they are likely host- rather than parasite-derived. Additionally, *in vivo* studies have shown that the "late" proteins persist in the hemolymph of the host for several days following parasite emergence from the host, further suggesting they may be of host origin. In contrast to the "early" induced proteins, they clearly are not inducible by the injection of polydnavirus into nonparasitized larvae (Beckage *et al.*, 1993), suggesting that they do not require activity of viral genes for their expression. In summary, although the "early" proteins represent protein products of polydnavirus genes, the origin of the "late" proteins detected in the host remains (frustratingly) unclear.

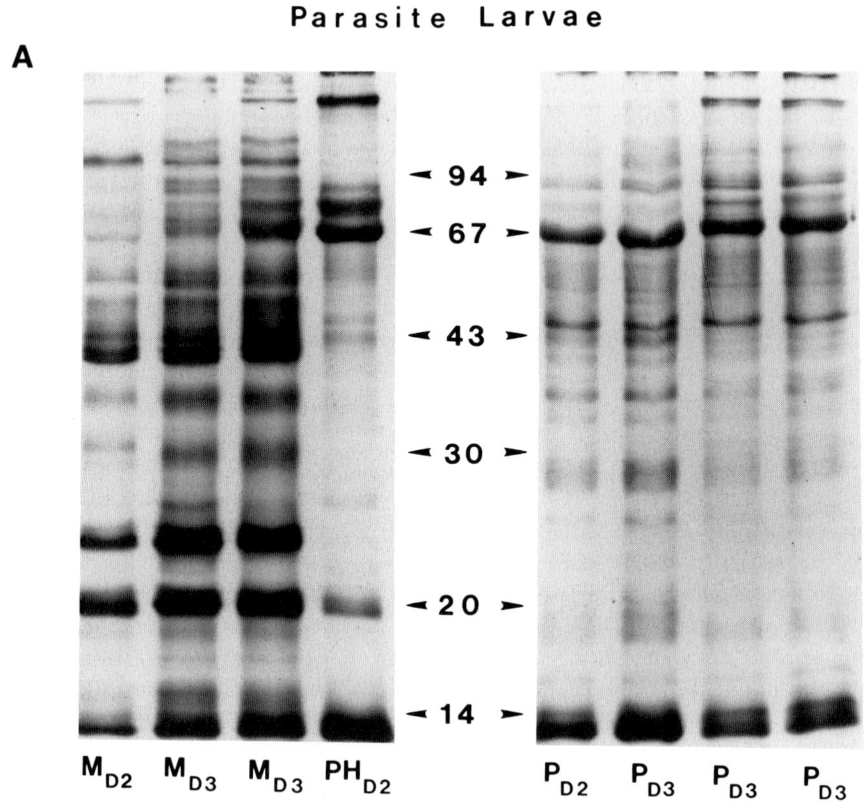

Figure 1 (A) Production of radiolabeled polypeptides by *Cotesia congregata* larvae dissected from host larvae on day 2 (D2) or 3 (D3) of the fifth instar, and cultured for 24 hr with sterile Grace's medium (formulated without methionine added) that was supplemented with [^{35}S]methionine to label proteins synthesized *de novo*. For culture procedures see de Buron *et al.* (1993) and Riddiford *et al.* (1979). Proteins recovered from the media (M), whole-body homogenates of the parasites (P), and parasite (*continues*)

Importantly, none of these proteins corresponds to any of the antibacterial proteins characteristically produced in nonparasitized tobacco hornworm larvae following injection of nonpathogenic bacteria (Beckage *et al.*, 1989; Kanost *et al.*, 1990; Faye and Hultmark, Volume 2, Chapter 2). Nor does injection of other DNA viruses induce their synthesis. For example, injection of lethal dosages of *Autographa californica* nuclear polyhedrosis virus (NPV) does not induce the same proteins (Fig. 2), indicating they are uniquely induced by parasitism. The tobacco hornworm represents a semipermissive host of the NPV virus and lives for an extended period following injection of

Figure 1 (*continued*) hemolymph (PH) were analyzed by SDS–PAGE and fluorography (for techniques, see Beckage *et al.*, 1987, 1989). Note that the media samples contain many radiolabeled polypeptides synthesized and secreted by the wasps; however, the major "parasitism-specific" bands detected in last-instar hosts (Beckage and Templeton, 1986) do not appear to be of wasp origin as determined by comparison to previously published gels (Beckage and Templeton, 1986; Beckage *et al.*, 1989). (B) Appearance of second-instar parasites similar to those used for the *in vitro* culture experiments described in Fig. 1a legend. Note the bulbous anal vesicle at the posterior end of each larva; these were punctured and hemolymph samples from several parasites were pooled and analyzed as described in Beckage *et al.* (1989). At this stage the mature second-instar parasites are ca. 3 mm in length and weigh 3 mg (Beckage and Riddiford, 1982). Photograph by Frances F. Tan.

this virus; injected larvae produce several novel proteins in response to NPV (Fig. 2) and ultimately die from the viral infection ca. 10–12 days later after molting to the fifth instar (Gretch *et al.*, 1991). A side-by-side comparison shows that the proteins detected in larvae following NPV injection clearly differ in MW from those induced following injection of the *C. congregata* PDV (Fig. 2) (Beckage *et al.*, 1987, 1989; Harwood, 1993; Harwood *et al.*, 1993a). Thus, these two classes of insect DNA viruses induce different sets of proteins to be expressed in the same "host." These differences are not surprising given that at least some of the novel proteins detected represent products of viral

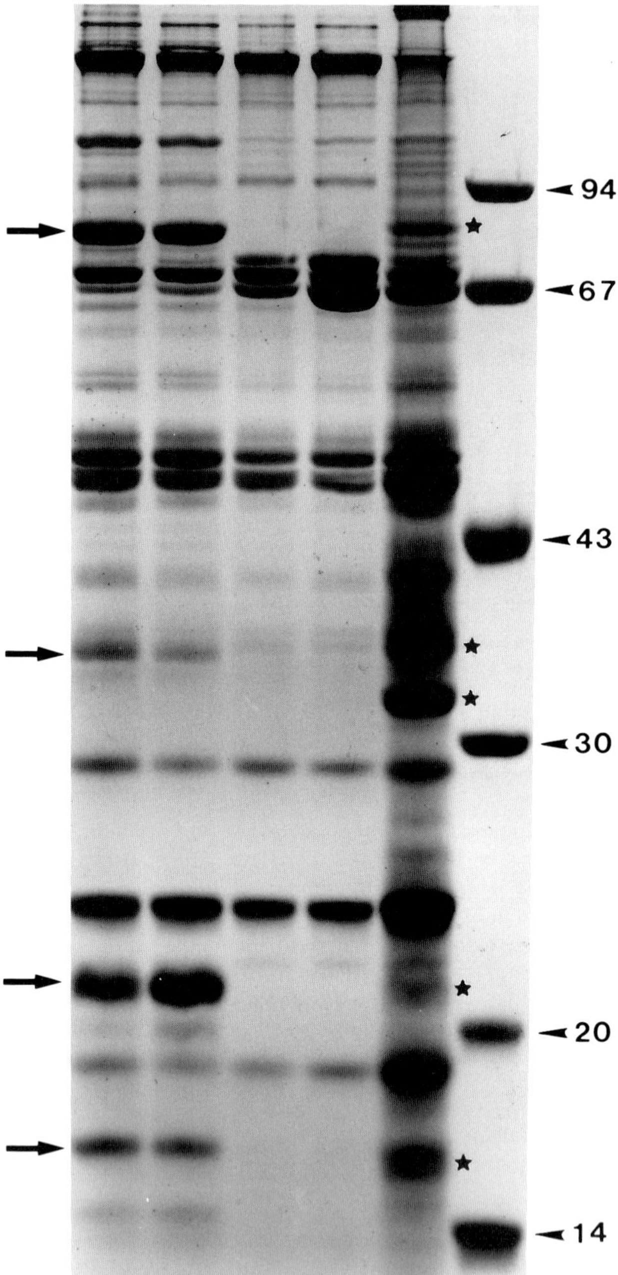

94
67
43
30
20
14
AcNPV
Control
PDV

genes (Harwood, 1993; Harwood *et al.*, 1993a, b) whose sequences may not be highly conserved among insect DNA viruses.

Unfortunately, what is currently lacking is a direct demonstration of the biological significance of "parasitism-specific" proteins and evidence that they play a critical role in the relationship between parasite and host. Since the "early" proteins appear prior to hatching of the first-instar parasites, the viruses and their gene products may serve as protective agents during this early vulnerable phase of the wasps' development, while during later stages the parasites themselves become capable of functioning in this capacity. Though this idea is attractive based on the relative time course of appearance of the proteins, strong experimental evidence for this hypothesis is lacking.

Conceivably at least some of the parasitism-induced proteins have a functional role in regulating some of the fundamental interactions characteristic of intimate or "habitual" host–parasite relationships during early phases of the relationship. Indeed, many of the parasitism-specific proteins characterized thus far are extensively glycosylated, similar to other surface-interacting molecules. Since glycosidated residues are now known to play major roles in cell–cell recognition and defense processes in other animal systems as well as plants, they may represent particularly critical molecular determinants of parasite–host interaction. For example, they could bind to parasite surfaces and prevent stimulation of hemocyte encapsulation; alternatively, they could interact with the cells themselves to elicit transformation to the nonencapsulative state (Stoltz, Volume 1, Chapter 8).

A reasonable corollary might be that required parasitism-specific proteins may not be produced at the normal rate or their function may be impaired in nonhabitual interactions culminating in the eventual death or demise of parasites in "alien" hosts. Cook *et al.* (1984) noted that production of a 55-

Figure 2 SDS–PAGE analysis of hemolymph samples collected from fourth-instar nonparasitized tobacco hornworm larvae 24 hr after injection of 10^6 pfu of the budded form of the *Autographa californica* nuclear polyhedrosis virus (AcNPV) diluted in sterile Grace's medium, control larvae injected with Grace's only, and larvae injected with ca. 2 wasp equivalents of filter-purified *Cotesia congregata* polydnavirus (Beckage *et al.*, 1993). The dose of AcNPV used is lethal for the larvae though they live for several days following virus injection (Gretch *et al.*, 1991). Hemolymph sampling and SDS–PAGE analysis were carried out as described previously (Beckage *et al.*, 1987). Arrows (left) denote novel proteins present only in samples collected from AcNPV-injected larvae; stars (right) denote the "parasitism-specific" proteins induced by injection of PDV (Beckage *et al.*, 1987, 1989; Harwood, 1993; Harwood *et al.*, 1993a). Note that the majority of the proteins induced by the two agents appear to have different mobilities; a few of the induced proteins in the AcNPV and PDV lanes appear to have roughly similar, though not identical, molecular weights. At least one of the novel proteins produced in the PDV-injected insects represents a viral transcript (see text).

kDa glycoprotein in host larvae parasitized by the ichneumon *Hyposoter exiguae* was correlated with the parasite's eventual successful development; moreover, the protein was not expressed in nonhabitual hosts in which the parasites were encapsulated. Their study further showed that the protein was inducible by injection of the wasp's polydnavirus into nonparasitized insects, whereas when the virus was inactivated this protein failed to be induced.

Similarly, in sphingids that are nonpermissive or partially permissive for the development of *C. congregata*, synthesis of the major early "polydnavirus-induced" protein appears altered (or markedly reduced) in cases where parasitism is not wholly effective and the parasites ultimately become encapsulated (Harwood, 1993; Harwood *et al.*, 1993b). However, the correlation of protein production with permissivity is not entirely clear-cut (as in a plus/minus correlation), as some production of the protein occurs in nonpermissive hosts, albeit at significantly reduced levels; moreover, Western blots show that their synthesis sometimes ceases prematurely in cases where the parasites are destined for encapsulation (Harwood, 1993; Harwood *et al.*, 1993b). Whether the reduced production of these parasitism-specific proteins or altered timing of their synthesis contributes to the parasites' encapsulation and failure to emerge from the host larva remains to be established; conceivably, the lack of protein production could represent an associated, but not causal, phenomenon. However, in species that have been determined to be fully permissive alternative host species, such as *Hyles lineata*, the full spectrum of both "early" and "late" proteins are produced (Harwood, 1993; Harwood *et al.*, 1993b). These observations support the hypothesis that their synthesis may be linked with successful parasitism of sphingid hosts by *C. congregata* but additional experimental evidence is required to yield definitive conclusions about their functional role(s). A future approach will be to inject antisense nucleotides into newly parasitized hosts, which would be expected to decrease parasite success if the complementary gene products are absolutely essential for successful parasitism.

Thus, our current working hypothesis in the *M. sexta* system is that the "early" proteins somehow act at the initial immunological interface between parasite and host. Where are they produced? Northern blots show the predominant "early" proteins are produced primarily by the fat body and hemocytes, although other tissues also synthesize the mRNAs to a lesser degree (Harwood, 1993; Harwood and Beckage, 1993). Experiments using *in vitro* culture of host fat body (Fig. 3) and hemocytes (data not shown) have verified that the "early" proteins are synthesized by fat body cells and secreted into the medium. Similarly, fat body and hemocytes represent the primary tissue sources of antibacterial proteins, while several others play a lesser role (Dunn, 1986; Faye and Hultmark, Volume 2, Chapter 2).

The next critical step will be to assay effects of purified parasitism-specific proteins on hemocyte behavior and other host physiological function to more

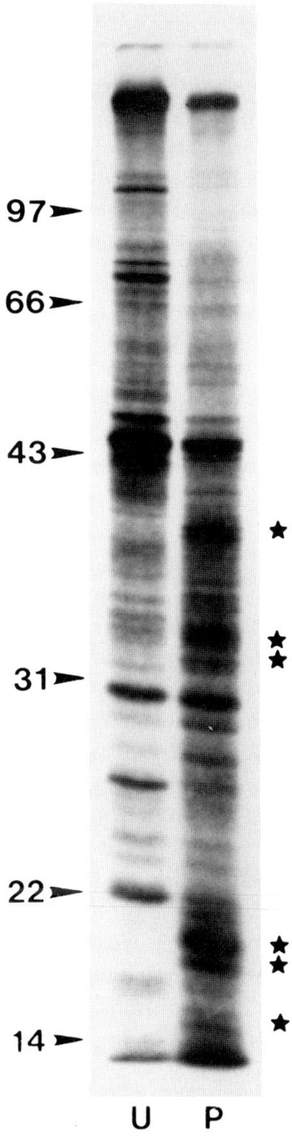

Figure 3 Results of SDS–PAGE and fluorograph analysis of proteins synthesized and secreted into Grace's medium by the fat bodies of unparasitized fourth-instar tobacco hornworm larvae (U) and newly parasitized (P) larvae 24 hr postparasitization. Culture techniques and fluorography were carried out as described for teratocyte cultures (de Buron *et al.*, 1993) and cultures of *M. sexta* tissues (Riddiford *et al.*, 1979). Stars (right) indicate bands present in the cultures of parasitized fat body that were undetected in the control media containing fat body from unparasitized larvae. Note multiple novel bands appearing in the media of the parasitized sample, in the <20 and 30–40 kDa MW ranges; these likely represent the early "parasitism-specific" proteins described in the text (Beckage *et al.*, 1987, 1989; Harwood, 1993).

directly address their biological role. Multicellular parasites obviously must utilize different mechanisms aside from direct cellular invasion to direct host biochemical changes, and these extracellular secreted molecules would appear to be one plausible mechanism of physiological interaction with the host. As mentioned earlier, some of the best clues about the nature of the requisite host–parasite molecular interactions may arise from examination of situations in which the host is completely nonpermissive, or exhibits only partial susceptibility to parasitism, so that the parasites begin developing but then become encapsulated before emerging. Several scenarios are plausible, and side-by-side comparisons of the "suite" of biochemical changes occurring in susceptible, nonpermissive, and semipermissive hosts may facilitate discrimination of the most critical genes whose expression invariably is linked to "successful" parasitism (or, alternatively, constitute an effective "antiparasite protection"). Another possibility is that the absence (or at least reduced synthesis) of "incompatible" host proteins may represent what is required. In the former case, presumably the normal expression of any critical genes would be hampered (possibly totally inhibited) in a nonpermissive host environment, whereas expression of noncritical genes would not show any modifying effect of endoparasitism. Nonpermissive hosts thus would appear similar to nonparasitized animals, whereas "successfully" parasitized individuals should show an attenuation or silencing of expression of any truly noncompatible protein whose presence might target the parasite for encapsulation. These might represent "assassin" genes, the expression of which must be knocked out by the parasite(s) to prevent a potentially lethal situation from developing.

In hornworm larvae injected with large amounts of PDV, the "early" proteins are induced, but other unique proteins eventually appear, which are never seen in naturally parasitized hosts or nonparasitized larvae. Specifically, administration of two or more wasp equivalents of purified *C. congregata* polydnavirus induces the "early" proteins described earlier, but also some proteins not normally detectable following natural exposure to wasp parasitoids; at 6–10 days postinjection *large* amounts of heavily glycosylated 18- to 22-kDa MW polypeptides are detectable in these PDV-treated "hosts" (Beckage *et al.*, 1993). Either these particular proteins are not inducible during natural parasitism, or for some reason their synthesis is suppressed in hosts whose parasites develop normally. Alternatively, they may be inducible only by injection of megadoses of polydnavirus not typically delivered by female wasps to hosts during parasitization. Intriguingly, these enigmatic proteins begin to appear when the larvae begin to show major pigmentation anomalies due to PDV injection, with the integument acquiring a distinctive rosy coloration (Beckage *et al.*, 1990), but whether the two phenomena, that is, the appearance of the proteins and pigments, are physiologically linked is unknown. Similar pigmentation changes occur in naturally parasitized larvae, but they are often exaggerated in larvae injected with large amounts of PDV. Additional studies are under way to characterize these unusual proteins and assess their possible relevance to other physiological changes induced by PDV injection.

Other possible roles of the "early" induced proteins include induction of host metabolic or behavioral effects, or modulation of the host's endocrine or neuroendocrine system culminating in induction of host arrest. Such action appears particularly plausible for the "late" proteins produced during the period when the host shows multiple metabolic and hormonal deviations from the normal nonparasitized state. However, experimental evidence for a role of peptides or proteins in altering the host's endocrine status, behavior, and developmental programming remains a "missing link" (Beckage, 1990) and their potential regulatory role in governing such changes remains purely speculative at this point.

Amazingly, the degree to which insect parasites themselves invoke changes in the hemolymph protein content of their host has received surprisingly little attention. The paucity of experimentation in this area sharply contrasts with rapid recent progress in identification of the spectrum of molecules secreted by parasites of vertebrate hosts; literally thousands of recent papers have described membrane-associated proteins and secretory/excretory antigens in many species, usually with a focus on exploiting that information for the purposes of disease diagnosis and vaccine development (see the following). Indeed, given the lack of research directed toward specific characterization of parasite- or parasitoid-secreted molecules in insect hosts, one might logically assume that endoparasitic wasps represent a "silent partner" or minor component in the physiological interaction between host and parasite. The comprehensive picture likely includes contributions from several interacting components, including the parasites as well as their associated teratocytes and polydnaviruses, plus the tissues of the host whose physiological functions are variously modified during parasitism.

For example, stage-specific proteins have been shown to be produced by embryos of *Microplitis croceipes* (Ferkovich and Dillard, 1986; Tilden and Ferkovich, 1987) and *Copidosoma floridanum* (Baehrecke *et al.*, 1992), but whether any embryo-derived proteins are secreted into the surrounding host tissues was not determined in the studies cited. In the tephritid fruitfly *Anastrepha suspensa,* the serosal cells (i.e., teratocytes) of the braconid wasp *Biosteres longicaudatus* produce a 24-kDa molecular weight parasitism-specific polypeptide detected in the pharate pupal host; the parasites also secrete the same protein, suggesting it could be produced by all parasite-derived cell lineages (Lawrence, 1990). Other teratocyte-secreted molecules are described by Dahlman (1991), Dahlman and Vinson (Volume 1, Chapter 7), and de Buron *et al.* (1993). In the latter paper, multiple (>30) proteins were observed to be synthesized and secreted by teratocytes of *C. congregata;* due to the large numbers (literally hundreds of thousands) of cells present in each host, their contribution to the host hemolymph milieu would be significant.

In one case, a peptide previously characterized as "parasitism-specific" was recently detected in nonparasitized individuals of the same species (Hayakawa, 1992). Thus, the previous designation of this peptide as "parasitism-specific" was a misnomer (Hayakawa, 1990, 1992) with the author now concluding that parasitism causes anomalous misexpression of this endogenous molecule at a stage later than it normally would be detected (Hayakawa, 1992) (see the following). In contrast, the proteins cited earlier appear to be truly "parasitism-specific" and cannot be "turned on" in nonparasitized insects subjected to various forms of physiological stress or manipulation, aside from parasitism.

Because of their unique association with parasitized insects, a future avenue for exploitation of parasitism-specific host hemolymph protein changes may be to utilize protein-specific probes to confirm parasitization in cases where the hosts are simply too small to be easily dissected. Western slot blots or ELISA tests then could be used to develop insect "squash blots" to estimate the percentage of parasitized insects in the host insect population. In agricultural entomology, an accurate assessment of the percentage of insects parasitized following release of biological control agents is critical to monitoring the success of biocontrol programs, yet testing is difficult to carry out, particularly with large populations of small insects. Just as parasite-associated antigens are now being utilized for detection of parasitic infection in mammalian hosts, and for detection of arthropod-borne parasites (e.g., malaria, leishmania) in vectors, so too might the parasitism-specific antigens described here be used for diagnostic purposes in screening insects for evidence of parasitism.

Another strategy for "technology transfer" may be to utilize gene transfer to exploit genes encoding parasitism-specific proteins for genetic engineering of parasites, possibly using polydnavirus promoters directing "early" expression of crucial genes needed for successful interaction. Site-directed mutagenesis first can be used to identify the most critical domains needed for appropriate interaction, which then would be transferred to other parasites or polydnaviruses; one goal might be to render a species having an extremely narrow host range more promiscuous, for example. Molecular "assasin" molecules might be formulated for targeting to host hemocytes, thereby rendering them incapable of mounting a successful defense. Although these strategies may be considered futuristic at this juncture, recent successes in gene transfer and transformation in other biological control agents such as beneficial predatory mites (Presnail and Hoy, 1992) are encouraging and lead to the optimistic prediction that this type of biotechnological approach might be made feasible for parasitic species in the next decade.

III. Parasitic Effects on Endogenous Proteins, Enzymes, and Peptides

Aside from stimulating *de novo* synthesis of new proteins, parasitism frequently also dramatically affects the titers of endogenous host hemolymph proteins. These changes may arise via alterations in the rate(s) of transcription of host genes or translational events. Not surprisingly, the levels of total hemolymph protein often are ultimately affected, with some species of host insects showing a decrease in total hemolymph protein whereas others display a significant increase. Aside from having direct effects on host protein expression, parasitism may also alter host blood volume, thereby indirectly affecting protein concentration; hemolymph volume may in turn be altered via effects on diuresis, food and fluid intake, or other processes affecting the dilution of endogenous proteins in the blood. Thus, host hemolymph protein concentration reflects the summation of action of multiple physiological factors.

How might the expression of host genes be affected by parasitism? Several hypothetical routes are plausible. Some alterations may arise via mechanisms mediated by effects on transcription factors, helicases, RNA processing enzymes, or other modifiers that ultimately impact the rate of host gene expression. Translation and protein processing may also be modified (see the following). Parasitic effects on host nutrition and the relative abundance of amino acids and other molecules required for synthesis of specific enzymes, peptides, and constitutive or transport proteins might also be expected to have a damping effect on the rate of host protein synthesis. Obviously one complicating factor is that the feeding behavior of the host frequently is altered during parasitism. This influences the rate of nutrient intake relative to that occurring in nonparasitized insects, and many species of parasitized insects show a decline or complete cessation of food consumption during the final stages of parasitism. Levels of amino acids are often reduced when the host insect is experiencing semistarvation and is feeding at a significantly slower rate, thus decreasing nutrient transfer from the gut to the hemolymph. An across-the-board decrease in all hemolymph proteins logically might ensue; however, in many cases the effects of parasitism are protein-specific, with some proteins showing no alterations or quantitative effects, whereas others show significant increases or decreases during parasitism. This selectivity strongly argues that a nonspecific lowering or shutdown in the rate of host protein synthesis is not responsible. Instead, the effects appear highly specific and targeted to certain genes whose levels of expression are variously modified during parasitism.

One class of proteins commonly affected during parasitism of larval stage hosts is the arylphorins. These high-molecular-weight storage proteins are synthesized during the larval feeding stage and function as an amino acid reservoir for use during developmental restructuring (Webb and Riddiford, 1988a,b). Labeling studies have shown that the amino acid components of arylphorin are periodically broken down and then incorporated into the cuticle and other tissues during larval molting and metamorphosis, providing evidence for their role in amino acid "reprocessing" (Webb and Riddiford, 1988a). During starvation and other nonfeeding periods such as molting, the synthesis of arylphorin ceases transiently, but is quickly resumed upon resumption of feeding (Webb and Riddiford, 1988a,b).

Two distinct trends are evident in parasitized insects. In a few instances parasitism causes precocious expression of high levels of arylphorins during what would normally be the penultimate instar, as exemplified by wasps belonging to the genus *Chelonus*, which invariably induce this phenomenon in host lepidopteran larvae (Jones *et al.*, 1985, 1986; Jones, 1989; Kunkel *et al.*, 1990). In others, hemolymph arylphorin levels are markedly depressed during parasitism, as seen in lepidopteran larvae parasitized by several species of *Cotesia*. For example, in fifth-instar larvae of *M. sexta* parasitized by *C. congregata*, the host's hemolymph shows a temporal decline in the concentration of arylphorin following host ecdysis to the fifth instar, whereas that of similarly aged nonparasitized animals exhibits a dramatic fourfold increase in arylphorin content (Beckage and Kanost, 1993). Interestingly, despite the lowered level of hemolymph arylphorin, the fat body of parasitized larvae contains roughly similar arylphorin mRNA levels (per mg tissue) as those detectable in tissues excised from unparasitized prewandering larvae, as determined by Northern hybridizations carried out using radiolabeled *M. sexta* arylphorin probes to analyze fat body mRNAs (Beckage and Webb, unpublished data). The latter study also showed that these mRNAs are translatable *in vitro* using the techniques of Webb and Riddiford (1988b). In such situations where normal mRNA levels continue to be produced although hemolymph titers of the protein are markedly depressed, one logical conclusion is that arylphorin expression may be regulated posttranscriptionally. Thus, similar to heat shock (and other stress-related) proteins, vertebrate transport proteins, and numerous viral gene products (Ilan, 1987; Thach, 1992; Hershey, 1992), arylphorin protein expression during parasitism may be regulated post-transcriptionally, rather than at earlier steps influencing the rate of accumulation of host mRNAs. Perhaps translational control is logical for the insect storage proteins, whose synthesis may be quickly up- or down-regulated synchronous with rapid changes in nutritional state; during nonfeeding periods, its accumulation stops (Webb and Riddiford, 1988a). Plausible mechanisms might involve fast-acting translation blocking agents or other factors. As other authors have noted a decrease in host hemolymph storage protein titers following parasitization by related species of *Cotesia* (Junnikkala, 1966), the mechanisms effecting a suppression of arylphorin accumulation may be wide-ranging among species.

Because storage proteins represent a large proportion of the total hemolymph protein, particularly during the prewandering feeding period immediately preceding the onset of metamorphosis, it is not surprising to find that parasitism frequently causes a lowering of the total protein content of the blood during this period. Hypoproteinemia has been reported in a wide range of parasitized insects, indicating that it likely is a common effect of parasitism (Vinson and Iwantsch, 1980). Effects are often complex, that is, in fourth-instar newly parasitized *M. sexta* larvae, the total protein concentration initially is elevated following oviposition (presumably due to induction of the parasitism-specific proteins) then is significantly decreased relative to the protein concentration measured in nonparasitized larvae, following molting to the fifth instar (Beckage *et al.*, 1989). This latter declining trend likely reflects the relative decrease in arylphorin titers (Beckage and Kanost, 1993). Thus, both transitory hyper- and hypo-regulatory effects can be observed in the same host.

Might another factor contributing to host hypoproteinuria be parasite uptake of host hemolymph proteins? Parasite eggs do not appear to absorb or utilize host apoproteins or their degradation products (Ferkovich and Dillard, 1986) though the situation might be different with larvae. Previously we had suspected that the rapidly developing second-instar *C. congregata* larvae sequestered arylphorin following uptake from the host hemolymph. However, Western blots using antibodies to *M. sexta* arylphorin showed that the latter protein was undetectable in hemolymph or whole-body homogenates of the wasps; although the wasps appeared to have endogenous storage proteinlike molecules with molecular weights similar to *M. sexta* arylphorin subunits, the wasp proteins did not cross-react with the *M. sexta* arylphorin antibodies (Beckage *et al.*, 1989). However, the possibility still exists that host proteins including arylphorin are consumed, rapidly degraded, then reprocessed by the wasps. Moreover, parasitoids that are cultured *in vitro* frequently require the presence of protein- or lipoprotein-rich supplements (in the form of fetal bovine serum, bovine serum albumin, etc.) in the medium to facilitate normal development (Greany, 1986; Ferkovich *et al.*, 1991; Pennachio *et al.*, 1992). It has thus been inferred that acquisition of host proteins might be critical in fulfilling the parasites' normal nutritional requirements, although the specific roles of host proteins or lipoproteins have yet to be established.

Arylphorin is not the only host hemolymph protein affected by parasitism. Hemolymph serpin levels also are dramatically lowered during the final stages of parasitism by *C. congregata* (Beckage and Kanost, 1993). Serine proteases are intimately involved in modulating activity of phenoloxidases and participate as well in other enzyme cascades; thus, induction of some modifications in host biochemical regulatory pathways might be expected to ensue concomitant with a reduction in serpin titers. Antibacterial proteins also frequently are lowered during parasitism, thus enhancing sensitivity to pathogenic infection in many species (Brooks, Volume 2, Chapter 11). Parasitized *M. sexta* larvae are more susceptible to bacterial infection as compared to nonparasitized larvae, and both cell- and humoral-mediated immune mechanisms likely are suppressed during parasitism (Ross and Dunn, 1989).

However, the vast majority of host hemolymph proteins in *M. sexta* remain unaffected by parasitism and effects of parasitism appear highly selective. In contrast to the depression in levels of arylphorin and serpins, insecticyanin and lipophorin are present at near-normal concentrations in the blood (Beckage and Kanost, 1993). The absence of any effect on lipophorin is surprising for two reasons. First, previous studies have noted a reduction in lipid and elevation in glycogen content of the fat body of parasitized larvae compared to nonparasitized larvae (Thompson *et al.*, 1990); this effect on stored lipids might logically be expected to alter the titers of lipid transport proteins circulating in the blood, yet the levels remain unaltered. The lack of effect on lipophorin was also unexpected given the reduced weights of parasitized larvae. Thus, parasitized larvae provide some clues as to its mode of regulation in that the lipophorin titer of 3- to 4-g host larvae with newly emerged wasps is the same as that reported for 8- to 10-g nonparasitized prewandering larvae, suggesting that larval weight per se has no major regulatory influence (Beckage and Kanost, 1993). Lipophorin has recently been shown to significantly reduce hemocyte adhesion behavior (Coodin and Caveney, 1992), leading this author to speculate: Might the relative abundance of this protein be related to the characteristic "transformations" in host hemocyte behavior occurring during parasitism?

Another host protein frequently affected by parasitism is the enzyme phenoloxidase (also see Christensen and Severson, Volume 1, Chapter 11; Dahlman and Vinson, Volume 1, Chapter 7). In insects attacked by ichneumonid (e.g., *Hyposoter exiguae*) and braconid (*Cotesia congregata*) wasps, polydnaviral invasion of host hemocytes elicits a reduction of phenoloxidase activity (Stoltz and Cook, 1983; Beckage *et al.*, 1990), possibly by affecting transcription of the host phenoloxidase gene(s), or subsequent translation of the mRNAs following virus uptake by the cells. Another interpretation was presented by Sroka and Vinson (1978), who attributed phenoloxidase inhibition in parasitized *Heliothis* larvae to tyrosine deficiency; in other species, phenoloxidase inhibitors may also be present in parasitized larvae (Führer and Willers, 1986; Kitano *et al.*, 1990).

Baculoviruses, which invade host hemocytes similar to polydnaviruses, also cause a suppression of hemolymph phenoloxidase activity and melanization (Andersons *et al.*, 1990), suggesting that a wide range of viruses may employ a common mechanism to down-regulate phenoloxidase production in infected insects. Possibly any type of viral or bacterial infection that has adverse effects on hemocytes, including parasitism by entomophagous nematodes and their associated bacteria (Yokoo *et al.*, 1992), may likewise inhibit phenoloxidase activity, since the enzyme normally is produced by these cells and released into the plasma. Hemocyte depletion or pathology might also contribute to a lowered capacity of the host to produce antibacterial proteins, as well as others of hemocyte origin.

Enzymes critical to endocrine regulation are also impacted by parasitism, thereby affecting the host's developmental program. The enzymes known to be affected by parasitism include juvenile hormone esterase, which is either depressed to near nondetectable levels (Beckage and Riddiford, 1982; Dahlman *et al.*, 1990; Hayakawa, 1990; Strand *et al.*, 1990; Zhang *et al.*, 1992) or precociously increased (Jones *et al.*, 1986), and ecdysone 20-monooxygenase may also be affected (Beckage and Templeton, 1986). Both of these enzymes are normally produced by the fat body, suggesting once again that this tissue represents a major site where effects of parasitism are expressed or "transduced," thereby modifying the host's physiology and development. In many species, the fat body of parasitized larvae often shows major debilitative effects of parasitism in that it fails to proliferate at the normal rate, perhaps thus accounting for some of the inhibitory effects of parasitism on enzyme production. Again, parasitic effects appear targeted to specific enzymes, as exemplified by the fact that in *M. sexta* hemolymph ketoreductase activity appears unaffected by parasitism (D. B. Gelman and N. E. Beckage, unpublished). This enzyme converts the 3-dehydroecdysone produced by the prothoracic glands to ecdysone (Kiriishi *et al.*, 1990) and appears to be aloof to any modifying effect of parasitism, in contrast to JH esterase titers in the same larvae, which are dramatically reduced. Thus, host hormonal changes may arise from parasitic effects on the activity of regulatory enzymes in the pathways leading from precursors to hormones to metabolites, but not all enzymes are affected equally.

Hayakawa (1990, 1991) reported the presence of a growth-blocking juvenile hormone esterase suppressive factor in the hemolymph of last-instar larvae of the armyworm *Pseudaletia separata* parasitized by the wasp *Apanteles* (=*Cotesia*) *kariyai*. Although initially he hypothesized that this factor of 4500 kDa MW was uniquely present in parasitized larvae (Hayakawa, 1990, 1991) and acted to prevent their metamorphosis by repressing JH esterase activity, a more recent study (Hayakawa, 1992) indicated that the factor is normally present in penultimate-stage and younger nonparasitized larvae but disappears in the last instar. Therefore, parasitism extends the period when the peptide is produced, presumably by suppressing metamorphosis.

This peptide is hypothesized to be required for maintenance of larval characteristics and speculated to be a hormonelike molecule (Hayakawa, 1992). However, another interpretation is plausible. Injection of purified peptide into nonparasitized larvae causes inhibition of food consumption and growth (Hayakawa, 1991), and lack of normal feeding behavior might be expected to depress levels of hemolymph juvenile hormone esterase. Previous studies have shown that starvation of nonparasitized larvae elevates the JH titer and depresses JH esterase, thereby delaying metamorphosis (Cymborowski *et al.*, 1982). Thus, effects of the peptide on JH esterase could be indirect, mediated via other mechanisms such as reduction of the host's feeding activity.

IV. Comparisons with Parasitism- and Virus-Induced Changes in Other Systems

Similar interactions involving proteins and peptides occur in a wide range of parasitized animals. In mammals, parasitism induces major alterations in the proteins and other components of the host's blood, frequently affecting levels of the host's endogenous proteins in conjunction with inducing new proteins (some of which are parasite-derived) to appear in the circulation. A variety of parasite-secreted molecules are released into the host milieu, and some are now utilized as diagnostic tools to document the occurrence of infection. Several of these so-called membrane-associated antigens, exoantigens, or excretory/secretory products have been proven to have a significant impact on their host's physiology (Turco, 1990; Blaxter *et al.*, 1992; Loker *et al.*, 1992; Noda, 1992). In addition to being extremely useful for diagnosis of acute and chronic parasitemias, some of these molecules are also being exploited in development of new methods of vaccine formulation based upon these parasite-derived antigens (see references in Coombs and North, 1991).

An exceptionally well-documented example of the induction of synthesis of novel peptides by parasitism that subsequently interfere with normal host function is illustrated by studies of schistosome-infected snails (summarized in de Jong-Brink, 1992). The host experiences parasitic castration, and the mechanisms causing cessation of reproduction in the fresh-water snail *Lymnaea stagnalis* have now been identified. The primary mediator of castration is a novel peptide termed schistosomin, which has a pronounced inhibitory effect on reproduction of the host snail (Schallig *et al.*, 1991; de Jong-Brink, 1992), and has both peripheral and central effects on the physiology of the host snail. This peptide is induced by the presence of schistosome cercariae in the host, conceivably by humoral factors secreted by the parasites. It has an inhibitory effect on reproduction by antagonizing the action of the snails' female gonadotropic hormones (caudodorsal cell hormone and calfluxin), suppressing their stimulatory action on, respectively, the gonad and the albumen gland, the female accessory sex gland that produces perivitellin fluid to nourish the developing embryo (de Jong-Brink *et al.*, 1988). Schistosomin also has central effects on the neuroendocrine centers regulating reproduction and growth, specifically by acting on the caudodorsal cells and light green cells (de Jong-Brink, 1992; de Jong-Brink *et al.*, 1992). Although it previously was speculated that ecdysteroids secreted by the parasites induce the synthesis of schistosomin by host tissues, experiments suggest this is not the case (Schallig, 1991). Thus, the proximate inducing factor(s), as well as the tissues producing them, await future isolation and characterization. Conceivably, the identification of these regulatory molecules, plus the modulatory factors influencing schistosomin production, ultimately may yield information relevant to development of new strategies to control proliferation of uninfected snails.

In parasitized crabs, proteins that appear "parasitism-specific" (Herberts, 1982) have been isolated; some of these appear to inhibit host spermatogenesis and reproduction (Rubiliani, 1983), but considerably less is known about those molecules compared to schistosomin. In insects, several species of insect parasitoids cause host castration by inhibiting normal development of the gonads during the larval stage (Junnikkala, 1985; Reed-Larsen and Brown, 1990). Though it is clear that the agents causing this effect are transferred to the host during a defined temporal window at parasitization, virtually nothing is known about the nature of the molecular mediators suppressing differentiation of the testes (Reed-Larsen and Brown, 1990; Brown and Kainoh, 1992). Proteins (or peptides) may mediate castration as occurs during suppression of spermatogenesis in crabs infected with rhizocephalan barnacle parasites; the latter parasites secrete molecules (estimated to be 25 to 30 kDa molecular weight) that act on the host gonad directly, as well as having indirect effects mediated via the sinus gland and the host neurosecretory system (Rubiliani, 1983).

Studies of other parasite taxa have confirmed that they secrete an astounding number of potent molecules that include, aside from hormones, a battery of proteases and other proteins (Coombs and North, 1991; Beier *et al.*, 1992), chitinases (Schlein *et al.*, 1992), cyclooxygenases (Kaiser *et al.*, 1992), growth-promoting factors (Phares and Watts, 1988), and other peptides (Thorndyke, 1990). These parasite-generated "cocktails" likely influence the host's physiology to varying degrees, most often for the parasite's ultimate benefit. Some integral membrane proteins appear to be particularly critical in the host–parasite interface; these are impregnated in the parasites' surface and act protectively to prevent immunorecognition processes from being mobilized to destroy the parasites (Blaxter *et al.*, 1992) or mediate cell–cell recognition during invasion of host cells. For example, surface lectins may mediate contact and cytoadherence with specific subsets of host cells (Thea *et al.*, 1992). Antigen shedding also occurs and allows the parasite to continuously renew the molecules exposed on its surface, allowing new gene products to be expressed during different stages of parasitism. Receptors for host hormones and growth factors may also be expressed on the parasite's surface, thus facilitating a response to host hormones or immune modulators such as tumor necrosis factor released during an immune response to parasitic infection (Amiri *et al.*, 1992). Unfortunately, our knowledge of the antigens associated with, and secreted by, parasites developing in insect hosts are considerably less well characterized in comparison to those produced by protozoan, helminth, and other parasites that invade and proliferate in mammalian hosts.

Similar to parasites, many viruses likewise affect transcription and/or translation of host gene products, and may do so without perturbing vital cell functions. For example, studies of lymphocytic choriomeningitis virus action in pituitary cells suggest the virus interferes with transcription of growth hormone genes, via an effect on growth hormone transactivator factor, but expression of prolactin and other required genes remains unaffected (de la Torre and Oldstone, 1992). The latter authors concluded that a variety of nonlytic viruses that persistently infect cells of the endocrine, immune, or nervous tissues may cause selective effects on hormone production, immune function, and neurotransmitter release, which heretofore were not recognized as having a viral etiology.

How might these findings relate to the phenomena occurring in parasitized insects? Many of the physiological changes observed in insect hosts may reflect action of these agents associated with insect parasites. Conceivably, the polydnaviruses have similar nonlytic, but nonetheless highly modulatory, roles in the target lepidopteran insects injected with these viruses. Moreover, several recent studies have verified that polydnavirus DNA "persists" in host insect cells for the duration of the association between parasite and host, as demonstrated using dot or slot blots for detection of "lingering" viral DNAs (Stoltz *et al.*, 1988; F. F. Tan and N. E. Beckage, unpublished; Strand *et al.*, 1992). Hence, "latent" infection of cells of the fat body and other tissues could influence host gene expression from the moment of the initial contact between parasite and host until the wasps successfully mature, so that small dosages of PDV may affect host endocrine regulation as well as immunity (Dushay and Beckage, 1993).

What remains to be unraveled are the potential mechanisms involved in modifying host gene expression. Clearly, the fact that some proteins are selectively depressed while others are maintained at titers approaching normal levels argues strongly that parasitism has complex effects upon the host genome. Viral transcription factors and translational control mechanisms affect host gene expression in a wide range of animals (not to mention plants), suggesting that polydnaviruses, acting independently or coordinately with the parasites carrying them, may have factors associated with them capable of influencing host gene expression for the duration of the parasitic relationship.

Thus, host genes may be regulated either at the transcriptional level (Ptashne, 1988) or at subsequent points in the protein synthetic pathway through alterations in the rate of translation of host mRNAs involving initiation, elongation, and processing of the nascent proteins. The latter may occur via direct interruption of the normal translation process, or indirectly via suppression of protein synthesis due to competition of the host and its parasites for available nutrients and the amino acids required for protein synthesis. Since the effects of parasitism often are highly specific, the latter phenomenon appears unlikely to be the major factor accounting for many of the observed changes in the hemolymph protein profile in host insects. Although the identification of the mechanisms of transcriptional/translational control and isolation of transcription factors, translational repressor proteins, and other modifying elements has progressed rapidly in mammalian and other systems, we unfortunately have only rudimentary information about these processes in insects. Obviously great challenges await us with regard to deciphering the mechanisms whereby parasitism alters the hemolymph proteins of insect hosts, and at present we can only speculate about the frequency of their occurrence and potential role(s) at the interface between parasite and host. We can only assume that many of the observed host changes likely have significant selective advantages for the parasites involved (as exemplified by the advantageous induction of insect host behavioral changes reported herein by Horton and Moore), and work from that point forward in assessing their roles. Although many points in this review have been obviously and deliberately highly speculative, hopefully they will stimulate new directions in research and provide a strong impetus for future molecular studies of these complex, though incredibly fascinating, interactions in insect hosts.

Acknowledgments

The author sincerely acknowledges Drs. Bruce Webb, Davy Jones, Steve Harwood, Mitch Dushay, and several other reviewers for their critical comments; Ms. Frances Tan, Ms. Therese Reignier, and Ms. Dorothy Nesbit for technical assistance. Many undergrads assisted with various aspects of this work; Anthony Nguyen, Lisa Cherubin, Roni Lane, and Trang Vo deserve special mention. Studies of the *Manduca sexta* system cited in this review were funded by grants from the NSF, USDA, and University of California-Riverside, all of which are gratefully acknowledged by the author.

The author dedicates this review, as well as her efforts in assembling these volumes, to her son Ross, in joyful celebration of his tenth birthday and his infectious enthusiasm for nature of all sorts; and acknowledges family members, especially her mother Pauline and brother Paul, and many friends for their enduring and much-appreciated support.

References

Amiri, P., Locksley, R. M., Parslow, T. G., Sadick, M., Rector, E., Ritter, D., and McKerrow, J. H. (1992). Tumour necrosis factor alpha restores granulomas and induces parasite egg-laying in schistosome-infected SCID mice. *Nature (London)* **356:**604–607.

Andersons, D., Gunne, H., Hellers, M., Johansson, H., and Steiner, H. (1990). Immune responses in *Trichoplusia ni* challenged with bacteria or baculoviruses. *Insect Biochem.* **20:**537–543.

Baehrecke, E. H., Strand, M. R., Williamson, J. L., and Aiken, J. M. (1992). Stage-specific protein and mRNA synthesis during morphogenesis of the polyembryonic parasitoid *Copidosoma floridanum* (Ashmead) (Hymenoptera: Encyrtidae). *Arch. Insect Biochem. Physiol.* **19:**81–92.

Barral-Netto, M., Barral, A., Bronell, C. E., Skeiky, Y.A.W., Ellingsworth, L. R., Twardzik, D. R., and Reed, S. G. (1992). Transforming growth factor-B in leishmanial infection: A parasite escape mechanism. *Science* **257:**545–548.

Barras, D. J., Kisner, R. T., Lewis, W. J., and Jones, R. T. (1972). Effects of the parasitoid, *Microplitis croceipes,* on the haemolymph proteins of the corn earworm, *Heliothis zea. Comp. Biochem. Physiol. B* **43B:**941–947.

Beckage, N. E. (1990). Induction of host endocrine and hemolymph protein alterations in tobacco hornworm larvae parasitized by *Cotesia congregata. Adv. Invertebr. Reprod.* **5:**149–155.

Beckage, N. E. (1993). In preparation.

Beckage, N. E., and Kanost, M. R. (1993). Effects of parasitism by the braconid wasp *Cotesia congregata* on host hemolymph proteins of the tobacco hornworm, *Manduca sexta. Insect Biochem. Mol. Biol.* (in press).

Beckage, N. E., and Riddiford, L. M. (1978). Developmental interactions between the tobacco hornworm *Manduca sexta* and its braconid parasite *Apanteles congregatatus. Entomol. Exp. Appl.* **23:**139–151.

Beckage, N. E., and Riddiford, L. M. (1982). Effects of parasitism by *Apanteles congregatus* on the endocrine physiology of the tobacco hornworm, *Manduca sexta. Gen. Comp. Endocrinol.* **47:**308–322.

Beckage, N. E., and Templeton, T. J. (1986). Physiological effects of parasitism by *Apanteles congregatus* in terminal-stage tobacco hornworm larvae. *J. Insect Physiol.* **32:**299–314.

Beckage, N. E., Templeton, T. J., Nielsen, B. D., Cook, D. I., and Stoltz, D. B. (1987). Parasitism-induced hemolymph polypeptides in *Manduca sexta* (L.) larvae parasitized by the braconid wasp *Cotesia congregata* (Say). *Insect Biochem.* **17:**439–455.

Beckage, N. E., Metcalf, J. S., Nielsen, B. D., and Nesbit, D. J. (1988). Disruptive effects of azadirachtin on development of *Cotesia congregata* in host tobacco hornworm larvae. *Arch. Insect Biochem. Physiol.* **9:**47–65.

Beckage, N. E., Nesbit, D. J., Nielsen, B. D., Spence, K. D., and Barman, M. A. (1989). Alterations of hemolymph polypeptides in *Manduca sexta:* A two-dimensional electrophoretic analysis and comparison with major bacteria-induced proteins. *Arch. Insect Biochem. Physiol.* **10:**29–45.

Beckage, N. E., Metcalf, J. S., Nesbit, D. J., Schleifer, K. W., Zetlan, S. T., and de Buron, I. (1990). Host hemolymph monophenoloxidase activity in parasitized *Manduca sexta* larvae and evidence for inhibition by wasp polydnavirus. *Insect Biochem.* **20:**285–294.

Beckage, N. E., Tan, F. F., Schleifer, K. W., Lane, R. D., and Cherubin, L. L. (1993). Characterization and biological effects of *Cotesia congregata* polydnavirus on host larvae of the tobacco hornworm, *Manduca sexta. Arch. Insect Biochem. Physiol.* (in press).

Beier, J. C., Vaughan, J. A., Madani, A., and Noden, B. H. (1992). *Plasmodium falciparum:* Release of circumsporozoite protein by sporozoites in the mosquito vector. *Exp. Parasitol.* **75:**248–256.

Bieber, A. J., Snow, P. M., Hortsch, M., Patel, N. H., Traquina, Z. R., Schilling, J., and Goodman, C. S. (1989). *Drosophila* neuroglian: A member of the immunoglobulin superfamily with extensive homology to the vertebrate neural adhesion molecule L1. *Cell (Cambridge, Mass.)* **59:**447–460.

Blaxter, M. L., Page, A. P., Rudin, W., and Maizels, R. M. (1992). Nematode surface coats: Actively evading immunity. *Parasitol. Today* **8:**243–247.

Blissard, G. W., Fleming, J.G.W., Vinson, S. B., and Summers, M. D. (1986). *Campoletis sonorensis* virus: Expression in *Heliothis virescens* and identification of expressed sequences. *J. Insect Physiol.* **32:**351–359.

Brewer, F. D., Glick, B., and Vinson, S. B. (1973). A comparative study of selected tissues from the parasitoid *Cardiochiles nigriceps,* its susceptible host *Heliothis virescens* and a resistant host *Heliothis zea* by polyacrylamide gel electrophoresis. *Comp. Biochem. Physiol. B* **46B:**567–574.

Brown, J. J., and Kainoh, Y. (1992). Host castration by *Ascogaster* spp. (Hymenoptera: Braconidae). *Ann. Entomol. Soc. Am.* **85:**67–71.

Coodin, S., and Caveney, S. (1992). Lipophorin inhibits the adhesion of cockroach (*Periplaneta americana*) haemocytes *in vitro. J. Insect Physiol.* **38:**853–862.

Cook, D. I., Stoltz, D. B., and Vinson, S. B. (1984). Induction of a new haemolymph glycoprotein in larvae of permissive hosts parasitized by *Campoletis sonorensis. Insect Biochem.* **14:**45–50.

Coombs, G., and North, M., eds. (1991). "Biochemical Protozoology." Taylor & Francis, Washington, DC.

Cymborowski, B., Bogus, M., Beckage, N. E., Williams, C. M., and Riddiford, L. M. (1982). Juvenile hormone titers and metabolism during starvation-induced supernumerary larval moulting of the tobacco hornworm, *Manduca sexta. J. Insect Physiol.* **28:**129–135.

Dahlman, D. L. (1991). Teratocytes and host/parasitoid interactions. *Biol. Control* **1:**118–126.

Dahlman, D. L., and Greene, J. R. (1981). Larval hemolymph protein patterns in tobacco hornworms parasitized by *Apanteles congregatus. Ann. Entomol. Soc. Am.* **74:**130–133.

Dahlman, D. L., Coar, D. L., Koller, N., and Neary, T. J. (1990). Contributing factors to reduced ecdysteroid titers in *Heliothis virescens* parasitized by *Microplitis croceipes. Arch. Insect Biochem. Physiol.* **13:**29–39.

de Buron, I., Nesbit, D. J., Tan, F. F., and Beckage, N. E. (1993). Development of *Cotesia congregata* teratocytes in host larvae of the tobacco hornworm, *Manduca sexta:* Correlation with morphological and biochemical changes in the cells. *Dev. Biol.* (submitted for publication).

de Jong-Brink, M. (1992). Interference of schistosome parasites with neuroendocrine mechanisms in their snail host causes physiological changes. *Adv. Neuroimmunol.* **2:**199–233.

de Jong-Brink, Elsaadany, M. M., and Boer, H. H. (1988). *Trichobilharzia ocellata:* Interference with endocrine control of female reproduction of *Lymnaea stagnalis. Exp. Parasitol.* **65:**91–100.

de Jong-Brink, M., Hordijk, P. L., Vergeest, D.P.E.J., Schllig, H.D.F.H., Kits, K. S., and ter Maat, A. (1992). The anti-gonadotropic neuropeptide schistosomin interferes with peripheral and central neuroendocrine mechanisms involved in the regulation of reproduction and growth in the schistosome-infected snail *Lymnaea stagnalis. Prog. Brain Res.* **92:**385–392.

de la Torre, J. C., and Oldstone, M.B.A. (1992). Selective disruption of growth hormone transcription by viral infection. *Proc. Natl. Acad. Sci. U.S.A.* **89:**9939–9943.

Dunn, P. E. (1986). Biochemical aspects of insect immunology. *Annu. Rev. Entomol.* **31:**321–339.

Dushay, M. S., and Beckage, N. E. (1993). Dose-dependent separation of *Cotesia congregata*-associated polydnavirus effects on *Manduca sexta* larval development and immunity. *J. Insect Physiol.* (in press).

Ferguson, E. L., and Anderson, K. V. (1992). Localized enhancement and repression of the activity of the TGF-B family member, *decapentaplegic,* is necessary for dorsal–ventral pattern formation in the *Drosophila* embryo. *Development (Cambridge, UK)* **114:**583–597.

Ferkovich, S. M., and Dillard, C. (1986). A study of the radiolabeled host proteins and protein synthesis during development of the eggs of the endoparasitoid, *Microplitis croceipes* (Cresson) (Braconidae). *Insect Biochem.* **16:**337–345.

Ferkovich, S. M., Greany, P. D., and Dillard, C. (1983). Changes in haemolymph proteins of the fall armyworm, *Spodoptera frugiperda* (J. E. Smith), associated with parasitism by the braconid parasitoid *Cotesia marginiventris* (Cresson). *J. Insect Physiol.* **29:**933–942.

Ferkovich, S. M., Dillard, C., and Oberlander, H. (1991). Stimulation of embryonic development in *Microplitis croceipes* (Braconidae) in cell culture media preconditioned with a fat body cell line derived from a nonpermissive host, gypsy moth, *Lymantria dispar. Arch. Insect Biochem. Physiol.* **18:**169–175.

Fisher, R. C., and Ganesalingam, V. K. (1970). Changes in the composition of host hemolymph after attack by an insect parasitoid. *Nature (London)* **277:**191–192.

Fittinghoff, C. M., and Riddiford, L. M. (1990). Heat sensitivity and protein synthesis during heat-shock in the tobacco hornworm, *Manduca sexta. J. Comp. Physiol.* **160:**349–356.

Flor, H. H. (1942). Inheritance of pathogenicity in *Melanmpsora lini. Phytopathology* **32:**653–669.

Führer, E., and Willers, D. (1986). The anal secretion of the endoparasitic larva *Pimpla turionellae:* Sites of production and effects. *J. Insect Physiol.* **32:**361–367.

Greany, P. D. (1986). *In vitro* culture of hymenopterous larval endoparasitoids. *J. Insect Physiol.* **32:**409–419.

Gretch, D. G., Sturley, S. L., Friesen, P. D., Beckage, N. E., and Attie, A. D. (1991). Baculovirus-mediated expression of human apolipoprotein E in *Manduca sexta* larvae generates particles that bind to the low density lipoprotein receptor. *Proc. Natl. Acad. Sci. U.S.A.* **88:**8530–8533.

Harwood, S. H. (1993). *A Cotesia congregata* polydnavirus gene: Cloning, characterization, and pattern of expression in host sphingid larvae. Ph.D. Dissertation, Department of Entomology, University of California-Riverside.

Harwood, S. H. and Beckage, N. E. (1993). Purification and characterization of an abundant polydnavirus-induced protein from the hemolymph of *Manduca sexta* larvae parasitized by *Cotesia congregata. Insect Biochem. Molec. Biol.* (submitted for publication).

Harwood, S. H., Grosovsky, A. J., Davis, J. W., and Beckage, N. E. (1993a). An abundantly expressed host hemolymph glycoprotein isolated from *Manduca sexta* is a *Cotesia congregata* polydnavirus gene product. *Virology* (submitted for publication).

Harwood, S. H., McElfresh, J. S., Nguyen, T. Q., and Beckage, N. E. (1993b). Pattern of expression of a *Cotesia congregata* polydnavirus gene transcript in different sphingid hosts: Correlation with successful parasitism? (submitted for publication).

Hayakawa, Y. (1990). Juvenile hormone esterase activity repressive factor in the plasma of parasitized insect larvae. *J. Biol. Chem.* **265:**10813–10816.

Hayakawa, Y. (1991). Structure of a growth-blocking peptide in parasitized insect hemolymph. *J. Biol. Chem.* **266:**7982–7984.

Hayakawa, Y. (1992). A putative new juvenile peptide hormone in lepidopteran insects. *Biochem. Biophys. Res. Commun.* **185:**1141–1147.

Herberts, C. (1982). Host–parasite relation between the shore crab *Carcinus maenas* and *Sacculina carcini* (Rhizocephala): Identification and characterization of a specific fraction correlated with parasitism. *J. Invertebr. Pathol.* **39:**60–65.

Hershey, J.W.B. (1992). Translational control in mammalian cells. *Annu. Rev. Biochem.* **60:**117–179.

Ilan, J., ed. (1987). "Translational Regulation of Gene Expression." Plenum, New York.

Jessell, T. M., and Melton, D. A. (1992). Diffusible factors in vertebrate embryonic induction. *Cell (Cambridge, Mass.)* **68:**257–270.

Jones, D. (1989). Protein expression during parasite redirection of host (*Trichoplusia ni*) biochemistry. *Insect Biochem.* **19:**445–455.

Jones, D., Jones, G., Rudnicka, M., and Click, A. (1985). Precocious expression of the final larval instar developmental program in larvae of *Trichoplusia ni* pseudoparasitized by *Chelonus* spp. *Comp. Biochem. Physiol. B* **83B:**339–346.

Jones, D., Jones, G., Rudnicka, M., Click, A., Reck-Malleczewen, V., and Iwaya, M. (1986). Pseudoparasitism of host *Trichoplusia ni* by *Chelonus* spp. as a new model system for parasite regulation of host physiology. *J. Insect Physiol.* **32:**315–328.

Joplin, K. H., Yocum, G. D., and Denlinger, D. L. (1990). Cold shock elicits expression of heat shock proteins in the flesh fly, *Sarcophaga crassipalpis*. *J. Insect Physiol.* **36:**825–834.

Junnikkala, E. (1966). Effect of braconid parasitization on the nitrogen metabolism of *Pieris brassicae* L. *Ann. Acad. Sci. Fenn., Ser. A4* **100:**1–83.

Junnikkala, E. (1985). Testis development in *Pieris brassicae* parasitized by *Apanteles glomeratus*. *Entomol. Exp. Appl.* **37:**283–288.

Kaiser, L., Lamb, V. L., Tithof, P. K., Gage, D. A., Chamberlin, B. A., Watson, J. T., and Williams, J. F. (1992). *Dirofilaria immitis:* Do filarial cyclooxygenase products depress endothelium-dependent relaxation in the *in vitro* rat aorta? *Exp. Parasitol.* **75:**159–167.

Kanai, A., and Natori, S. (1989). Cloning of gene cluster for sarcotoxin I, antibacterial proteins of *Sarcophaga peregrina*. *FEBS Lett.* **258:**199–202.

Kanost, M. R., Kawooya, J. K., Law, J. H., Ryan, R. O., Van Huesden, M. C., and Ziegler, R. (1990). Insect hemolymph proteins. *Adv. Insect Physiol.* **22:**299–396.

King, P. E., and Rafai, J. (1970). Host discrimination in a gregarious parasitoid *Nasonia vitripennis* (Walker) (Hymenoptera: Pteromalidae). *J. Exp. Biol.* **53:**245–254.

Kiriishi, S., Rountree, D. B., Sakurai, S., and Gilbert, L. I. (1990). Prothoracic gland synthesis of 3-dehydroecdysone and its hemolymph 3B-reductase mediated conversion to ecdysone in representative insects. *Experientia* **46:**716–721.

Kitano, H., Wago, H., and Arakawa, T. (1990). Possible role of teratocytes of the gregarious parasitoid *Apanteles glomerata* in the suppression of phenoloxidase activity in the larval host, *Pieris rapae crucivora*. *Arch. Insect Biochem. Physiol.* **13:**177–185.

Kunkel, J. G., Grossniklaus-Buergin, C., Karpells, S. T., and Lanzrein, B. (1990). Arylphorin of *Trichoplusia ni:* Characterization and parasite-induced precocious increase in titer. *Arch. Insect Biochem. Physiol.* **13:**117–125.

Lawrence, P. O. (1990). Serosal cells of *Biosteres longicaudatus* (Hymenoptera: Braconidae): Ultrastructure and release of polypeptides. *Arch. Insect Biochem. Physiol.* **13:**199–216.

Loker, E. S., Cimino, D. F., and Hertel, L. A. (1992). Excretory-secretory products of *Echinostoma paraensei* sporocysts mediate interference with *Biomphalaria glabrata* hemocyte functions. *J. Parasitol.* **78:**104–115.

Natori, S. (1990). Dual functions of insect immunity proteins in defence and development. *Res. Immunol.* **141:**938–939.

Noda, S. (1992). Effects of excretory-secretory products of *Echinostoma paraensei* larvae on the hematopoietic organ of M-line *Biomphalaria glabrata* snails. *J. Parasitol.* **78:**512–517.

Pennachio, F., Vinson, S. B., and Tremblay, E. (1992). Preliminary results on *in vitro* rearing of the endoparasitoid *Cardiochiles nigriceps* from egg to second instar. *Entomol. Exp. Appl.* **64:**209–216.

Phares, C. K., and Watts, D. J. (1988). The growth hormone-like factor produced by the tapeworm *Spirometra mansonoides* specifically binds receptors on culture human lymphocytes. *J. Parasitol.* **74:**896–898.
Presnail, J. K., and Hoy, M. A. (1992). Stable transformation of a beneficial arthropod, *Metaseiulus occidentalis* (Acari: Phytoseiidae), by a microinjection technique. *Proc. Natl. Acad. Sci. U.S.A.* **89:**7732–7736.
Ptashne, M. (1988). How eukaryotic transcriptional activators work. *Nature* (*London*) **335:**683–689.
Reed-Larsen, D. A., and Brown, J. J. (1990). Embryonic castration of the codling moth, *Cydia pomonella,* by an endoparasitoid, *Ascogaster quadridentata*. *J. Insect Physiol.* **36:**111–118.
Riddiford, L. M., Curtis, A. T., and Kiguchi, K. (1979). Culture of the epidermis of the tobacco hornworm *Manduca sexta*. *Tissue Cult. Assoc. Man.* **5:**975–985.
Ross, D. R., and Dunn, P. E. (1989). Effect of parasitism by *Cotesia congregata* on the susceptibility of *Manduca sexta* to bacterial infection. *Dev. Comp. Immunol.* **13:**205–216.
Rubiliani, C. (1983). Action of a rhizocephalan on the genital activity of host male crabs: Characterization of a parasitic secretion inhibiting spermatogenesis. *Int. J. Invertebr. Reprod.* **6:**137–147.
Sayles, P. C., and Wassom, D. L. (1988). Immunoregulation in murine malaria. Susceptibility of inbred mice to infection with *Plasmodium yoelli* depends on the dynamic interplay of host and parasite genes. *J. Immunol.* **141:**241–248.
Schallig, H.D.F.H. (1991). Neuro-endocrine interactions between schistosome parasites and their intermediate hosts, freshwater snails, studied in the combination *Trichobilharzia ocellata–Lymnaea stagnalis*. Ph.D. Thesis, Vrije Universiteit, Amsterdam, The Netherlands.
Schallig, H.D.F.H., Sassen, M.J.M., Hordijk, P. L., and de Jong-Brink, M. (1991). *Trichobilharzia ocellata:* Influence of infection on the fecundity of its intermediate snail host *Lymnaea stagnalis* and cercarial induction of the release of schistosomin, a snail neuropeptide antagonizing female gonadotropic hormones. *Parasitology* **102:**85–91.
Schlein, Y., Jacobson, R. L., and Messer, G. (1992). *Leishmania* infections damage the feeding mechanism of the sandfly vector and implement parasite transmission by bite. *Proc. Natl. Acad. Sci. U.S.A.* **89:**9944–9948.
Schmidt, O., and Schuchmann-Feddersen, I. (1989). Role of virus-like particles in parasitoid–host interaction of insects. *Subcell. Biochem.* **15:**91–119.
Schmidt, O., Andersson, K., Will, A., and Schuchmann-Feddersen, I. (1990). Viruslike particle proteins from a hymenopteran endoparasitoid are related to a protein component of the immune system in the lepidopteran host. *Arch. Insect Biochem. Physiol.* **13:**107–115.
Sibley, L. C., and Boothroyd, J. C. (1992). Virulent strains of *Toxoplasms condii* comprise a single clonal lineage. *Nature* (*London*) **359:**82–85.
Smilowitz, Z. (1973). Electrophoretic patterns in haemolymph protein of cabbage looper during development of the parasitoid *Hyposoter exiguae*. *Ann. Entomol. Soc. Am.* **66:**93–99.
Smilowitz, Z., and Smith, C. L. (1977). Hemolymph proteins of developing *Pieris rapae* larvae parasitized by *Apanteles glomeratus*. *Ann. Entomol. Soc. Am.* **70:**447–454.
Soldevila, A. I., and Jones, D. (1991). Immunoanalysis of unique protein in *Trichoplusia ni* larvae parasitized by the braconid wasp *Chelonus* near *curvimaculatus*. *Insect Biochem.* **21:**845–856.
Soldevila, A. and Jones, D. (1993). Expression of a parasitism-specific protein in lepidopteran hosts of *Chelonus* sp. *Arch. Insect Biochem. Physiol.* (in press).
Sroka, P., and Vinson, S. B. (1978). Phenyloxidase activity in the hemolymph of parasitized and unparasitized *Heliothis virescens*. *Insect Biochem.* **8:**399–402.
Stoltz, D. B., and Cook, D. I. (1983). Inhibition of host phenoloxidase activity by parasitoid hymenoptera. *Experientia* **39:**1022–1024.
Stoltz, D. B., and Cook, D. I. (1986). Apparent haemocytic transformations associated with parasitoid induced inhibition of immunity in *Malacosoma disstria* larvae. *J. Insect Physiol.* **32:**377–388.

Stoltz, D. B., Guzo, D., Belland, E. R., Lucarotti, C. J., and MacKinnon, E. A. (1988). Venom promotes uncoating *in vitro* and persistence *in vivo* of DNA from a braconid polydnavirus. *J. Gen. Virol.* **69:**903–907.

Strand, M. R., Dover, B. A., and Johnson, J. A. (1990). Alterations in the ecdysteroid and juvenile hormone esterase profiles of *Trichoplusia ni* parasitized by the polyembryonic wasp *Copidosoma floridanum. Arch. Insect Biochem. Physiol.* **13:**41–51.

Strand, M. R., McKenzie, D. I., Grassl, V., Dover, B. A., and Aiken, J. M. (1992). Persistence and expression of *Microplitis demolitor* polydnavirus in *Pseudoplusia includens. J. Gen. Virol.* **73:**1627–1635.

Sun, S. C., Lindström, I., Boman, H. G., Faye, I., and Schmidt, O. (1990). Hemolin: An insect-immune protein belonging to the immunoglobulin superfamily. *Science* **250:**1729–1732.

Thach, R. E. (1992). Cap recap: The involvement of EIF-4F in regulating gene expression. *Cell (Cambridge, Mass.)* **68:**177–179.

Thea, D. M., Pereira, M.E.A., Kotler, D., Sterling, C. R., and Keusch, G. T. (1992). Identification and partial purification of a lectin on the surface of the sporozoite of *Cryptosporidium parvum. J. Parasitol.* **78:**886–893.

Theilmann, D. A., and Summers, M. D. (1988). Identification and comparison of *Campoletis sonorensis* virus transcripts expressed from four genomic segments in the insect hosts *Campoletis sonorensis* and *Heliothis virescens. Virology* **167:**329–341.

Thompson, J. N., and Burdon, J. J. (1992). Gene-for-gene coevolution between plants and parasites. *Nature (London)* **360:**121–125.

Thompson, S. N., Lee, R.W.K., and Beckage, N. E. (1990). Metabolism of parasitized *Manduca sexta* examined by nuclear magnetic resonance. *Arch. Insect Biochem. Physiol.* **13:**127–143.

Thorndyke, M. C. (1990). Neurohormonal peptides in parasitic worms: A new frontier in host–parasite pathophysiology. *In* "Progress in Comparative Endocrinology," pp. 716–721. Wiley-Liss, New York.

Tilden, R. L., and Ferkovich, S. M. (1987). Regulation of protein synthesis during egg development of the parasitic wasp, *Microplitis croceipes* (Cresson) (Braconidae). *Insect Biochem.* **17:**783–792.

Turco, S. J. (1990). The leishmanial lipophosphoglycans: A multifunctional molecule. *Exp. Parasitol.* **70:**241–245.

Vinson, S. B., and Barras, D. J. (1970). Effects of the parasitoid, *Cardiochiles nigriceps,* on the growth, development, and tissues of *Heliothis virescens. J. Insect Physiol.* **16:**1329–1338.

Vinson, S. B., and Iwantsch, G. F. (1980). Host regulation by insect parasitoids. *Q. Rev. Biol.* **55:**143–165.

Webb, B. A., and Riddiford, L. M. (1988a). Synthesis of two storage proteins during larval development of the tobacco hornworm, *Manduca sexta. Dev. Biol.* **130:**671–681.

Webb, B. A., and Riddiford, L. M. (1988b). Regulation of expression of arylphorin and female-specific protein mRNAs in the tobacco hornworm. *Manduca sexta. Dev. Biol.* **130:**682–692.

Yokoo, S., Tojo, S., and Ishibashi, N. (1992). Suppression of the prophenoloxidase cascade in the larval haemolymph of the turnip moth, *Agrotis segetum,* by an entomopathogenic nematode, *Steinernema carpocapsae,* and its symbiotic bacterium. *J. Insect Physiol.* **38:**915–924.

Zhang, D., Dahlman, D. L., and Gelman, D. B. (1992). Juvenile hormone esterase activity and ecdysteroid titer in *Heliothis virescens* larvae injected with *Microplitis croceipes* teratocytes. *Arch. Insect Biochem. Physiol.* **20:**231–242.

Zitnan, D., Kramer, S. J., and Beckage, N. E. (1993). Accumulation of neuropeptides in the cerebral neurosecretory system of *Manduca sexta* larvae parasitized by the braconid wasp *Cotesia congregata. Develop. Biol.* (submitted for publication).

Chapter 3

Hormonal Interactions between Insect Endoparasites and Their Host Insects

Pauline O. Lawrence
Department of Zoology
University of Florida
Gainesville, Florida

Beatrice Lanzrein
Division of Developmental Biology
Zoological Institute
University of Berne
Berne, Switzerland

I. Introduction and Terminology

Parasitic organisms living within their hosts necessarily encounter a range of physiological and biochemical conditions that reflect the developmental, maturational, and stress-related changes the hosts undergo. The immunological, nutritional, and hormonal conditions of the host influence and are themselves influenced by such changes. Endoparasites therefore react to the prevailing host environment to minimize damage to themselves and to capitalize on those nutritional and physiological conditions that will ultimately enhance their own fitness. Insect parasites have essentially the same family of hormones as their insect hosts. Consequently, some are likely to utilize these hormones directly to regulate their own developmental and metabolic processes whereas others may utilize the fluctuations in host hormone titers and/or hormone-mediated metabolic and nutritional status as cues to initiate their own endocrine and metabolic machinery. Some endoparasites may require conditions of homeostasis that are similar to their hosts' and may readily utilize these conditions with minimal disruption of the host's endocrine or metabolic pathways. Consequently, at parasite loads optimal for their development, these conformers do not significantly disrupt their host's behavior or development or cause macroecological niche alterations of their hosts (Lawrence, 1986a, 1990b). Still other parasites (regulators) disrupt the normal metabolic and endocrine processes of their hosts whose microenvironmental conditions are suboptimal for parasite growth and development (Lawrence, 1986a, 1990b). Consequently, these regulators alter the development, behavior, and ultimately, the niche of their hosts. A large cadre of endoparasitic insects may utilize a combined conformer–regulator strategy depending on the stage(s) of the host they attack and in which they develop, on the changes in their nutritional requirements (Lawrence, 1990b), and probably on the extent of the coevolutionary relationship between the particular host and parasite species (see Lawrence, 1988c).

The process of parasitism stimulates reciprocal responses from the host. For example, feeding of parasite larvae depletes the host's nutrients and these nutritional changes likely alter the endocrine milieu of the host, thereby impacting the parasite (Lawrence, 1990b). Thus, a necessary underpinning of discussions of endocrine interactions between insect hosts and their endoparasites is the appreciation of the intricate interplay among metabolism, nutritional status, and endocrine feedback mechanisms that occur in insects (Nijhout, 1975; Cymborowsky *et al.*, 1982; Downer and Laufer, 1983; Kerkut and Gilbert, 1985).

Approximately 10% of all species of animals are parasitic insects (Askew, 1971). A large number of these insects are parasitic as adults but an even greater number are parasitic during their immature stages and are therefore termed protelean parasites (Askew, 1971). Protelean insect parasites that ultimately kill their insect hosts and later emerge as free-living adults are termed parasitoids (Doutt, 1964). (However, since parasitoids behave essentially like other multicellular parasites during their development within their hosts, the term parasites will be utilized throughout this chapter.)

Insect parasites are categorized on the basis of the host stage they attack and from which they exit as mature larvae or adults, as well as on the optimal number of progeny that can develop per host. Thus, egg-larval parasites attack the host egg and emerge from the larval stage, and so on. Solitary and gregarious parasites develop one and more than one progeny per host, respectively (Doutt, 1964). In polyembryonic species, several offspring develop from a single egg. A host is said to be superparasitized if it has more progeny of a single species than can develop. It is multiply parasitized with progeny from two or more parasite species (Doutt, 1964).

The majority of parasitic species occur in the orders Hymenoptera and Diptera, and exhibit a variety of developmental and physiological interactions with their hosts. These interactions presumably occur along a continuum, with conformers and regulators at the extremes (Lawrence, 1986a). However, the paucity of endocrine studies has restricted our ability to better understand these interactions. The availability of sensitive biochemical and molecular techniques has facilitated separate and direct evaluations of the endogenous hormone fluctuations and related biochemical events in insect hosts and their endoparasites. However, many host–parasite taxa remain to be studied in order to facilitate a holistic appreciation of their endocrine interactions. The purpose of this paper is to (1) review our current knowledge on the hormonal interactions between hymenopteran endoparasites (the most studied group of parasitic insects) and their hosts and (2) identify problems that restrict our understanding of these interactions and suggest possible solutions to them.

II. Hormones Controlling Insect Development: An Overview

Three types of hormones play major roles in insect development: peptides produced by the brain or other neural tissues, ecdysteroids (Fig. 1) produced by the prothoracic gland (or ring gland in Diptera), and juvenile hormones (JHs) (Fig. 2), which are sesquiterpenoids produced by the corpora allata (CA). The reader is referred to reviews by Lafont and Koolman (1984) on ecdysteroids and Baker (1990) on JHs. The most important neuropeptides are the prothoracicotropic hormones (PTTHs), which stimulate the synthesis and release of ecdysone by the prothoracic gland; allatotropins and allatostatins, which stimulate and inhibit, respectively, the production of JH by the CA; and eclosion hormone, which elicits eclosion behavior.

Data from *Manduca sexta* indicate that the ecdysteroids 2- and 3-dehydroecdysone are produced by the prothoracic glands and are immediately converted to ecdysone (reviewed by Gilbert, 1989). It appears that, depending on the developmental stage of the insect, ecdysone acts as a hormone by itself or as a precursor (or prohormone) of 20-hydroxyecdysone, the most commonly occurring molting hormone. 20,26-Dihydroxyecdysone is generally considered a metabolite of 20-hydroxyecdysone but may exert its own morphogenetic effects. It has been suggested that in embryos of *M. sexta*, 26-hydroxyecdysone and/or 20,26-dihydroxyecdysone may act morphogenetically (reviewed in Gilbert, 1989). Makisterone A is unique among the ecdysteroids in that it contains a methyl substituent at C-24 of the side chain

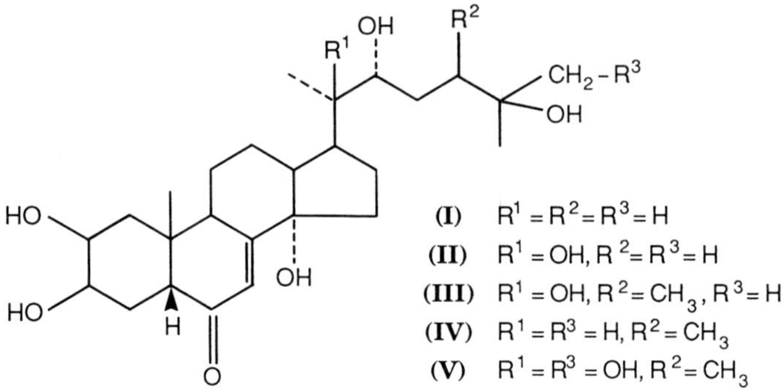

Figure 1 Structures of ecdysteroids of insects. (I) ecdysone; (II) 20-hydroxyecdysone; (III) makisterone A; (IV) 20-deoxymakisterone A; (V) 26-hydroxymakisterone A.

and is not synthesized from cholesterol as are all the other ecdysteroids. It has been identified in several Hemiptera, as well as in the honeybee (Hymenoptera) and *Drosophila* (Diptera) (Kelly *et al.*, 1981, 1984; Feldlaufer, 1989).

A major function of JHs in larval insects is to dictate the character of the molt: the presence of JH during a certain period in the molting cycle leads to

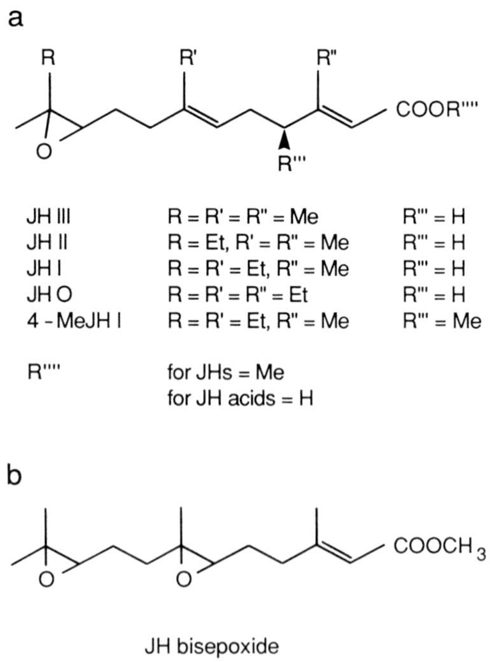

Figure 2 Structures of juvenile hormones in insects. (a) JHs found *in vivo*. (b) JH bisepoxide synthesized by cyclorraphous dipteran ring glands *in vitro*.

larval molts, and its absence leads to metamorphic molts (for reviews, see Downer and Laufer, 1983; Kerkut and Gilbert, 1985; Sehnal, 1989; Gupta, 1990). JH III appears to be the most ubiquitous JH as it was found in all orders investigated, including Lepidoptera, Hymenoptera, and Diptera, whereas the higher homologs JH II, JH I, JH 0, and 4-methyl-JH I have so far only been found in larval and/or embryonic Lepidoptera (reviewed in Baker, 1990). In prepupal stages of *M. sexta* and possibly other Lepidoptera, the CA switch from secreting JH to secreting JH acid, which can be methylated to JH by imaginal disks (Bhaskaran *et al.*, 1986). JH III bisepoxide has so far only been observed *in vitro* upon incubation of ring glands of larval *Drosophila melanogaster* and related Diptera and may represent a biologically active form of JH in cyclorraphous Diptera (Richard *et al.*, 1989). Although insect development is punctuated by molts, it is a continuous process in which both ecdysteroids and JHs have functions in addition to those given here (see Gilbert, 1989; Sehnal, 1989).

The fluctuations in the titer of JHs and ecdysteroids are brought about by variations in both their rates of biosynthesis and degradation. In the hemolymph JH is metabolized by ester hydrolysis to JH acid (reviewed by Hammock, 1985), but in the whole insect epoxide hydrolysis leading to JH diol and JH acid-diol appears to be equally important (Klages and Emmerich, 1979; Halarnkar and Schooley, 1990). The major routes of degradation of 20-hydroxyecdysone are 26-hydroxylation leading to 20,26-dihydroxyecdysone and the subsequent formation of 20-hydroxy-ecdysonoic acid. It is, however, important to note that ecdysone metabolism pathways vary with species, developmental stage, and tissue (reviewed by Lafont and Connat, 1989).

During larval molts and metamorphosis, complex interendocrine interactions occur and are essentially similar in dipterans and lepidopterans (see Richards, 1981a,b; Redfern, 1984; Gilbert, 1989; Baker, 1990; Rauschenbach, 1991), but lepidopteran systems have been studied more extensively. Current models for lepidopteran metamorphosis are that a critical size must first be achieved to allow the JH titer to fall and PTTH to be released. The absence of JH appears to be necessary for both PTTH release to occur and for prothoracic glands to gain steroidogenic competence, that is, to respond to PTTH by producing ecdysone (see reviews by Bollenbacher, 1988; Watson and Bollenbacher, 1988; Watson *et al.*, 1989). A slight increase in ecdysone and/or 20-hydroxyecdysone in the absence of JH induces pupal commitment (Riddiford, 1980). Thereafter a second release of PTTH induces a second, much higher increase in ecdysone and 20-hydroxyecdysone, and JH increases again along with JH acids (see review by Bollenbacher, 1988). Presumably this increase in JH is necessary for further activation of the prothoracic gland (Gruetzmacher *et al.*, 1984; Watson *et al.*, 1989) as well as for the proper formation of the pupa (Hiruma, 1980, 1986). In some species this prepupal increase in JH is also presumed to induce juvenile hormone esterase (JHE) (Jones and Hammock, 1983) but in other species this seems not to be the case (Baker *et al.*, 1987). In still other species, allatectomized larvae undergo normal pupation (Sehnal and Granger, 1975), indicating species-specific importance of the prepupal JH peak. The hormonal changes associated with the initiation of insect metamorphosis also drastically influence the synthesis and storage of proteins, lipids, and glycogen in the fat body (Collins, 1974; Locke, 1980).

III. Parasite Hormones Relative to Those of Their Hosts

The qualitative and quantitative changes of JHs and ecdysteroids in the course of development of lepidopteran and dipteran hosts create a variety of biochemical conditions that endoparasites encounter. The host milieu clearly varies with the species, the environmental and nutritional conditions, the instar, and the time point within an instar. Endoparasites have different nutritional and physiological requirements (see review by Lawrence, 1990b). Consequently, particular parasites will select particular host species, whether eggs, early instars, or late instars (Lawrence, 1990b). The range of host species (niche breadth) selected depends on whether parasites are mono- or polyphagous. Therefore, a parasite, especially a polyphagous species, will likely encounter hosts of various endocrine and physiological states. Thus, either regulation of or conformation to the existing host conditions or a combined conformer–regulator coevolutionary relationship with the host will ensue.

JH and ecdysteroid titers have been measured in only very few parasitic hymenopterans and their hosts. First-instar larvae of the braconid *Diachasmimorpha* (*Biosteres*) *longicaudata,* a larval-pupal parasite of *Anastrepha suspensa* (Diptera, Tephritidae), contained high quantities of JH III (Lawrence *et al.*, 1990). The corresponding late third-instar and pharate pupal hosts had an abnormally high level of JH III compared with the unparasitized controls (Lawrence *et al.*, 1990). It is not known whether this elevated JH titer is a consequence of increased JH production by the host or whether it represents JH that was released from the parasite or its serosa (extraembryonic membrane). Interestingly, the high JH titer in pharate pupal hosts co-occurred with a decrease in JHE, while JH and JHE levels in unparasitized individuals were low and high, respectively (Lawrence *et al.*, 1990), as expected in healthy dipterans (Rauschenbach, 1991). Ecdysteroids in *D. longicaudata* first instars within pharate pupae of *A. suspensa* increased with each decline of the pharate pupal ecdysteroid peaks of the host (Lawrence, 1988b, 1991). This relationship is further indicated in *in vitro* experiments; parasites picked up [^{3}H]ecdysone from culture medium and converted it to [^{3}H]-20-hydroxyecdysone (Lawrence and Hagedorn, 1986), which they require to synthesize a second larval cuticle (Lawrence, 1988d). Studies *in vitro* indicated that the addition of 20-hydroxyecdysone at various intervals to culture media resulted in ecdysteroid increases in first- and second-instar *D. longicaudata* and stimulated molting to subsequent instars, although a small percentage of these later instars have been observed to molt without continued hormone supplement (Lawrence, 1991, and unpublished). The ecdysteroid requirement by *D. longicaudata in vitro* suggests a possible endocrine dependence on the host.

This appears not to be the case in the braconid *Chelonus* sp., an egg-larval parasite of *Trichoplusia ni* (Lepidoptera, Noctuidae). In this parasite JH III was identified with physicochemical methods in late second-instar larvae whereas JH II was identified in the host (Jones *et al.*, 1990). This is the only parasite/host system where fluctuations in JH II, JH III, and various ecdysteroids were measured in parasites and corresponding host hemolymph (Grossniklaus-Bürgin and Lanzrein, 1990b). In late first-instar parasites a very high titer (60 ng/g) of JH III was seen followed by a smaller peak (30 ng/g) in the middle of the second instar while the titer of JH II in host hemolymph fluctuated between 0 and 3 ng/ml (Grossniklaus-Bürgin and Lanzrein, 1990b). Ecdysteroids in the parasite increased before the molt to the second instar and fluctuated at high levels in second- and third-instar larvae while ecdysone and 20-hydroxyecdysone were low in corresponding host hemolymph; in third-instar parasites ecdysteroids were tentatively identified by the use of high-performance liquid chromatography (HPLC) combined with radioimmunoassay (RIA) (two different antibodies) as ecdysone, 20-hydroxyecdysone, 20,26-dihydroxyecdysone, and unknowns, and gas chromatography–mass spectrometry (GC-MS) analyses confirmed the presence of 20-hydroxyecdysone, 20,26-dihydroxyecdysone, ecdysone, and possibly 26-hydroxyecdysone in a proportion of 100 : 14.8 : 11.3 : 2.0 (Grossniklaus-Bürgin and Lanzrein, 1990b). To our knowledge this represents the only identification of ecdysteroids in an endoparasitic insect. As neither JHs nor ecdysteroids fluctuated in a consistent pattern in the parasite relative to those in the host and were in general higher in the former, and as the parasite contains JH III while the host contains JH II, this parasite clearly makes its own hormones at stages after the late first instar (Grossniklaus-Bürgin and Lanzrein, 1990b).

Taken together the few data available so far suggest that in endoparasitic as in free-living hymenopterans the major JH is JH III. Its concentration in parasites is very high, particularly in the first and second instars. No data on metabolism of JH by parasites have been reported to date but preliminary *in vitro* data obtained with homogenates of first- and second-instar larvae of the parasite *Chelonus inanitus* indicate that labeled JH III is degraded only at a very low rate and that the major metabolite is JH III diol (B. Lanzrein and B. D. Hammock, unpublished). Thus the high JH titers seen in parasites might reflect low rates of JH degradation rather than high rates of JH biosynthesis. So far there exists no evidence that parasites release hormones into the host. This possibility appears to be attractive as it would offer to the parasite multiple possibilities of influencing the host's endocrine system, particularly in view of the complex interendocrine regulatory mechanisms. However, two aspects have to be considered. (1) It is not clear by which mechanism hormones produced inside the parasite could reach the host's hemolymph without being degraded. (2) Although the concentration of JH and ecdysteroids was found to be high in the parasites investigated so far, the quantity in absolute terms and in comparison to the host hemolymph volume is very small. Consequently, parasites would have to produce very large amounts of hormones in order to substantially alter the titer in the host. The possibility of release of hormone metabolites seems more likely.

Experiments with *Chelonus* sp./*T. ni,* where labeled ecdysone was injected into the host, followed by subsequent analysis of labeled ecdysteroids in parasite and host, suggested that this parasite makes apolar ecdysteroids that are released into the host (Grossniklaus-Bürgin *et al.*, 1989). Also the presence of JH III acid in late second-instar *Chelonus* sp. and corresponding carcasses of *T. ni* (Jones *et al.*, 1990) could be interpreted as release of JH III acid from the parasite into the host. Therefore there is indication that some parasites use and metabolize host hormones and/or release hormone metabolites into the host but no coherent picture emerges yet as to the endocrine exchanges between parasites and hosts.

IV. Direct or Indirect Host Endocrine Effects on Endoparasites

The impact of host quality on parasite development has been termed host suitability (Doutt, 1964), which is influenced by the nutritional, endocrine, biochemical, and immunological factors within the host (Vinson and Iwantsch, 1980a). These factors in turn may impact each other through feedback loops. Environmentally induced stress conditions may also alter the dynamic interplay of the various physiological factors that modulate host suitability (Lawrence, 1988c).

There have been numerous studies on parasitism-induced endocrine changes in the insect host (see Section V) but very few on the host's impact on the parasite (see reviews by Lawrence, 1986a, 1990b). This paucity probably reflects the inherent difficulty in separating host-related versus parasite-related factors and identifying their respective causes and effects *in vivo*.

Host endocrine effects on endoparasites were demonstrated indirectly in the *Opius concolor* (Hymenoptera, Braconidae)–*Ceratitis capitata* (Diptera, Tephritidae) larval-pupal parasite–host system. The parasite's first larval molt was induced by the injection of ecdysone into the host after their developmental arrest by γ irradiation (Cals-Usciati, 1969, 1975). Since pupariation/pupation of the injected host preceded the parasite's larval molt, it is not clear whether the hormone effects directly on the parasite or indirectly by mediating important biochemical changes associated with host pupariation and pupation. In the related Hymenoptera–Diptera parasite–host systems *Opius* (*Diachasmimorpha*) *tryoni*–*C. capitata* and *Opius melleus*–*Rhagoletis pomonella,* the parasites' first larval molt appears to be influenced by and occurs in relation to the pupariation/pupation of their hosts (Pemberton and Willard, 1918; Lathrop and Newton, 1933). Though significant endocrine changes are known to be necessary for molting, pupariation, pupation, and diapause, their role in influencing endoparasites needs clarification. Only correlations between parasite behavior and development and host endocrine-mediated events (see Mellini, 1975; Lawrence, 1986a, 1990b) and host hormone titers have so far been established (Plantevin *et al.*, 1986; Grossniklaus-Bürgin and Lanzrein, 1990b; Strand *et al.*, 1991) in the majority of studies to date. *In vitro* culture and hormone therapy of a few parasite species have demonstrated a direct effect of host hormones on parasite development. For example, Nenon (1972) first demonstrated the necessity of JH and ecdysone for growth and development of the hymenopteran parasite *Ageniaspis fusciollis in vitro*. More recently, Lawrence (1988d) demonstrated that 20-hydroxyecdysone supplementation of culture medium was necessary for cuticulogenesis of *D. longicaudata* second instars preceding the first larval molt. Furthermore, 20-hydroxyecdysone cell-free culture medium containing *D. longicaudata* larvae effected increases in parasite endogenous ecdysteroid titers and subsequent larval molts (Lawrence, 1991). The development of new *in vitro* culture media must necessarily precede studies on a larger number of endoparasites to determine host endocrine effects. It should be noted that the effects of 20-hydroxyecdysone, JH, and other insect hormones on endoparasites *in vitro* are not surprising but the absence of parasite development in hormone-free medium provides strong evidence for a direct host hormonal role—providing the medium is optimal for parasite growth.

In the absence of empirical evidence of various host endocrine effects on endoparasites, we will briefly address selected host factors and the possible ways in which they likely impact the developing parasite directly or via the endocrine system. The endocrine and nutritional status of host insects is known to impact the development and viability of their endoparasites. For example, Salt (1937) demonstrated that *Trichogramma evanescens* reared on a neuropteran host (*Sialis*) emerged as apterous adults, whereas those reared on the usual lepidopteran host eggs were normal (winged). He argued that the nutrients in *Sialis* were suboptimal for alate adult development. The size and quality of the host in influencing parasite development were further demonstrated by the rearing of *Trichogramma* on eggs of different-sized lepidopteran hosts. Salt (1941) obtained parasite adults that were stunted or had malformed wings from small host eggs whereas normal-sized individuals emerged from the usual lepidopteran host eggs. Some parasites contain the same types of amino acids as their hosts (Thompson, 1986) and probably sequester all or some of these from their hosts. Consequently, a deficiency in such nutrients within the host could be deleterious to the developing parasite. Cals-Usciati (1969, 1975) and several others have shown that larval endoparasites feed on host hemolymph during the first larval stadium and may even puncture host cells with their mandibles to utilize cell contents.

The nutrients available in a host are influenced by the food eaten by the host itself as well as by the endocrine and developmental status of the host (see Lawrence, 1990b, for review). Barbosa (1988) has reviewed the effects of substances such as nicotine, rutin, and other allelochemicals within the host's food on endoparasites. They inhibit parasite growth, prolong larval development, and even cause mortality in a large percentage of endoparasites. Some plant-derived substances like coumarins can cross the insect gut wall and bind to storage proteins in the hemolymph (Shapiro *et al.*, 1988) and are therefore encountered by the parasite larvae. Phytohormones, such as juvadecene (Nishida *et al.*, 1983) and phytoecdysteroids, plus antihormones such as precocene, can directly or indirectly impact the developing parasite. Azadiractin, a neem tree derivative, was found to disrupt vitellogenesis as well as JH and ecdysteroid levels in the exposed insects (Rembold and Schmutterer, 1981). Thus, endoparasites that attack actively feeding larvae of phytophagous insects are the most likely to be affected by plant-derived substances from the host's food. Such effects may occur within a short time after the host feeds or at a later date in cases where substances like coumarins may be bound to arylphorins (Shapiro *et al.*, 1988), lipoproteins, or other storage molecules. Second-order effects of allelochemicals on beneficials have been studied in a few cases (Barbosa, 1988). However, the impact of these allelochemicals on host and parasite endocrine systems, though likely, has not been evaluated.

The level of hemolymph nutrients in insects is known to be influenced by their endocrine status (Locke, 1980; Wyatt, 1980). No studies have been done on the immediate and ultimate effects of endocrine-induced nutritional changes in hosts on their developing endoparasites. One can only speculate on the possible effects. For example, actively feeding lepidopteran larvae may have a higher level of free hemolymph trehalose than nonfeeding larvae. Consequently, hormones that stimulate the metabolism of glycogen and the production of trehalose (Collins, 1974) may differ in titer between mature larvae and younger stage feeding larvae with correspondingly different effects on their respective endoparasites. Several of these hormones that regulate insect metabolism have been reviewed (Goldsworthy and Gäde, 1983; Kerkut and Gilbert, 1985). Their possible direct and indirect effects on developing endoparasites have been discussed (Lawrence, 1990b) but empirical studies are needed to determine their impact on endoparasites. Results from such studies could be particularly valuable in developing artificial media for rearing endoparasites.

Several hormonelike substances may also work via the host to adversely impact the developing parasite. These include hormone analogs (Wright and Spates, 1972; McNeil, 1975; Beckage, 1985), antihormones (Bowers, 1983), hormone agonists (Wing, 1988), and chitin synthesis inhibitors (Lawrence, 1981). Although much effort is being made to develop pesticides with minimal deleterious environmental effects, little effort is being made to address the impact of these substances in comprehensive pest control programs in which beneficial insects, primarily endoparasites, are employed. For example, mortality of the endoparasite *D. longicaudata,* treated with the chitin synthesis inhibitor diflubenzuron, as first instars within pharate pupae of the dipteran host, *A. suspensa,* was lower than that of parasites that were either fourth instar or pupae when the host was treated (Lawrence, 1981). However, greater than 70% of those parasites (treated as first instars) that emerged as adults were malformed due to abnormal sclerite development (Lawrence, 1981, 1983). Presumably, the inhibitory effect of diflubenzuron on *N*-acetylglucosamine uptake during chitin synthesis (Mayer *et al.*, 1980) may have impacted the parasites and prevented the normal formation of pupal and adult cuticles.

The newly developed ecdysone agonist RH 5849 is a potential insecticide that binds ecdysone receptors and has similar effects on molting as 20-OHE (Wing, 1988). This agonist is more effective *in vivo* than natural ecdysteroids because it is not as readily metabolized by endogenous enzymes as are the ecdysteroids. Since some parasites such as *C. congregata* apparently utilize increased ecdysone titers as a signal to exit their treated hosts (Beckage, 1985), to what extent could this pesticide influence premature emergence of endoparasite larvae from their treated hosts? In the case of those parasites requiring host ecdysteroids to initiate cuticle synthesis and subsequent development, what impact will substances like RH 5849 have on parasite development when applied at sublethal dosages to disrupt the pest host's development or reproduction?

V. Direct or Indirect Effects of Endoparasites on Host Endocrine System

The effects of endoparasites on host development are numerous and diverse (reviewed in Vinson and Iwantsch, 1980b; Beckage, 1985, 1990a,b; Lawrence, 1986a, 1988a). In most cases it is not known whether parasites directly interfere with the host's endocrine system or whether parasite effects are indirect and caused by stress and/or influences on the nutritional status of the host. In the following we shall discuss various types of developmental and endocrine effects of parasites on their hosts by focusing on selected and well-investigated parasite/host systems. Table 1 summarizes the effects of several hymenopterous parasites on their lepidopterus hosts. It shows the developmental effects, the endocrine manifestations, and where available, the putative factors involved. It includes parasites that prolong larval life of the host and/or cause its developmental arrest in the last instar, or that induce precocious onset of metamorphosis, or that prolong the feeding phase in the last instar.

The braconids *C. congregata, Apanteles kariyai, Cardiochiles nigriceps,* and *Microplitis croceipes* are all larval parasites that cause developmental arrest of the host in the last instar. This is reflected in lowered JHE and ecdysteroid levels as well as altered sensitivity of prothoracic glands to PTTH and/or altered sensitivity of target tissues to ecdysteroids, depending on the system. Calyx (which contains virus), venom, and teratocytes appear to be involved in causing some of the endocrine changes in the host but their precise function has not yet been elucidated (see below). The ichneumonid *Campoletis sonorensis* also causes developmental arrest of the host and polydnaviruses induce degeneration of the prothoracic gland and suppress ecdysone production but only if last-instar larvae are attacked (Dover and Vinson, 1990). This appears to be exceptional under natural conditions, and this effect might thus rather represent a safety measure that would allow some parasites to survive when no younger larvae are available. *C. sonorensis* is a habitual parasite of third-instar *H. virescens* (Davies *et al.,* 1987).

While the larval parasites discussed in the foregoing prolong larval life of the host and/or prevent host pupation, egg-larval parasites of the genus *Chelonus* (Braconidae) induce precocious cocoon spinning of the host in the penultimate instar followed by parasite emergence (Jones *et al.,* 1981). These solitary parasites oviposit into eggs of various lepidopteran species. The first instar hatches when the host is a late embryo or a newly hatched first-instar larva. The parasites remain in their first stadium for several days and molt into the second instar only after the host's ecdysis into its precocious last instar (Grossniklaus-Bürgin and Lanzrein, 1990b).

Various endocrine parameters have been measured in the course of development in nonparasitized and parasitized *T. ni* larvae. The pattern of JHE (Bühler *et al.*, 1985), the variations in JH II and various ecdysteroids (Grossniklaus-Bürgin and Lanzrein, 1990b), and the metabolism of ecdysone (Grossniklaus-Bürgin *et al.*, 1989) were found to be similar in parasitized precocious last-instar larvae and in nonparasitized last-instar larvae (Table 1). Also the storage protein arylphorin was seen to appear precociously (Kunkel *et al.*, 1990). All of these data indicate that this parasite induces a true precocious initiation of metamorphosis. Experiments carried out with pseudoparasitized larvae (larvae developing from stung eggs but containing no live parasite) that become developmentally arrested in the prepupal stage indicated suppression of ecdysone production and metabolism (Jones, 1986) but no indication of either suppression of ecdysone production or conversion to 20-hydroxyecdysone was seen in parasitized *T. ni* up to the stage when the parasites exit the host (Grossniklaus-Bürgin *et al.*, 1989; Grossniklaus-Bürgin and Lanzrein, 1990b). It is not clear which factors are responsible for induction of the host's precocious metamorphosis.

The observation that pseudoparasitized larvae go into precocious metamorphosis and the fact that removal of second-instar parasites from host larvae does not prevent initiation of precocious metamorphosis (Bühler *et al.*, 1985) indicate that the second-instar parasite alone cannot induce precocious metamorphosis. From indirect approaches it has been claimed that material from the female wasp alone is responsible for inducing precocious metamorphosis (Jones, 1987) but so far direct evidence is lacking and an involvement of embryonic or first-instar parasites or teratocytes cannot yet be excluded. Injection of calyx fluid plus venom from *Chelonus insularis* into *Spodoptera ornithogalli* was seen to imitate the growth-reducing effect of parasitism but not the precocious initiation of metamorphosis (Ables and Vinson, 1981). Injection of calyx fluid, purified virus, or venom from *Chelonus inanitus* also did not induce precocious metamorphosis in either *S. littoralis* or *S. exigua* (C. Grossniklaus-Bürgin and B. Lanzrein, unpublished). This suggests that factors in addition to virus and venom are necessary to induce precocious metamorphosis.

A comparison of JHs in nonparasitized eggs of *T. ni* and eggs parasitized by *Chelonus* sp. revealed that before hatching parasitized eggs contained high quantities of JH (JH I, JH II, and JH III) whereas nonparasitized ones contained only little JH (JH I and JH II). At this stage the parasites hatch and release the serosal membrane, and it was suggested that the latter produces JH III (Grossniklaus-Bürgin and Lanzrein, 1990b). The difference in

Table 1
Effects of Parasites on Host Development and Endocrinology, and Putative Parasite Factors Involved in These Effects

Parasite	Host	Category	Developmental effects	Endocrine manifestation in host	Factors	Reference
Cotesia congregata Braconidae	*Manduca sexta* Sphingidae	Gregarious larval	Developmental arrest in last instar, occasionally supernumerary larval instar	JH titer elevated, JH esterase lowered, ecdysone 20-monooxygenase suppressed, sensitivity to ecdysteroids lowered		Reviewed in Beckage and Templeton (1986); Beckage (1990b)
Apanteles kariyai Braconidae	*Pseudaletia separata* Noctuidae	Gregarious larval	Prolongation of larval life, developmental arrest in last stadium	Inhibition of PTTH synthesis or secretion, lowered sensitivity to 20-hydroxyecdysone, prothoracic gland sensitivity to PTTH suppressed; JH esterase suppressing factor	Calyx + venom + teratocytes	Tanaka *et al.* (1987); Tanaka (1987); Wani *et al.* (1990); Hayakawa (1990)
Cardiochiles nigriceps Braconidae	*Heliothis virescens* Noctuidae	Gregarious larval	Prolongation of larval life, developmental arrest in last stadium	Function of prothoracic gland and its sensitivity to PTTH depressed	Calyx + venom	Tanaka and Vinson (1991)

Microplitis croceipes Braconidae	*Heliothis virescens* Noctuidae	Solitary larval	Developmental arrest in last stadium	Ecdysteroids lowered, JH esterase lowered	Teratocytes, calyx	Webb and Dahlman (1986); Zhang and Dahlman (1989); Dahlman *et al.* (1990a,b)
Campoletis sonorensis Ichneumonidae	*Heliothis virescens* Noctuidae	Solitary larval	Developmental arrest when last instar larvae are attacked	Degeneration of prothoracic glands, ecdysteroids lowered	Calyx or polydnavirus	Dover *et al.* (1987, 1988a,b, 1989); Dover and Vinson (1990)
Chelonus sp. Braconidae	*Trichoplusia ni*, Noctuidae	Solitary egg-larval	Precocious onset of metamorphosis, emergence of parasite from precocious prepupa	Precocious appearance of premetamorphic changes in JH, ecdysteroids, and JH esterase; elevated JH titer in parasitized eggs	Polydnavirus? Teratocytes?	Jones *et al.* (1981); Bühler *et al.* (1985) Grossniklaus-Bürgin and Lanzrein (1990b)
Copidosoma floridanum Encyrtidae	*Trichoplusia ni*, Noctuidae	Polyembryonic egg-larval	Prolongation of feeding in last stadium	JH titer elevated, JH esterase and ecdysteroid increase delayed		Strand *et al.* (1990, 1991)

JH titer appears not to be due to different rates of metabolism of JH as labeled JH III was seen to be degraded to the same extent by homogenates of parasitized and nonparasitized eggs in the closely related system *C. inanitus/S. exigua* (B. Lanzrein, unpublished).

The polyembryonic egg-larval parasite *Copidosoma floridanum* (Encyrtidae) parasitizes *T. ni* and other plusiine Lepidoptera. *C. floridanum* oviposits into the eggs and during the first through fourth stadia the parasite embryo proliferates to form an average of 1200 morula stage embryos. Approximately 40 of these embryos initiate morphogenesis during this period to form a morph referred to as the precocious larva (Strand, 1989). These larvae cease to be formed in the fourth stadium of *T. ni,* do not molt, and die in the fifth stadium. The remaining embryos develop into a second morph, referred to as the reproductive larva. Differentiation of morulae begins with the *T. ni* molt to the fifth stadium (Baehrecke and Strand, 1990) and the reproductive larvae eclose on day 3 of the host's fifth stadium. Parasitized larvae attain a larger final weight and initiate wandering one day later than nonparasitized larvae. JH titer measurements showed that parasitized *T. ni* contained three times more JH than nonparasitized ones during the molt to the fourth stadium (Strand *et al.*, 1991). As the identity of the JH was not determined, it is not clear whether it is of parasite or host origin. In the fifth stadium the JH titer of parasitized larvae fell one day later compared to that of nonparasitized larvae (Strand *et al.*, 1991) and correspondingly ecdysteroid and JHE increases were also delayed (Strand *et al.*, 1990). It is not known whether the delay in the initiation of metamorphosis and thus the increase in length of the host's feeding phase are caused directly by hormonal signals from the parasite or whether it is a consequence of altered host physiology.

In the following we discuss the putative endocrine effects of the various parasite-associated factors.

A. Substances Secreted by Serosal Membranes/Teratocytes

Teratocytes are cells derived from the serosal membrane of Braconidae, Trichogrammatidae, and Scelionidae; these cells apparently assume different roles (trophic, immunosuppression, secretory) in different species (reviewed by Dahlman, 1990). Endocrinologically, the link between the serosal membrane/teratocytes and the JH system is most intriguing. Some observations suggest that serosal membranes and young teratocytes of some species contain and/or release JH. In the system *D. longicaudata–A. suspensa,* comparative JH III measurements in nonparasitized and parasitized *A. suspensa* and dissected parasites suggested that serosal membranes probably contain JH III (Lawrence *et al.*, 1990). The same hypothesis was put forward in the system *Chelonus* sp.–*T. ni,* where parasitized eggs were seen to contain much higher quantities of JH than nonparasitized ones at the stage of release of serosal membranes into the host (Grossniklaus-Bürgin and Lanzrein, 1990b). In the system *C. nigriceps–H. virescens,* injection of teratocytes was seen to have JH-like effects, with younger teratocytes being more effective (Vinson, 1970), and JH analyses made by *Galleria* bioassay revealed high concentrations of JH in 7-day-old teratocytes and much lower concentrations in teratocytes of mixed ages (Joiner *et al.,* 1973). Interestingly, serosal epithelia of the locust *Locusta migratoria* have been shown to produce JH III (Hartmann *et al.,* 1987), which demonstrates the capability of this tissue to synthesize JH.

Older teratocytes of some species appear to interfere with the JH system and to increase the JH titer in the host by acting on the JHE. In the system *M. croceipes–H. viresecens,* injection of teratocytes was seen to inhibit larval–pupal transformation in the host in a dose- and age-dependent manner, with younger teratocytes being more active and younger hosts being more sensitive (Zhang and Dahlman, 1989). It is interesting to note that *C. congregata* teratocytes collected from *M. sexta* had no effect when injected into *H. virescens;* measurement of JHE in hemolymph of *H. virescens* larvae injected with teratocytes from *M. croceipes* revealed that JHE was dramatically depressed when the normal number (750) of teratocytes from a single parasite was injected, but not when teratocytes from *C. congregata* were injected (Zhang and Dahlman, 1989). In the system *Apanteles karyiai–Pseudaletia separata,* injection of teratocytes delayed larval–larval and larval–pupal ecdysis (Wani *et al.,* 1990). In this system hemolymph of larvae collected 3–6 days after parasitization was seen to contain a factor that suppresses JHE; this factor was purified, found to be a protein, and partially sequenced (Hayakawa, 1990), but its origin (host, teratocytes, polydnavirus) has not been investigated. In all of these systems, parasitization causes developmental arrest before pupation (Table 1). In the endocrine changes associated with metamorphosis an absence of JH is necessary not only for pupal commitment but also for the regulation of PTTH and steroidogenic competence of the prothoracic glands (see Section V.B), and consequently an interference with the removal of JH by esterases has multiple effects.

B. Calyx Fluid, Viruses, and Venom Gland Secretions

Viruses have been implicated in immunosuppression and in altering the nutritional and hormonal balance of the host (Stoltz and Vinson, 1979). In the system *C. congregata–M. sexta,* polydnavirus has been shown to induce the appearance of parasitism-specific proteins in the host (Beckage, 1990a). In the system *C. sonorensis/H. virescens,* injection of calyx fluid or polydnavirus into fifth-instar larvae led to degeneration of the prothoracic gland and reduced ecdysteroid titers (Dover *et al.,* 1987, 1988a,b, 1989). Hosts stung in the fourth stadium showed no signs of prothoracic gland degeneration and injection of calyx into early fourth-stadium larvae did not cause prothoracic gland degeneration, suggesting that induction of prothoracic gland degeneration by polydnavirus is specific to the last stadium of the host (Dover and Vinson, 1990). In the system *M. croceipes–H. virescens,* injection of calyx fluid inhibited growth and development as well as JHE activity, effects also seen upon the injection of teratocytes (see Table 1 and earlier discussion), and it was suggested that teratocytes contain polydnavirus or some of its genomic information (Dahlman *et al.,* 1990b).

Venom gland secretions exhibit a wide range of activities. Those produced by ectoparasites are often paralytic, those of some endoparasites have been shown to contribute to the abrogation of the host's immune system, and others have been implicated in the perturbation of the endocrine system of the host (Stoltz and Vinson, 1979; Stoltz, 1986). An important function of venoms appears to be their role as synergists of viruses in that they facilitate penetration and survival of virus in host cells (Stoltz *et al.,* 1988). In the system *A. karyiai–P. separata* (see Table 1), calyx fluid plus venom have been shown to suppress the sensitivity of prothoracic glands to PTTH, and in the system *C. nigriceps–H. virescens* (see Table 1), the two components together have been shown to depress the function of the prothoracic gland and its competence to respond to PTTH (Tanaka and Vinson, 1991).

From these descriptions it is obvious that the mechanisms and factors through which parasites alter their host's endocrine system vary with the parasite species involved and with the host species and stage attacked. Some effects may be due to parasite-derived substances acting on endocrine glands, on hormone metabolism, or on the sensitivity of target glands and tissues, or interfering with interendocrine axes in the PTTH/ecdysone and JH system and feedback loops. Interactions with metabolic hormones have not yet been investigated but are also a likely target for parasite action. At least in one system it has been shown that the three factors calyx, venom, and teratocytes injected in a sequence that simulated the *in vivo* situation simulated the developmental effect of parasitism (Wani *et al.,* 1990). This indicates that the various parasite-associated factors act in concert and might explain why application of only one factor usually does not induce the full spectrum of parasitismlike alterations.

VI. Limitations

The foregoing review clearly demonstrates that our study of parasite–host endocrine interactions is in its infancy and much needs to be done before we can attempt to understand this very complicated association. Among the factors that contribute to our limited progress are the following: (1) our inability to separate (successfully) parasite from host over a prolonged period in order to study the physiology, nutritional biochemistry, and endocrinology separately; (2) the lack of appropriate technology that allows the study of very small organisms such as insect parasites, particularly the Hymenoptera, and the minute volumes of hemolymph available for accurate hormone analysis, both qualitative and quantitative; (3) our interpretations of the effects of hormones on hosts and parasites, for example, increases in hormone titer may facilitate certain cellular events not readily detectable but a decline in the same hormone may then result in some measurable change; and (4) numerous peptide hormones occur in insects but most parasite–host studies have been confined to JH and ecdysteroids. What roles do peptide hormones play in host–parasite interactions?

The major limitations in doing endocrinological studies with insect parasites arise from the small size of parasite eggs and larvae (fractions of a microgram to a few milligrams). For measurement of JHs for instance, the most reliable method is GC-MS, which allows precise quantification of individual JHs (for details, see Baker, 1990), but this method requires quantities of biological material that are often not available and only a few laboratories have the necessary costly equipment and the expertise.

Radioimmunoassays of biological extracts do not provide absolute information about the concentration of individual ecdysteroids. This is primarily because the antibodies commonly used cross-react with several ecdysteroids to varying degrees (for details, see Warren and Gilbert, 1986; Reum and Koolman, 1989). Consequently, a combination of RIA with chromatographic methods such as thin-layer chromatography or preferably HPLC allows quantification of individual ecdysteroids, provided the cross-reactivity of the various ecdysteroids with the antibody is known. For measuring JHs, several approaches are being used: physicochemical methods (GC-MS), bioassays, RIAs, and combinations of HPLC with bioassay or RIA. Information about the fluctuation of individual JHs can be best obtained by GC-MS. The combination of HPLC with bioassay or RIA can also fulfill this goal, provided standard curves for the different JHs are available (bioassays) or cross-reactivities of the antibody with the individual JHs are known (RIA).

In cases where specific methods were used and measurements taken at short intervals (6–8 hr), hormone titers were seen to fluctuate rapidly. For example, PTTH was seen to be released in bursts (reviewed in Gilbert, 1989), individual JHs displayed short and sharp peaks (Baker *et al.*, 1987; Grossniklaus-Bürgin and Lanzrein, 1990a), and also ecdysone and 20-hydroxyecdysone were seen to fluctuate rapidly (Lafont *et al.*, 1975; Grossniklaus-Bürgin and Lanzrein, 1990a). In addition, since very minor hormone increases and decreases can be biologically significant, it would be most desirable to analyze precisely staged insects at short intervals.

The small size of parasites makes it impossible to measure hormone titers in the hemolymph or to routinely measure rates of hormone synthesis by incubating glands (CA, prothoracic glands, or brains) *in vitro*. Consequently, many approaches used to investigate the endocrine status of parasites are indirect. In the case of hormone application to the host, it is not possible to know if a parasite's reaction is caused by the hormone itself or by hormone-induced changes in the host. Likewise, analysis of hormone metabolism by parasites *in vivo* is very difficult as the labeled hormone has to be injected into the host, which also metabolizes it. If the same hormone metabolites are identified in parasite and host it is not possible to know whether they were made by the parasite or simply taken up as metabolites. If parasites use as yet unidentified, parasite-specific hormones, their analysis would be extremely difficult because of the lack of specific bioassays.

VII. Possible Solutions

To date, the physiology/endocrinology of only very few host–parasite systems have been studied. Consequently, further analyses of parasite hormones in relation to those of their hosts should be carried out. Emphasis on hosts that have already been endocrinologically analyzed would facilitate the task. However, the majority of these host species are lepidopterans. Therefore, additional efforts should be made to include host groups such as the hemipterans and dipterans since reasonable endocrinological studies have been done on selected members of these groups. Furthermore, several species of hemipterans are major agricultural pests and numerous dipteran species are either themselves parasites or vectors of human and animal diseases. Consequently, studies on the hormonal interactions of parasites of these hosts would be valuable in pest control efforts.

When analyses of JHs are being undertaken, they should be measured individually. In the case of several lepidopterous hosts, this would allow discrimination between the hymenopteran parasite-produced JH III and host JH II and JH I, or in the case of lepidopteran host eggs, JH 0. However, the problems will be more difficult in hymenopteran–dipteran parasite–host systems in which JH III is common to both groups. One approach that could alleviate this problem and simultaneously elucidate the effect of parasites on the endocrine glands of the host and vice versa is the analysis of the production *in vitro* of hormones or other substances by intact parasites, or by glands of hosts and their nonparasitized counterparts. This approach has been successfully used to demonstrate parasite uptake of labeled proteins (Beckage *et al.*, 1989) and amino acids (Ferkovich and Dillard, 1986), labeled hormones (Lawrence and Hagedorn, 1986), the release of proteins by parasites (Lawrence, 1990a), and host prothoracic gland activity (Tanaka *et al.*, 1987). Other *in vitro* studies on parasites have included the successful rearing of pupal parasites and a study of the nutritional requirements of others (for reviews, see Greany, 1986; Nettles, 1990) and the requirements of some parasites for exogenous hormones (Neon, 1972; Lawrence, 1986b, 1988d, 1991).

Other as yet unresolved questions in host–parasite interactions relate to those parasite-derived factors that disrupt the host's endocrinology (see Section V). Their influence on suspected target tissues and glands of the host could be evaluated *in vitro*.

VIII. Summary

In the preceding discussion, we have attempted to focus primarily on hymenopterous protelean endoparasites that attack eggs and larvae of lepidopterans and larvae of dipterans. We have utilized existing information on the general endocrine systems and interendocrine interactions known in insects as the basis of our discussion and have complemented this with the limited data available from a few host–parasite systems.

The information available so far indicates that the dependence of parasites on the host's endocrine system varies from strong to marginal and is characteristic of each parasite–host system. It appears that the endocrine interactions between parasites and their hosts vary from those that require the host's hormones (especially those governing growth, development, and metamorphosis) for their own development (conformers) to those that significantly modify their host's endocrine system in order to retain the host for some period of time in a particular physiological and likely, nutritional state (regulators). A third category of parasites appears to utilize a combined strategy. In this latter group, early stages (embryos to first-instar larvae) appear to display a higher degree of dependence than later stages. Parasite effects become manifest in later stages of host development and often involve a disturbance of the PTTH/ecdysone axis and/or the JH axis. In causing these effects, all factors, namely, virus, venom, and teratocytes (if present), appear to be involved but their interrelationship and impact on the host's endocrine system are not yet understood. Some parasites have been shown to produce their own hormones (at least after the late first instar) and this might be generally true, but it is not clear whether parasite hormones or metabolites thereof are ever released into the host and used to influence its development.

Acknowledgments

Financial support from the National Science Foundation, Grant DCB 9005514, and the USDA, CRGO Grant 9001300, to P.O.L., and the Swiss National Science Foundation, Grant 31-27723.89, to B. L. is gratefully acknowledged.

References

Ables, J. R., and Vinson, S. B. (1981). Regulation of host larval development by egg-larval endoparasitoid *Chelonus insularis*. *Entomophaga* **26:**453–458.

Askew, R. R. (1971). "Parasitic Insects." C.F.R. Woodward, London.

Baehrecke, E. H., and Strand, M. R. (1990). Embryonic morphology and growth of the embryonic parasitoid *Copidosoma floridanum* (Ashmead) (Hymenoptera: Encyrtidae). *Int. J. Insect Morphol. Embryol.* **19:**165–175.

Baker, F. C. (1990). Techniques for identification and quantification of juvenile hormones and related compounds in arthropods. *In* "Morphogenetic Hormones of Arthropods" (A. P. Gupta, ed.), pp. 389–453. Rutgers Univ. Press, New Brunswick, NJ and London.

Baker, F. C., Tsai, L. W., Reuter, C. C., and Schooley, D. A. (1987). *In vivo* fluctuation of JH, JH acid, and ecdysteroid titer, and JH esterase activity, during development of fifth stadium *Manduca sexta*. *Insect Biochem.* **17:**989–996.

Barbosa, P. (1988). Natural enemies and herbivore–plant interactions: Influence of plant allelochemicals and host specificity. *In* "Novel Aspects of Insect–Plant Allelochemicals and Host Specificity" (P. Barbosa and D. Letourneau, eds.), pp. 201–229. Wiley, New York.

Beckage, N. E. (1985). Endocrine interactions between endo-parasitic insects and their hosts (review). *Annu. Rev. Entomol.* **30:**371–413.

Beckage, N. E. (1990a). Induction of host endocrine and hemolymph protein alterations in tobacco hornworm larvae parasitized by *Cotesia congregata*. *Adv. Invertebr. Reprod.* **5:**149–155.

Beckage, N. E. (1990b). Parasitic effects on host development. *In* "New Directions in Biological Control: Alternatives for Suppressing Agricultural Pests and Diseases" (R. Baker and P. R. Dunn, eds.), pp. 497–515. Liss, New York.

Beckage, N. E., and Templeton, T. J. (1986). Physiological effects of parasitism by *Apanteles congregatus* in terminal-stage tobacco hornworm larvae. *J. Insect Physiol.* **32:**299–314.

Beckage, N. E., Nesbit, D. J., Nielsen, B. D., Spence, K. D., and Barman, M.A.E. (1989). Alteration of hemolymph polypeptides in *Manduca sexta* larvae parasitized by *Cotesia congregata:* A two-dimensional electrophoretic analysis and comparison with major bacteria-induced proteins. *Arch. Insect Biochem. Physiol.* **10:**1–17.

Bhaskaran, G., Sparagana, S. P., Barrera, P., and Dahm, K. H. (1986). Change in corpus allatum function during metamorphosis of the tobacco hornworm *Manduca sexta:* Regulation at the terminal step in juvenile hormone biosynthesis. *Arch. Insect Biochem. Physiol.* **3:**321–338.

Bollenbacher, W. E. (1988). The interendocrine regulation of larval-pupal development in the tobacco hornworm, *Manduca sexta:* A model. *J. Insect. Physiol.* **34:**941–947.

Bowers, W. S. (1983). The precocenes. *In* "Endocrinology of Insects" (R.G.H. Downer and H. Laufer, eds.), pp. 517–523. Liss, New York.

Bühler, A., Hanzlik, T. N., and Hammock, B. D. (1985). Effects of parasitization of *Trichoplusia ni* by *Chelonus* sp. *Physiol. Entomol.* **10:**383–394.

Cals-Usciati, J. (1969). Influence de l'état physiologique de l'hôte *Ceratitis capitata* Weid. (Diptera) sur le développement du parasite *Opius concolor* Szepl. (Hymenoptera). *C. R. Hebd. Séances/Acad. Sci., Sér. D* **269:**342–344.

Cals-Usciati, J. (1975). Répercussion de la modification du cycle normal de *Ceratitis capitata* Weid. (Diptère: Trypetidae), par irradiation Y et injection d'ecdysone, sur le développment de son parasite *Opius concolor* Szepl. (Hymenoptère: Braconidae). *C. R. Hebd. Séances/Acad. Sci., Sér. D* **281**:275–278.

Collins, J. V. (1974). Hormonal control of protein sequestration in the fat body of *Calpodes ethlius*. *Can. J. Zool.* **52**:639–642.

Cymborowsky, B., Bogus, M., Beckage, N. E., Williams, C. M., and Riddiford, L. M. (1982). Juvenile hormone titers and metabolism during starvation-induced supernumerary larval moulting of the tobacco hornworm, *Manduca sexta* L. *J. Insect Physiol.* **28**:129–135.

Dahlman, D. L. (1990). Evaluation of teratocyte functions: An overview. *Arch. Insect Biochem. Physiol.* **13**:159–166.

Dahlman, D. L., Coar, D. L., Koller, C. N., and Neary, T. J. (1990a). Contributing factors to reduced ecdysteroid titers in *Heliothis virescens* parasitized by *Microplitis croceipes*. *Arch. Insect Biochem. Physiol.* **13**:29–39.

Dahlman, D. L., Zhang, D., and Neary, T. D. (1990b). Alteration of juvenile hormone esterase activity by parasitoid-derived factors [abstract]. *In* "Molecular Insect Science" (H. H. Hagedorn, J. G. Hildebrand, M. G. Kidwel, and J. H. Law, eds.), pp. 295. Plenum, New York.

Davies, D. H., Strand, M. R., and Vinson, S. B. (1987). Changes in differential hemocyte count and in vitro behavior of plasmatocytes from host *Heliothis virescens* caused by *Campoletis sonorensis* polydnavirus. *J. Insect Physiol.* **33**: 143–153.

Doutt, R. L. (1964). Biological characteristics of entomophagous adults. *In* "Biological Control of Insect Pests and Weeds" (P. DeBach and E. I. Schlinger, eds.), pp. 145–167. Reinhold, New York.

Dover, B. A., and Vinson, S. B. (1990). Stage-specific effects of *Campoletis sonorensis* parasitism on *Heliothis virescens* development and prothoracic glands. *Physiol. Entomol.* **15**:405–414.

Dover, B. A., Davies, D. H., Strand, M. R., Gray, R. S., Keeley, L. L., and Vinson, S. B. (1987). Ecdysteroid-titre reduction and developmental arrest of last-instar *Heliothis virescens* larvae by calyx fluid from the parasitoid *Campoletis sonorensis*. *J. Insect Physiol.* **33**:333–338.

Dover, B. A., Davies, D. H., and Vinson, S. B. (1988a). Dose-dependent influence of *Campoletis sonorensis* polydnavirus on the development and ecdysteroid titers of last-instar *Heliothis virescens* larvae. *Arch. Insect Biochem. Physiol.* **8**:113–126.

Dover, B. A., Davies, D. H., and Vinson, S. B. (1988b). Degeneration of last instar *Heliothis virescens* prothoracic glands by *Campoletis sonorensis* polydnavirus. *J. Invertebr. Pathol.* **51**:80–91.

Dover, B. A., Strand, M. R., Davies, D. H., and Vinson, S. B. (1989). Ultrastructure of host tissues exposed to the calyx fluid of the parasitoid, *Campoletis sonorensis* (Cameron) (Hymenoptera, Ichneumonidae). *Int. J. Insect Morphol. Embryol.* **18**:47–57.

Downer, R.G.H., and Laufer, H., eds. (1983). "Endocrinology of Insects." Liss, New York.

Feldlaufer, M. F. (1989). Diversity of molting hormones in insects. *In* "Ecdysone" (J. Koolman, ed.), pp. 308–311. Thieme, Stuttgart.

Ferkovich, S. M., and Dillard, C. R. (1986). A study of radiolabeled host proteins and protein synthesis during development of eggs of the endoparasitoid, *Microplitis croiceipes* (Cresson) (Braconidae). *Insect Biochem.* **16**:337–343.

Gilbert, L. I. (1989). The endocrine control of molting: The tobacco hornworm, *Manduca sexta*, as a model system. *In* "Ecdysone" (J. Koolman, ed.), pp. 448–471. Thieme, Stuttgart.

Goldsworthy, G. J., and Gäde, G. (1983). The chemistry of hypertrehalosemic factors. *In* "Endocrinology of Insects" (R.G.H. Downer and H. Laufer, eds.), pp. 109–119. Liss, New York.

Greany, P. D. (1986). In vitro culture of hymenopterous larval endoparasitoids. *J. Insect Physiol.* **32**:409–419.

Grossniklaus-Bürgin, C., and Lanzrein, B. (1990a). Qualitative and quantitative analysis of juvenile hormone and ecdysteroids from the egg to the pupal molt in *Trichoplusi ni*. *Arch. Insect Biochem. Physiol.* **14:**13–30.

Grossniklaus-Bürgin, C., and Lanzrein, B. (1990b). Endocrine interrelationship between the parasitoid *Chelonus* sp. and its host *Trichoplusi ni*. *Arch. Insect Biochem. Physiol.* **14:**201–216.

Grossniklaus-Bürgin, C., Connat, J. L., and Lanzrein, B. (1989). Ecdysone metabolism in the host–parasitoid system *Trichoplusia ni*. *Arch. Insect Biochem. Physiol.* **11:**79–92.

Gruetzmacher, M. C., Gilbert, L. I., Granger, N. A., Goodman, W., and Bollenbacher, W. E. (1984). The effect of juvenile hormone on prothoracic gland function during the larval-pupal development of *Manduca sexta:* An in situ and in vitro analysis. *J. Insect Physiol.* **30:**331–340.

Gupta, A. P., ed. (1990). "Morphogenetic Hormones of Arthropods." Rutgers Univ. Press, New Brunswick, NJ and London.

Halarnkar, P. P., and Schooley, D. A. (1990). Reversed-phase liquid chromatographic separation of juvenile hormone and its metabolites, and its application for an in vivo juvenile hormone metabolism study in *Manduca sexta*. *Entomol. Exp. Appl.* **188:**394–397.

Hammock, B. D. (1985). Regulation of juvenile hormone titer: Degradation. *In* "Comprehensive Insect Physiology, Biochemistry and Pharmacology" (G. A. Kerkut and L. I. Gilbert, eds.), Vol. 7, pp. 431–472. Pergamon, New York.

Hartmann, R., Jendrsczok, C., and Peter, M. G. (1987). The occurrence of a juvenile hormone binding protein and in vitro synthesis of juvenile hormone by the serosa of *Locusta migratoria* embryos. *Roux's Arch. Dev. Biol.* **196:**347–355.

Hayakawa, Y. (1990). Juvenile hormone esterase activity repressive factor in the plasma of parasitized insect larvae. *J. Biol. Chem.* **265:**10813–10816.

Hiruma, K. (1980). Possible roles of juvenile hormone in the prepupal stage of *Mamestra brassicae*. *Gen. Comp. Endocrinol.* **41:**392–399.

Hiruma, K. (1986). Regulation of prothoracicotropic hormone release by juvenile hormone in the penultimate and last instar larvae of *Mamestra brassicae*. *Gen. Comp. Endocrinol.* **63:**201–211.

Joiner, R. L., Vinson, S. B., and Benskin, J. B. (1973). Teratocytes as source of juvenile hormone activity in a parasitoid–host relationship. *Nature* (*London*) *New Biol.* **246:**120–121.

Jones, D. (1986). *Chelonus* sp.: Suppression of host ecdysteroids and developmentally stationary pseudoparasitized prepupae. *Exp. Parasitol.* **61:**10–17.

Jones, D. (1987). Material from adult female *Chelonus* sp. directs expression of altered developmental programme of host Lepidoptera. *J. Insect Physiol.* **33:**129–134.

Jones, D., Jones, G., and Hammock, B. D. (1981). Developmental and behavioral responses of larval *Trichoplusia ni* to parasitization by an imported braconid parasite *Chelonus* sp. *Physiol. Entomol.* **6:**387–394.

Jones, G., and Hammock, B. D. (1983). Prepupal regulation of juvenile hormone esterase through direct induction by juvenile hormone. *J. Insect Physiol.* **29:**471–475.

Jones, G., Hanzlik, T., Hammock, B. D., Schooley, D. A., Miller, C. A., Tsai, L. W., and Baker, F. C. (1990). The juvenile hormone titer during the penultimate and ultimate larval stadia of *Trichoplusia ni*. *J. Insect Physiol.* **36:**77–83.

Kelly, T. J., Woods, C. W., Redfern, R. E., and Borkevec, A. B. (1981). Makisterone A: The moulting hormone of larval *Oncopeltus*. *J. Exp. Zool.* **218:**127–132.

Kelly, T. J., Aldrich, J. R., Woods, C. W., and Borkevec, A. B. (1984). Makisterone A: Its distribution and physiological role as the moulting hormone of true bugs. *Experientia* **40:**996–997.

Kerkut, G. A., and Gilbert, L. I., eds. (1985). "Comprehensive Insect Physiology, Biochemistry and Pharmacology." Pergamon, New York.

Klages, G., and Emmerich, H. (1979). Juvenile hormone metabolism and juvenile hormone esterase titer in hemolymph and peripheral tissues of *Drosophila hydei. J. Comp. Physiol.* **132:**319–325.

Kunkel, J. G., Grossniklaus-Bürgin, C., Karpells, S. T., and Lanzrein, B. (1990). Arylphorin of *Trichoplusia ni:* Characterization and parasite-induced precocious increase in titer. *Arch. Insect Biochem. Physiol.* **13:**117–125.

Lafont, R., and Connat, J. L. (1989). Pathways of ecdysone metabolism. *In* "Ecdysone" (J. Koolman, ed.), pp. 167–173. Thieme, Stuttgart.

Lafont, R., and Koolman, J. (1984). Ecdysone metabolism. *In* "Biosynthesis, Metabolism and Mode of Action in Invertebrate Hormones" (J. A. Hoffman and M. Porchet, eds.), pp. 196–226. Springer-Verlag, Heidelberg.

Lafont, R., Mauchamp, B., Pennetier, J. L., Tarroux, P., De Hys, L., and Delbecque, J. P. (1975). α-ecdysone and β-ecdysone levels in insect hemolymph—correlation with developmental events. *Experientia* **31:**1241–1242.

Lathrop, F. H., and Newton, R. C. (1933). The biology of *Opius melleus* Gahan, a parasite of the blueberry maggot. *J. Agric. Res.* **46:**143–160.

Lawrence, P. O. (1981). Developmental and reproduction biologies of *Biosteres longicaudatus* reared on hosts treated with a chitin synthesis inhibitor. *Insect Sci. Appl.* **1:**403–406.

Lawrence, P. O. (1983). Age-specific fecundity and offspring survivorship in the Caribbean fruit fly, *Anastrepha suspensa,* after diflubenzuron treatment. *Insect Sci. Appl.* **4:**285–290.

Lawrence, P. O. (1986a). Host–parasite hormonal interactions: An overview. *J. Insect Physiol.* **32:**295–298.

Lawrence, P. O. (1986b). The role of 20-hydroxyecdysone in the moulting of *Biosteres longicaudatus,* a parasite of the Caribbean fruit fly, *Anastrepha suspensa. J. Insect Physiol.* **32:**329–337.

Lawrence, P. O. (1988a). Superparasitism of the Caribbean fruit fly *Anastrepha suspensa* (Diptera: Tephritidae), by *Biosteres longicaudatus* (Hymenoptera: Braconidae): Implications for host regulation. *Ann. Entomol. Soc. Am.* **81:**233–239.

Lawrence, P. O. (1988b). Ecdysteroid titers and integument changes in superparasitized puparia of *Anastrepha suspensa. J. Insect Physiol.* **34:**603–608.

Lawrence, P. O. (1988c). Hormonal interactions between parasitoids and hosts: Adaptation to stress? *In* "Endocrinological Frontiers in Physiological Insect Ecology" (F. Sehnal, A. Zabza, and D. L. Denlinger, eds.), pp. 423–435. Wroclaw Tech. Univ. Press, Wroclaw, Poland.

Lawrence, P. O. (1988d). In vivo and in vitro development of first instars of the parasitic wasp *Biosteres longicaudatus* (Hymenoptera: Braconidae). *In* "Advances in Parasitic Hymenoptera Research" (V. Gupta, ed.), pp. 351–366. E. J. Brill, New York.

Lawrence, P. O. (1990a). Serosal cells of *Biosteres longicaudatus* (Hymenoptera:: Braconidae): Ultrastructure and release of polypeptides. *Arch. Insect Biochem. Physiol.* **13:**199–216.

Lawrence, P. O. (1990b). The biochemical and physiological effects of insect hosts on the development and ecology of their insect parasites: An overview. *Arch. Insect Biochem. Physiol.* **13:**217–228.

Lawrence, P. O. (1991). Hormonal effects on insects and other endoparasites in vitro. *In Vitro Cell. Dev. Biol.* **27:**487–496.

Lawrence, P. O., and Hagedorn, H. H. (1986). Relationship between the ecdysteroid titres of a host and those of its parasite. *Insect Biochem.* **16:**163–168.

Lawrence, P. O., Baker, F. C., Tsai, L. W., Miller, C. A., Schooley, D. A., and Geddes, L. G. (1990). JH III levels in larvae and pharate pupae of *Anastrepha suspensa* (Diptera: Tephritidae) and in larvae of the parasitic wasp *Biosteres longicaudatus* (Hymenoptera: Braconidae). *Arch. Insect Biochem. Physiol.* **13:**53–62.

Locke, M. (1980). The cell biology of fat body development. *In* "Insect Biology in the Future—VBW 80" (M. Locke and D. S. Smith, eds.), pp. 227–252. Academic Press, New York.

Mayer, R. T., Meola, S. M., Coppage, D. L., and DeLoach, J. R. (1980). Utilization of imaginal tissues from pupae of the stable fly for the study of chitin synthesis and screening of chitin synthesis inhibitors. *J. Econ. Entomol.* **73:**76–80.

McNeil, J. (1975). Juvenile hormone and analogues: Detrimental effects on the development of an endoparasitoid. *Science* **189:**640–642.

Mellini, E. (1975). Studi sui Ditteri larvevoridi. XXV. Sul determinismo ormondale delle influenze esercitate dagli ospiti sui loro parassiti. *Boll. Ist. Entomol. Univ. Bologna* **31:**165–203.

Nenon, J.-P. (1972). Culture in vitro des embryos d'un Hyménoptère endoparasite polyembryonnaire: *Ageniaspis fuscicollis*—Rôle des hormones de synthesis. *C. R. Hebd. Séances/ Acad. Sci., Sér. D* **274:**3409–3412.

Nettles, W. C., Jr. (1990). In vitro rearing of parasitoids: Role of host factors in nutrition. *Arch. Insect Biochem. Physiol.* **13:**167–175.

Nijhout, H. F. (1975). Dynamics of JH action in larvae of the tobacco hornworm, *Manduca sexta. Biol. Bull.* (*Woods Hole, Mass.*) **149:**568–579.

Nishida, R., Bowers, W. S., and Evans, P. H. (1983). Juvadecene: Discovery of a juvenile hormone mimic in the plant *Macropiper excelsum. Arch. Insect Biochem. Physiol.* **1:**17–24.

Pemberton, C. E., and Willard, H. F. (1918). A contribution to the biology of fruit fly parasites in Hawaii. *J. Agric. Res.* **15:**419–465.

Plantevin, G., Grenier, S., Richard, G., and Nardon, C. (1986). Larval development, developmental arrest, and hormone levels in the couple *Galleria mellonella* (Lepidoptera–Pyralidae)–*Pseudoperichaeta nigrolineata* (Diptera–Tachinidae). *Arch. Insect Biochem. Physiol.* **3:**457–470.

Rauschenbach, I. Y. (1991). Changes in juvenile hormone and ecdysteroid content during insect development under heat stress. *In* "Hormones and Metabolism in Insect Stress" (J. Ivanovič and M. Jankovič-Hladni, eds.), pp. 115–148. CRC Press, Boca Raton, FL.

Redfern, C. P. (1984). Evidence for the presence of makisterone A in *Drosophila* larvae and the secretion of 20-deoxymakisterone A by the ring gland. *Proc. Natl. Acad. Sci. U.S.A.* **81:**5643–5647.

Rembold, H., and Schmutterer, D. (1981). Disruption of insect growth by neem seed components. *In* "Regulation of Insect Development and Behavior" (F. Sehnal, A. Zabza, J. J. Menn, and B. Cymborowski, eds.), pp. 1087–1090. Wroclaw Tech. Univ. Press, Wroclaw, Poland.

Reum, L., and Koolman, J. (1989). Radioimmunoassay of ecdysteroids. *In* "Ecdysone" (J. Koolman, ed.), pp. 131–143. Thieme, Stuttgart.

Richard, D. S., Applebaum, S. W., Sliter, T. J., Baker, F. C., Schooley, D. A., Reuter, C. C., Henrich, V. C., and Gilbert, L. I. (1989). Juvenile hormone bisepoxide biosynthesis in vitro by the ring gland of *Drosophila melanogaster:* a putative juvenile hormone in the higher Diptera. *Proc. Natl. Acad. Sci. U.S.A.* **86:**1421–1425.

Richards, G. (1981a). The radioimmunoassay of ecdysteroid titers in *Drosophila. Mol. Cell. Endocrinol.* **21:**181–197.

Richards, G. (1981b). Insect hormones in development. *Biol. Rev. Cambridge Philos. Soc.* **56:**501–549.

Riddiford, L. M. (1980). Interaction of ecdysteroids and juvenile hormone in the regulation of larval growth and metamorphosis of the tobacco hornworm. *In* "Progress in Ecdysone Research" (J. A. Hoffmann, ed.), pp. 409–430. Elsevier/North-Holland, Amsterdam.

Salt, G. (1937). The egg parasite of *Sialis lutaria:* A study of the influence of the host upon a dimorphic parasite. *Parasitology* **29:**539–553.

Salt, G. (1941). The effects of hosts upon their insect parasites. *Biol. Rev. Cambridge Philos. Soc.* **16:**239–264.

Sehnal, F. (1989). Hormonal role of ecdysteroids in insect larvae and during metamorphosis. *In* "Ecdysone" (J. Koolman, ed.), pp. 271–277. Thieme, Stuttgart.

Sehnal, F., and Granger, N. A. (1975). Control of corpora allata function in larvae of *Galleria mellonella*. *Biol. Bull.* (*Woods Hole, Mass.*) **148:**106–116.

Shapiro, J. P., Mayer, R. T., and Schroeder, W. J. (1988). Absorption and transport of natural and synthetic toxins mediated by hemolymph proteins. *In* "Endocrinological Frontiers in Physiological Insect Ecology" (F. Sehnal, A. Zabza, and D. L. Denlinger, eds.), Wroclaw Tech. Univ. Press, Wroclaw, Poland.

Stoltz, D. B. (1986). Interactions between parasitoid-derived products and host insects: An overview. *J. Insect Physiol.* **32:**347–350.

Stoltz, D. B., and Vinson, S. B. (1979). Viruses and parasitism in insects. *Adv. Virus Res.* **24:**125–171.

Stoltz, D. B., Guzo, E. R., Belland, E. R., Lucarotti, C. J., and Mackinnon, E. A. (1988). Venom promotes uncoating in vitro and persistence in vivo of DNA from a braconid polydnavirus. *J. Gen. Virol.* **69:**903–907.

Strand, M. R. (1989). Development of the polyembryonic parasitoid *Copidosoma floridanum* in *Trichoplusia ni*. *Entomol. Exp. Appl.* **50:**37–46.

Strand, M. R., Dover, B. A., and Johnson, J. A. (1990). Alterations in the ecdysteroid and juvenile hormone esterase profiles of *Trichoplusia ni* parasitized by the polyembryonic wasp *Copidosoma floridanum*. *Arch. Insect Biochem. Physiol.* **13:**41–51.

Strand, M. R., Goodman, W. G., and Baehrecke, E. H. (1991). The juvenile hormone titer of *Trichoplusia ni* and its potential role in embryogenesis of the polyembryonic wasp *Copidosoma floridanum*. *Insect Biochem.* **21:**205–214.

Tanaka, T. (1987). Calyx and venom fluids of *Apanteles kariyai* (Hymenoptera, Braconidae) as factors that prolong larval period of the host, *Pseudaletia separata* (Lepidoptera, Noctuidae). *Ann. Entomol. Soc. Am.* **80:**530–533.

Tanaka, T., and Vinson, S. B. (1991). Depression of prothoracic gland activity of *Heliothis virescens* by venom and calyx fluids from the parasitoid, *Cardiochiles nigriceps*. *J. Insect Physiol.* **37:**139–144.

Tanaka, T., Agui, N., and Hiruma, K. (1987). The parasitoid *Apanteles kariyai* inhibits pupation of its host *Pseudaletia separata,* via disruption of prothoracicotropic hormone release. *Gen. Comp. Endocrinol.* **67:**364–374.

Thompson, S. N. (1986). The metabolism of insect parasites (parasitoids): An overview. *J. Insect Physiol.* **32:**421–423.

Vinson, S. B. (1970). Development and possible function of teratocytes in the host–parasite association. *J. Invertebr. Pathol.* **16:**93–101.

Vinson, S. B., and Iwantsch, G. F. (1980a). Host suitability for insect parasitoids. *Annu. Rev. Entomol.* **25,** 397–419.

Vinson, S. B., and Iwantsch, G. F. (1980b). Host regulation by insect parasitoids. *Q. Rev. Biol.* **55:** 143–165.

Wani, M., Yagi, S., and Tanaka, T. (1990). Synergistic effect of venom, calyx and teratocytes of *Apanteles kariyai* on the inhibition of larval-pupal ecdysis of the host, *Pseudaletia separata*. *Entomol. Exp. Appl.* **57:**101–104.

Warren, J. T., and Gilbert, L. I. (1986). Ecdysone metabolism and distribution during the pupal-adult development of *Manduca sexta*. *Insect Biochem.* **16:**65–82.

Watson, R. D., and Bollenbacher, W. E. (1988). Juvenile hormone regulates the steroidogenic competence of *Manduca sexta* prothoracic glands. *Mol. Cell. Endocrinol.* **57:**251–259.

Watson, R. D., Spaziani, E., and Bollenbacher, W. E. (1989). Regulation of ecdysone biosynthesis in insects and crustaceans: A comparison. *In* "Ecdysone" (J. Koolman, ed.), pp. 188–203. Thieme, Stuttgart.

Webb, B. A., and Dahlman, D. L. (1986). Ecdysteroid influence on the development of the host *Heliothis virescens* and its endoparasite *Microplitis croceipes*. *J. Insect Physiol.* **32:**339–345.

Wing, S. (1988). RH 5849, a nonsteroidal ecdysone agonist: Effects on a *Drosophila* cell line. *Science* **241:**467–469.

Wright, J. E., and Spates, G. E. (1972). A new approach in integrated control: Insect juvenile hormone plus a hymenopteran parasite against the stable fly. *Science* **178:**1292–1293.

Wyatt, G. R. (1980). The fat body as a protein factory. *In* "Insect Biology in the Future—VBW 80" (M. Locke and D. S. Smith, eds.), pp. 201–225. Academic Press, New York.

Zhang, D., and Dahlman, D. L. (1989). *Microplitis croceipes* teratocytes cause developmental arrest of *Heliothis virescens* larvae. *Arch. Insect Biochem. Physiol.* **12:**51–61.

Chapter 4

Reproductive Disturbances Induced by Parasites and Pathogens of Insects

Hilary Hurd
Department of Biological Sciences
University of Keele
Staffordshire, United Kingdom

I. Introduction

Parasites and pathogens play a major role as regulators of host life histories and population dynamics (see, e.g., Dobson, 1988; Holmes and Zohar, 1990; May and Anderson, 1990) and their influence upon the evolution and maintenance of host sexual behavior has been reviewed recently (Read, 1990). It is my intention to confine this review to organisms that affect insect populations, not as lethal agents, but by reducing reproductive output or causing host sterility. Parasitized invertebrates commonly exhibit some form of reproductive curtailment (Hurd, 1990a) and diverse aspects of physiology and behavior can be affected, ranging from irreversible destruction of reproductive tissue (true castration) to changes in sexual behavior or juvenilization. Although examples of reproductive dysfunction in infected insects are abundant, attempts to elucidate the underlying mechanisms have rarely been made. Wherever possible I will focus upon these studies but material is arranged to present a taxonomic rather than mechanistic survey. Thus, the effects of microparasites [defined by Anderson and May (1981) as small pathogens with a high rate of reproduction within the host and a short generation time], nematodes, cestodes, and endoparasitic insects upon insect reproduction will be examined.

II. Microparasites

The impact of nonlethal viral, fungal, and bacterial infections upon insect population dynamics is very great, both host reproductive output and survivorship of offspring being affected. However, a survey of this field would require more space than is available and I shall therefore concentrate upon those microparasites generally regarded as being within the remit of parasitology, the Protista. Although many insects act as hosts for protozoa with single-stage life cycles, some are vectors or intermediate hosts for parasites with vertebrate definitive hosts, often causing diseases of medical or economic importance. Our knowledge of the interaction between the Protista and insect-host reproductive output is surprisingly scanty, especially in view of the importance of, for example, the Diptera in the epidemiology of parasitic disease. The majority of studies have concentrated upon microsporidial infections and these will be dealt with separately.

A comparison of reproductive output in six strains of *Aedes aegypti* (Hacker, 1971) showed a marked difference in the reduction in fecundity caused by *Plasmodium gallinaceum*, with the ROCK strain being the most affected. The possible impact of the parasite upon the genetic structure of a polymorphic host population, and its role in the maintenance of a balanced population structure, was discussed. Further studies (Hacker and Kilama, 1974) showed that a reduction in mosquito fecundity occurred at very low levels of parasitemia, measured in terms of both number of parasites per 100 chicken erythrocytes in the blood meal and oocyst density in the mosquito gut. Host fecundity was not related to parasite burden in this study.

Freier and Friedman (1976) confirmed that *P. gallinaceum*-infected *A. aegypti* laid fewer eggs (an average of 35% less) than those fed on uninfected chickens and also demonstrated that egg viability was unaffected by infection. The size of the blood meal taken by *A. aegypti* is known to influence the number of eggs laid (see discussion in Freier and Friedman, 1976). Female mosquitoes feeding on parasitized chickens were shown to take 43% less blood than those feeding on uninfected blood and decrease in blood meal size and fecundity were both associated with a rise in blood parasitemia (Freier and Friedman, 1976), although considerable variability existed between individual insects. Infected mosquitoes are known to exhibit altered probing behavior, probably associated with the presence of sporozoites in the salivary glands (see e.g., Molyneux and Jefferies, 1976). Freier and Friedman (1976) found that probing behavior of uninfected females did not differ when feeding on infected chickens but blood ingestion rates were reduced from 1.23 to 0.66 mg/min. Furthermore, feeding choice experiments indicated that females preferred uninfected chickens and could distinguish between birds suspended 1 cm from the cage. This report indicates that factors other than, or in addition to, quality of blood meal are involved in feeding choice and the authors conclude that olfaction and gustation may be noninteracting factors affecting feeding behavior.

There is clearly much work still to be done to clarify the relationship between blood intake and fecundity. Investigations need to be performed to determine the effect of malaria on egg production in the second and subsequent gonotrophic cycles, when the parasite is established in the mosquito. In addition there have been no detailed studies of the effect of *Plasmodium* spp. upon egg laying in *Anopheles* mosquitoes, the vectors of human malaria. Shieh and Rossignol (1992) discussed the influence of the anaemia associated with blood-borne parasites upon mosquito feeding and subsequent fecundity. They confirmed the Daniel and Kingsolver model predictions by demonstrating that *Aedes aegypti* feed faster on anaemic blood and also demonstrated the negative influence of anaemic blood on egg laying. The significance of these findings to the field situation, where fast feeding may be important when feeds are frequently interrupted due to host defences, was discussed.

The amoeba *Malamoeba locustae* infects the Malpighian tubules of the desert locust and causes a progressive attrition of the host fat body. Changes in fat body protein metabolism result in a reduction in hemolymph protein titer, which may account for the significant delay in oocyte development observed in infected females (Papillon and Cassier, 1978). Fat body attrition is also associated with *M. locustae* infection of grasshoppers.

Using the *M. locustae*/grasshopper model Jackson *et al.* (1968) made an examination of egg lipids. Although no change in total lipid content or lipid composition occurred in infected grasshoppers, the relative abundance of fatty acids was altered by infection. The authors were unable to determine whether this difference accounted for the decrease in egg viability associated with infection.

A. Microsporidia

The majority of microsporidian–insect associations produce chronic, nonlethal infections (Weiser, 1974) of which a common feature is a reduction in host reproductive output. This reproductive curtailment is engendered in a number of ways and generalization is hampered by the variation in assessment parameters adopted by investigators. However, a reduction in mating success, fecundity, and egg hatch have been described by many authors as shown in Table 1. Mercer and Wigley (1987) measured mating success in *Nosema* sp. (NCS)-infected poroporo stem borers, *Sceliodes cordalis,* as the ability to lay viable eggs. In addition, Gaugler and Brooks (1975) observed a higher incidence of locked matings (failure to separate after copulation) in

Table 1

The Effect of Microsporidial Infections upon the Reproductive Success of Their Insect Hosts

Parasite	Host	Mating success[a]	Fecundity	Egg hatch	Vertical transmission	Venereal transmission	Source
Nosema pyraustra	*Ostrinia nubilalis*	?	-ve	-ve	?	?	Windels *et al.* (1976)
		?	-ve	-ve	+	?	Siegel *et al.* (1986)
		?	?	?	++ (ho)	?	Sajup and Lewis (1988)
Nosema sp. (NSC)	*Sceliodes cordalis*	-ve	-ve	-ve	++ (ss)	?	Mercer and Wigley (1987)
Nosema whitei	*Triboluim* spp.	?	-ve	0	No	?	Milner (1972)
	T. castaneum	-ve	-ve[b]	?	?	?	Armstrong and Bass (1986)
Nosema heliothidis	*Heliothus zea*	-ve	-ve[c]	0	+	?	Gaugler and Brooks (1975)
Nosema algerae	*Aedes aegypti*	?	-ve	-ve	++ (ho)	?	Nnakumusana (1986)
	Culex fatigans	?	-ve	-ve	++ (ho)	?	
Nosema carpocapsae	*Cydia pomonella*	-ve	-ve	-ve	++ (ho) (ss)	No	Malone and Wigley (1981)
Nosema kingi	*Drosophila willistoni*	?	?	?	+	Yes	Armstrong (1977)
Nosema plodiae	*Plodia interpunctella*	?	-ve	-ve	+	Yes	Kellen and Lindegren (1971)
Nosema bombysis	*Bombyx mori*	?	?	-ve	++ (ss)	No	Han and Watanabe (1988)
Nosema sp. M11		?	?	-ve	++ (ss)	No	
Nosema stegomyiae	*Culex fatigans*	?	-ve	-ve	No	Possible	Reynolds (1971)
Nosema fumiferana	*Choristoneura fumiferana*	0	-ve	0	+	?	Bauer and Nordin (1989)
Microsporidium itiiti	*Listronotus bonariensis*	?	-ve	0	?	?	Malone (1987)

[a]-ve, parasite adversely affects the host; 0, no change occurs with infection; ?, feature not investigated; +, transmission occurs; ++, transovarial transmission demonstrated by surface sterilization; of eggs (ss) or by histological observation (ho).
[b]Fecundity measured as offspring emergence.
[c]Only occurs when insects are infected *per os*.

Nosema heliothidis infections of *Heliothis zea*, and Armstrong and Bass (1986) measured mating frequency in *N. whitei*-infected *Tribolium castaneum*. Fecundity is usually measured as the number of eggs laid (e.g., Kellen and Lindegren, 1971; Malone and Wigley, 1981; Malone, 1987) and egg viability is determined by egg hatch rate. However, Mercer and Wigley (1987) distinguished infertile eggs from those undergoing incomplete embryogenesis (and therefore not hatching), though they did not determine whether the significant decrease in successful embryo development was due to depletion of food reserves or interference with embryogenesis.

The mechanisms underlying this reduced fecundity are diverse. Several authors have suggested that parasite-induced nutritional deprivation is the cause of reduced egg production (Malone, 1987). Depletion of fat and protein reserves is associated with microsporidial infection of insect fat bodies (e.g., Thompson and Sikorowski, 1979) and may result in reduced vitellogenin production (Gaugler and Brooks, 1975).

The involvement of the host endocrine system in the interaction between microsporidia and larval insect hosts was proposed by Fisher and Sanborn (1962, 1964) to explain the disturbance of larval molts and occurrence of adultoids that they observed in these associations. They demonstrated that *Nosema* spp. secretory products restored the nymphal or larval molts to allatectomized insects when the protozoa were applied externally and separated from the hemocoel by a 0.45-μm millipore filter (Fisher and Sanborn, 1964). Juvenile hormone-like activity was demonstrated in spore extracts from *Nosema* (Fisher and Sanborn, 1964), *Farinocystis tribolii* (Rabindra *et al.*, 1981), and *N. bombycis* (Karavaeva, 1987). However, Milner (1972) was unable to demonstrate the disturbance of larval molts described by Fisher and Sanborn (1962, 1964) and several authors have attributed the parasite associated disturbance of insect development to fat body invasion (e.g., Listov, 1977). It has been suggested that destruction of fat body tissue would reduce the production of juvenile hormone degradation enzymes and lead to an accumulation of the hormone, thus resulting in abnormal larval development (Rabindra *et al.*, 1981).

Ultrastructural studies of corpora allata from *Nosema apis*-infected worker bees demonstrated that neurosecretory axon granules were rarely present in infected bees (Liu, 1986). The author suggested that gland activity may be affected but no histolysis or degradation was observed. Later studies of the effect of juvenile hormone III on the honeybee hypopharangeal gland showed that the hormone caused a depletion of ribosomes in the cytoplasm of the cells, resulting in inhibition of protein synthesis. The observation that *Nosema* infection did not cause ribosome depletion led to the conclusion that the action of *N. apis* may not be synonymous with JH action (Liu, 1989, 1990). Clearly, further studies are required to elucidate the role of endocrine factors in microsporidial–insect associations and in particular their participation, if any, in the reduction of host reproductive output.

Histological studies of the ovaries of *N. pyrausta*-infected *Ostrinia nubilalis* (Sajup and Lewis, 1988) demonstrated massive germ cell destruction, which would account for reduced fecundity. However, ovarian invasion by microsporidia does not always decrease egg viability as infected larvae are produced in many associations. Sajup and Lewis (1988) postulated that the concentration of microsporidia in the center of oocytes may prevent interference with developing embryos.

Although many microsporidia are confined to host tissues such as fat body or gut, those that invade the reproductive system are provided with an avenue for transmission. Thus, both horizontal transmission of parasites between hosts and vertical transmission to the next generation can occur within the same associations. Some authors have distinguished between transovarial and transovum transmission (see Table 1). In the former, infection occurs while the egg is developing within the ovary, and in the latter the pathogen is adsorbed onto the egg surface and infection of the larvae occurs during or after hatching. Thus, Sajup and Lewis (1988) observed that *N. pyrausta* may pass into developing oocytes with nutrients, from the trophocytes of *O. nubilalis* and, in addition, infection of the epithelial cells of the oviduct may result in transovum transmission. If histological studies have not been performed, surface sterilization of eggs is an essential step in the determination of whether true transovarial transmission has occurred when infected larvae emerge. Venereal transmission, resulting in infected larvae being produced by noninfected females mated to infected males has also been described. This may be due to contamination of the egg surface (Kellen and Lindegren, 1971) or true infection of the female reproductive tissues (Armstrong, 1977).

The host reproductive system thus appears to play a major role in microsporidial transmission (see Table 1) although the exact mode (transovarial or transovum) has not always been determined. Nnakumusana (1986) suggested that this is the sole method of transmission for more than 30 species of microsporidial larval pathogens of dipteran vectors. Smith and Dunn (1991) recently discussed cytoplasmic inheritance of intracellular symbionts and stressed the selection pressures faced by parasites causing pathology in host offspring. In view of this, it is surprising that this is the sole route of transmission for so many microsporidia.

Few ecological studies of the effects of microsporidial infections have been performed (but see Andreadis, 1986). However, it is clear that a knowledge of the degree of importance of horizontal and vertical transmission is an essential factor in the construction of models of the influence of microsporidia upon host population dynamics (Anderson and May, 1981; Onstad and

Maddox, 1989). The potential for the use of insect pathogens as microbial control agents of pest species has long been recognized. Although microsporidial infections are usually sublethal, they result in severe reduction in fecundity and, in the case of *N. pyrausta–O. nubilalis* infections, have been shown to down-regulate the host population below the carrying capacity of the environment (Onstad and Maddox, 1989). For example, *N. locusta* spores have been shown to be effective population regulators both alone (Raina *et al.*, 1987) and in combination with insecticides (Germida *et al.*, 1987; Johnson and Henry, 1987). Moreover, the potential for microsporidia as control agents of stored product Coleoptera has recently been emphasized by Khan and Selman (1989).

III. Nematodes

The ratio of parasite to host tissue usually attains far greater proportions in entomogenous nematode infections than in the insect–microparasite associations discussed in the foregoing. Thus, the potential drain on host resources is likely to be greater. Although little direct gonadal tissue invasion has been observed, some authors report examples of nematode-induced reproductive curtailment ranging from slight reductions in egg laying to male and/or female sterilization. In many cases parasite burden has been correlated with loss of reproductive output and evidence suggesting that nutritional deprivation is linked to host-resource reallocation is discussed in the following.

Allantonematids infect a variety of bark beetles (Scolytidae) and alter several aspects of their reproduction. Behavioral patterns such as swarming were shown to be affected (Lieutier, 1981) and many of the infected bark beetles that still laid eggs were found to excavate aberrant galleries, which had no egg niches. Examples of the effect of allantonematids on bark beetle reproduction are given in Table 2 and range from parasite-induced castration of male and female hosts to associations with no observable effects upon host gametogenesis. In the majority of cases female insects are more severely affected than males.

Histological examination of host gonads in several host–parasite associations led Tomalak *et al.* (1990) to observe that the scale of damage inflicted was related to the timing of infection, release of juvenile nematodes into the hemocoel, and their relative abundance. For example, in *Sulphuretylenchus pseudoundulatus*-infected *Polygraphus rufipennis*, large numbers of juveniles were released synchronously when the maternal nematode cuticle was ruptured. Development of germ cells in the testes was inhibited, tropharia in the ovaries were reduced, and no follicular development occurred in infected females. However, *Contortylenchus reversus* females deposited juveniles into the

Table 2

The Effect of Allantonematid Nematodes on Bark Beetle Reproduction

Species	Host	Oocyte formation[a]	Oviposition	Spermatogenesis	Genitalia malformed	Nuptial flight	Gallery construction	Source
Contortylenchus reversus	*Dendroctonus pseudotsugae*	+	+	0	?	?	*	Thong and Webster (1975a,b)
C. brevicomi	*D. fontalis*	?	+	?	?	?	*	MacGuidwin *et al.* (1980)
Sulphuretylenchus elongalus	*Scolytus ventralis*	++	++	++	?	?	?	Massey (1964)
	S. elongatus	++	++	++	?	?	?	Schvester (1957)
	S. ventralis	?	+	0	?	Yes	*	Ashraf and Berryman (1970)
Parasitaphelenchus papillatus	*Tomicus piniperda*	?	?	?	Males	?	?	Tomalak *et al.* (1984)
S. pseudoundulatus	*Polygraphus rufipennis*	++	++	++	?	?	**	Tomalak *et al.* (1990)
S. nopimingi	*Pityokteines sparsus*	++	++	+	?	?	**	Tomalak *et al.* (1990)
S. posteruteri	*Ips perroti*	++	++	0	?	?	**	Tomalak *et al.* (1990)
Parasitylenchus sp.	*I. perturbatus*	+	+	0	?	?	*	Tomalak *et al.* (1990)
Neoparasitylenchus ipinius	*I. perturbatus*	+	+	0	?	?	*	Tomalak *et al.* (1990)
Allantonema paramorosum	*Dryocoetes autographus*	+	0	0	?	?	*	Tomalak *et al.* (1990)
C. reversus	*D. simplex*	+	+	0	?	?	*	Tomalak *et al.* (1990)
Parasitaphelenchus sp.	*I. sexdentatus*	+	+	?	?	Yes	0	Lieutier (1981, 1982a,b)
P. sexdentati		+	+	?	?	Yes	0	Lieutier (1984a)
C. diplogaster		+	+	?	?	Yes	0	Lieutier (1984a)
Parasitorhabditis ipsophila	*I. sexdentatus*	0	+	?	?	Yes	+	Lieutier (1984b)

[a] ++, complete cessation; +, reduction; 0, unaffected/hardly affected; ?, not investigated; *, egg gallery length altered; **, very short egg galleries and no egg niches.

host hemocoel over a long time period and in this association the authors found that oogenesis and vitellogenesis proceeded in host ovaries, although at a reduced rate, and egg laying occurred (see Table 2).

The lack of tissue damage associated with allantonematid infections has led to the suggestion that host reproductive curtailment may be due to parasite-induced nutrient deprivation resulting in increased fat body catabolism and decreased anabolism (Tomalak *et al.*, 1990). Little evidence has been produced to support this hypothesis, although Lieutier (1984a,b) demonstrated a reduction in the protein content of fat bodies taken from nematode-infected *Ips sexdentatus*. The author suggested that the delay in ovarian maturation was a result of reduced protein sequestration, although rate of fat body synthesis and hemolymph protein content were not examined. Using polyacrylamide gel electrophoresis (PAGE), Thong and Webster (1975a) demonstrated that *Contortylenchus reversus* also depleted the hemolymph proteins of its host, *Dendroctonus pseudotsugae*. Importantly, a link *has* been established between fat body protein synthesis, hemolymph vitellogenin titers, and ovarian sequestration in a cestode–insect association (see the following).

Mermithid infections also result in host reproductive disturbances (see review by Gordon, 1981). Investigations into the underlying mechanisms have focused upon locust and blackfly infections. Gordon and his associates (1973) investigated the physiological mechanisms involved in the cessation of egg production in *Mermis nigrescens*-infected *Schistocerca gregaria*. Ovaries from female locusts infected 3 days after the imaginal molt exhibited impaired sequestration of vitellogenic proteins in the second week of infection and, by week 3, terminal and penultimate oocytes were resorbed. Determination of quantitative and qualitative changes in fat body and hemolymph proteins revealed that termination of vitellogenesis was not due to depletion of hemolymph vitellogenins (which was not significant until 1 week later) and was thought to result from an impairment of the ability of the follicular epithelium to sequester proteins. Absorption of oocytes may have resulted in the accumulation of vitellogenic proteins in the host hemocoel, observed in the fourth week of infection, however, the authors did not make this association. The effect of the nematodes on fat body protein turnover was ascribed to the provision of a dietary source of amino acids for the nematodes within the hemolymph. In addition, the authors suggested that parasite-induced changes in host neurosecretory activity may affect fat body metabolism and that a reduction in circulatory JH may inhibit ovarian sequestration of yolk proteins. Although there is little evidence to support or refute this view, *Neomesomermis flumenalis* was shown to increase the nuclear DNA/RNA activity (measured in terms of nuclear/cytoplasmic ratios), the volume of the host corpora allata (CA), and the amount of neurosecretory material accumulated in the cerebral neurosecretory cells of *Simulium venustum* (Condon and Gordon, 1977).

A reduction in CA volume was also associated with *Sphaerularia bombis* infections of *Bombus terrestris*. No ovarian development occurred in infected queen bees, and the hemolymph vitellogenin titer was reduced (Röseler and Röseler, 1973). As both biologically active juvenile hormone and juvenile hormone analogues (Rogers, 1973) and ecdysteroids (see Barker and Rees, 1990) occur in nematodes, there is clearly a need to reassess the potential for interaction between insect and nematode endocrine systems.

Reduced egg production in Diptera infected with filarial nematodes has been correlated with parasite burden. Using refractory and susceptible stocks of *Aedes aegypti*, Javadian and Macdonald (1974) demonstrated a negative linear correlation between parasite load and egg production in infections with older larvae of *Brugia pahangi* and in *A. aegypti* infected with *Dirofilaria repens*, in both early and late infections. A negative linear correlation between parasite burden and mean number of basal follicles developed per female was also demonstrated in the *B. pahangi*/*A. aegypti* association (Gaaboub, 1976). In *D. immitis*-infected *A. trivittatus*, reduced egg production has also been related to parasite burden, with those exceeding 15 juveniles per mosquito severely limiting reproduction (Christensen, 1981). *Dirofilaria immitis* and *B. pahangi* are Malpighian tubule and flight muscle dwellers, respectively, and no evidence of direct damage to host ovaries has been described.

The fecundity of wild-caught *Simulium damnosum* was found to be reduced by *Onchocerca* sp. infections (Cheke *et al.*, 1982) and two species of United Kingdom blackflies exhibited reductions in fecundity related to *O. lienalis* burden when infected by blood feeding or by intrathoracic injections (Ham and Banya, 1984). Further experiments demonstrated an approximately 50% reduction in egg laying in *S. ornatum* injected with 30 microfilarae (Ham and Gale, 1984). Intrathoracic injections ruled out the possible effect of pathology caused by gut penetration on fecundity. However, both the quality and quantity of blood ingested by infected vectors may alter nutrient availability for the symbionts. Using radioisotope estimations of blood meal size, Courtney *et al.* (1985) demonstrated a reduced intake when *A. aegypti* were feeding on *D. immitis*-infected blood and suggested that nutrient deprivation was the cause of reduced fecundity in infected mosquitoes. Ham and Banya (1984) found a reduction in feeding rate in blackflies fed on infected blood.

IV. Cestodes

Although cestode parasitization of insects does not result in host castration, reproductive output is reduced in hymenolepid infections of Tenebrionidae (Coleoptera). Embryonated "eggs" of the rat tapeworm *Hymenolepis diminuta* are passively transmitted to intermediate hosts when these stored grain pests feed on feces from infected definitive hosts. The oncosphere is released in the gut, penetrates the midgut wall, and develops into a metacestode in 8–12 days at 26°C. The parasites absorb nutrients from the hemolymph and no tissue invasion occurs.

A marked long-term depression of host populations was observed in *Hymenolepis diminuta* infections of *Tribolium castaneum* (Keymer, 1981) and reduction of host fecundity was related to total parasite biomass (Keymer, 1980). Maema (1986) associated host egg output with age of infection, finding a significant reduction 14 or more days postinfection. Hurd and Arme (1986a) also demonstrated reduced oviposition in the larger *Tenebrio molitor* 13–16 days postinfection with *H. diminuta,* although only when the hosts were maintained in very crowded conditions was this reduction still evident 30 days postinfection. Egg viability was depressed by 9.4% in parasitized *Tenebrio molitor* kept as single pairs and by 12% in beetles kept in crowded conditions (Hurd and Arme, 1986b). In addition, a significant reduction in the number of larvae produced during a 6-week period was recorded (Hurd, 1990b).

The reproductive physiology of *Tenebrio molitor*-infected with *Hymenolepis diminuta* has been extensively investigated by the author. These findings have been reviewed recently (Hurd, 1990a,b) and thus will not be dealt with in great detail in this chapter. However, perturbance of host egg production in this system appears to have some parallels with the mermithid infections of locusts, addressed earlier, and thus these aspects will be discussed. Host reproductive curtailment probably results from a retardation of oocyte maturation and is undoubtedly the result of impaired vitellogenesis. Despite a significant reduction in yolk volume in eggs from beetles infected with *H. diminuta* (Hurd and Arme, 1986b), hemolymph protein titers were not depleted, but increased by 46.7%, in females harboring mature parasites. Densitometric analysis of hemolymph separated on SDS-PAGE gels revealed that this increase was entirely due to increased titers of the vitellogenin subunits (Hurd and Arme, 1984). Thus, in common with the mermithid/locust association, vitellogenin is present in abundance and reduced egg output is not a direct result of nutrient deprivation. Ovarian sequestration of yolk was investigated *in vivo* by monitoring the incorporation of ^{14}C-labeled vitellogenin and shown to be decreased by 51.5% in parasitized females (Hurd and Arme, 1986a). The terminal follicles of these telotropic ovaries exhibit a significant reduction in patency (Hurd and Arme, 1987) and it is proposed that this reduction results in an impairment of yolk uptake and hence a reduction in fecundity.

The mechanisms underlying this parasite-induced perturbance of vitellogenesis are not known. It is thought that juvenile hormone III regulates vitellogenesis in *Tenebrio molitor* (Laverdure, 1970). However, CA secretion, circulating titers, and rate of degradation of juvenile hormone are not affected by infection (Hurd and Weaver, 1987; Hurd *et al.*, 1990). Attempts to demonstrate changes in ecdysteroid titers in infected beetles have also failed (Hurd *et al.*, 1987) and, although ecdysteroids have been demonstrated in adult *H. diminuta* (Mercer *et al.*, 1987), no free ecdysteroid activity was found in the metacestodes using RIA methods (Mercer *et al.*, 1987) or immunochemistry (H. D. F. H. Schallig and H. Hurd, personal observations). Despite the apparent normality of hormone titers in infected females, a juvenile hormone analogue, methoprene, applied to beetles at the time of infection, produced a rescue effect by preventing the elevation of hemolymph protein titers associated with infection (Hurd *et al.*, 1990). In addition, other juvenile hormone-controlled events such as fat body vitellogenin synthesis (Hurd and Arme, 1986a) and sex pheromone production (Hurd and Parry, 1991) are also down-regulated by *H. diminuta* metacestodes. Is it apparent that although none of the investigations performed so far directly supports the proposition that metacestodes manipulate host endocrine control of reproduction, new avenues need to be explored. Parasites may not change hormone titer but could, for example, induce a change in host response to mediator substances (i.e., JH) in the target tissues.

V. Endoparasitic Insects

Insect parasitoids usually attack immature insects (Kuris, 1974) and kill their hosts before they reach sexual maturity. Host castration is thus rarely observed in insect–insect associations. However, Junnikkala (1985) noted a reduction in the volume and wall thickness of developing testes in *Pieris brassicae* larvae parasitized by *Apanteles glomeratus*, with the degree of retardation not being related to parasite burden. Bundles of sperm were, however, produced before the parasitoid killed its host at the prepupal stage, indicating that the gonad was still functional at this stage. The timing of attack determined the degree of gonadal damage in the codling moth, *Cydia pomonella*, parasitized by *Ascogaster quadridentata* (Reed-Larson and Brown, 1990). Castration occurred in larvae developing from eggs exposed to adult wasps during the first 36 h; however, some larvae developed small testes if attacked between 36 and 48 h and those parasitized at 64 h had testes that, at the fourth stadium, exhibited precocious development. Reed-Larson and Brown (1990) discussed the potential benefit to parasitoids of host castration and suggested that the elimination of host target organs may make more ecdysone available to act as a cue for *A. quadridentata* development, and also remove potential competition for glycogen and lipid reserves.

Two well-documented cases in which the infected female host reaches adulthood are outlined in the following. In both examples, interaction between the parasite and host endocrine system has been investigated.

Triungulinid larvae of the parasitic wasp *Xenos vesparum* invade *Polistis* spp. (Hymenoptera) larvae in the spring. Parasite growth proceeds slowly in host larvae and the last larval instars and nymphal and imaginal molts occur in young adult wasp hosts. Male parasites emerge from their hosts, mate, and die, whereas females, infecting future queen wasps, overwinter in their hosts (Nassonov, 1910; Hoffmann, 1914). Parasitic castration (stylopization syndrome) is induced by suppression of oocyte maturation and the occurrence of intersexes in stylopized bees is well known (Wülker, 1975).

Strambi *et al.* (1982) demonstrated that, at the time of host emergence from diapause, terminal oocytes were significantly reduced in infected females. Control *X. vesparum* females had higher blood protein titers at this time (Strambi *et al.*, 1982) and, although present, vitellogenin titers were reduced in parasitized females (Roux, 1973). Infected adults were found to have reduced fat bodies, ovaries, and CA volume. After overwintering, synthesis of juvenile hormone by the CA reached 4 pmoles per pair of glands per hour in control wasps but only 0.2 pmoles were produced by CA from infected wasps (Röseler *et al.*, 1980). In addition, hemolymph juvenile hormone titers were significantly reduced in infected females (A. Strambi, personal observations) and CA implantation into parasitized females induced vitellogenesis and ovarian development (Strambi and Girardie, 1973). Thus, Strambi *et al.* (1982) maintained that it was a juvenile hormone deficiency that inhibited egg maturation in infected insects. Cytological evidence suggest that, unlike in uninfected counterparts, large amounts of neurosecretory material are released from infected wasps and that this may control CA functioning (e.g., Strambi, 1966; Strambi and Strambi, 1973). The method by which the parasite exerts its influence on the neurosecretory cells remains to be elucidated.

In *Metacemyia calloti* (Diptera, Tachinidae) infections of the grasshopper *Anacridium aegyptium*, impairment of host oocyte development has been associated with a depletion of hemolymph proteins. Synthesis of vitellogenic proteins was reduced and vitellogenins were shown to be involved in the nutritional requirements of the parasites (Girardie, 1977). Unlike the strepsipteran infection outlined earlier, host juvenile hormone production was not reduced, but elevated in infected females (Girardie and Joly, 1975). Ultrastructural studies confirmed the status of CA from infected insects and CA implantation failed to initiate oocyte development in parasitized females (Girardie and Granier, 1974). However, a combination of radioactive tracer studies and light and electron microscopy demonstrated the lack of activity in the median neurosecretory cells of infected grasshoppers. Electrical stimulation of the pars intercerebralis enhanced ovarian development (Girardie and

Girardie, 1977). Girardie (1977) postulated that hemolymph protein deficiency may be the result of both parasite nutritional requirements and hypoactivity of the median neurosecretory cells resulting in decreased protein synthesis, possibly via JH control.

VI. Conclusions

The associations selected for discussion in this chapter aptly illustrate the widespread occurrence of parasite-induced reduction in host fecundity. Dawkins (e.g., 1982, 1990) regarded parasitic castration as an example of parasite manipulation of hosts via an extension of the influence of parasite genes. The potential that this strategy provides for enhancing symbiont survival by diversion of metabolic resources away from host reproduction has been discussed by many authors (e.g., Baudoin, 1975; Hurd, 1990a).

The majority of methods whereby host reproduction is suppressed fall into one of three categories: the inhibition of host mating due to morphological or behavioral changes, the destruction of host reproductive tissue, and the reduction of egg output due to malfunction of ovarian yolk sequestration. In all three cases, nutrients, which in control insects would be utilized by developing oocytes, are made available to sustain the host and developing parasites. Our understanding of the mode of parasite action is rudimentary but in most associations it appears to involve a more subtle interplay than the simple scenario of parasite utilization of nutrients destined for egg production. Indirect diversion of nutrients via alteration of fat body metabolism and/or perturbance of neuroendocrine control mechanisms are undoubtedly involved in some associations and putative parasite secretion of mediator substances, which may alter gonadal sensitivity or affect hormone metabolism, has been discussed recently (Holmes and Zohar, 1990). However, at the moment there is no example of parasite-induced disturbance of insect reproduction in which the underlying mechanisms have been fully elucidated and this remains a challenge for the future.

References

Anderson, R. M., and May, R. M. (1981). The population dynamics of microparasites and their invertebrate hosts. *Philos. Trans R. Soc. London* **291**:451–524.

Andreadis, T. G. (1986). Dissemination of *Nosema pyrausta* in feral populations of the European corn borer, *Ostrinia nubilalis*. *J. Invertebr. Pathol.* **48**:335–343.

Armstrong, E. (1977). Transmission of *Nosema kingi* to offspring of *Drosophila willistoni* during copulation. *Z. Parasitenkd.* **53**:311–315.

Armstrong, E., and Bass, L. K. (1986). Effects of infection by *Nosema whitei* on the mating frequency and fecundity of *Tribolium castaneum*. *J. Invertebr. Pathol.* **47**:310–316.

Ashraf, M., and Berryman, A. A. (1970). Biology of *Sulphuretylenchus elongatus* (Nematoda: Sphaerulariidae), and its effect on its host, *Scolytus ventralis* (Coleoptera: Scolytidae). *Can. Entomol.* **102:**197–213.

Barker, G. C., and Rees, H. H. (1990). Ecdysteroids in nematodes. *Parasitol. Today* **6:**384–387.

Baudoin, M. (1975). Host castration as a parasitic strategy. *Evolution (Lawrence, Kans.)* **29:**335–352.

Bauer, L. S., and Nordin, G. L. (1989). Effect of *Nosema fumiferanae* (Microsporida) on fecundity, fertility, and progeny performance of *Choristoneura fumiferana* (Lepidoptera: Torticidae). *Environ. Entomol.* **18:**261–265.

Cheke, R. A., Garms, R., and Kerner, M. (1982). The fecundity of *Simulium damnosum* s.l. in northern Toga and infections with *Onchocerca* spp. *Ann. Trop. Med. Parasitol.* **54:**561–568.

Christensen, B. M. (1981). Effect of *Dirofilaria immitis* on the fecundity of *Aedes trivittatus. Mosq. News* **41:**78–81.

Condon, W. J., and Gordon, R. (1977). Some effects of mermithid parasitism on the larval blackflies *Prosimulium mixtum fuscum* and *Simulium venustum. J. Invertebr. Pathol.* **29:**56–62.

Courtney, C. C., Christensen, B. M., and Goodman, W. G. (1985). Effect of *Dirofilaria immitis* on blood meal size and fecundity in *Aedes aegypti* (Diptera: Culicidae). *J. Entomol.* **22:**389–400.

Dawkins, R. (1982). "The Extended Phenotype." Oxford Univ. Press, Oxford.

Dawkins, R. (1990). Parasites, desiderata lists and the paradox of the organism. *Parasitology* **100:**S63–S75.

Dobson, A. P. (1988). The population biology of parasite-induced changes in host behaviour. *Q. Rev. Biol.* **63:**139–165.

Fisher, F. M., Jr., and Sanborn, R. C. (1962). Production of insect juvenile hormone by the microsporidian parasite *Nosema. Nature (London)* **194:**1193.

Fisher, F. M., Jr., and Sanborn, R. C. (1964). *Nosema* as a source of juvenile hormone in parasitised insects. *Biol. Bull. (Woods Hole, Mass.)* **126:**235–252.

Freier, J. E., and Friedman, S. (1976). Effect of host infection with *Plasmodium gallinaceum* on the reproductive capacity of *Aedes aegypti. J. Invertebr. Pathol.* **28:**161–166.

Gaaboub, I. A. (1976). Observations on the basal follicle numbers developed per female of two strains of *Aedes aegypti* after being fed on hosts with different levels of microfilariae of *Brugia pahangi. J. Invertebr. Pathol.* **28:**203–207.

Gaugler, R. R., and Brooks, W. M. (1975). Sublethal effects of infection by *Nosema heliothis zea. J. Invertebr. Pathol.* **26:**57–63.

Germida, J. J., Onofriechuk, E. E., and Ewen, A. B. (1987). Effect of *Nosema locustae* (Canning) microsporidia and three chemical insecticides on microbial activity in soil. *Can. J. Soil Sci.* **67:**631–638.

Girardie, J. (1977). Contrôle neuroendocrine des protéines sanguines vitellogènes *d'Anacridium aegyptium* sains et parasités. *J. Insect Physiol.* **23:**569–577.

Girardie, J., and Granier, S. (1974). Rôle des corps allates dans la castration parasitaire *d'Anacridium aegyptium* (Insecte, Orthoptère) infesté par *Metacemya calloti* (Insecte, Diptöre). *Arch. Anat. Microsc. Morphol. Exp.* **63:**269–280.

Girardie, J., and Girardie, A. (1977). Intervention des cellules neurosécrétrices médiances dans la castration parasitaire *d'Anacridium aegyptium* (Orthoptöres). *J. Insect Physiol.* **23:**461–467.

Girardie, J., and Joly, L. (1975). Dosage biologique de l'hormone juvénile dans l'hémolymphe d'adultes *d'Anacridium aegyptium* (Insectes, Orthoptöres) sains et parasites, pendant et après rupture expérimentale de la diapause. *C.R. Hebd. Seances Acad. Sci., Ser. D* **281:**719–722.

Gordon, R. (1981). Mermithid nematodes: Physiological relationships with their insect hosts. *J. Nematol.* **13:**266–274.

Gordon, R., Webster, J. M., and Hislop, T. G. (1973). Mermithid parasitism, protein turnover and vitellogenesis in the desert locust, *Schistocerca gregaria. Comp. Biochem. Physiol. B* **46B:**575–593.

Hacker, C. S. (1971). The differential effect of *Plasmodium gallinaceum* on the fecundity of several strains of *Aedes aegypti*. *J. Invertebr. Pathol.* **18:**373–377.

Hacker, C. S., and Kilama, W. L. (1974). The relationship between *Plasmodium gallinaceum* density and the fecundity of *Aedes aegypti*. *J. Invertebr. Pathol.* **23:**101–105.

Ham, P. J., and Banya, A. J. (1984). The effect of experimental *Onchocerca* infections on the fecundity and oviposition of laboratory reared *Simulium* sp. (Diptera, Simuliidae). *Tropenmed. Parasitol.* **35:**61–66.

Ham, P. J., and Gale, L. (1984). Blood meal enhanced *Onchocerca* development and its correlation with fecundity in laboratory reared blackflies (Diptera, Simuliidae). *Tropenmed. Parasitol.* **35:**212–216.

Han, M.-S., and Watanabe, H. (1988). Transovarial transmission of two microsporidia in the silkworm, *Bombyx mori*, and disease occurrence in the progeny population. *J. Invertebr. Pathol.* **51:**41–45.

Hoffmann, R. W. (1914). Die embryonalen Vorgänge bei der Strepsiteren and ihre Deutung. *Verh. Dtsch. Zool. Ges.* **24:**192–216.

Holmes, J. C., and Zohar, S. (1990). Pathology and host behaviour. In "Parasitism and Host Behaviour" (C. J. Barnard and J. M. Behnke, eds.), pp. 34–64. Taylor & Francis, London.

Hurd, H. (1990a). Physiological and behavioural interactions between parasites and invertebrate hosts. *Adv. Parasitol.* **29:**271–318.

Hurd, H. (1990b). Parasite induced modulation of insect reproduction. *Adv. Invertebr. Reprod.* **5:**163–168.

Hurd, H., and Arme, C. (1984). Pathology of *Hymenolepis diminuta* infections in *Tenebrio molitor:* Effect of parasitism on haemolymph proteins. *Parasitology* **89:**253–262.

Hurd, H., and Arme, C. (1986a). *Hymenolepis diminuta:* Influence of metacestodes upon synthesis and secretion of fat body protein and its ovarian sequestration in the intermediate host, *Tenebrio molitor. Parasitology* **93:**111–120.

Hurd, H., and Arme, C. (1986b). *Hymenolepis diminuta:* The effect of metacestodes upon egg production and viability in the intermediate host, *Tenebrio molitor. J. Invertebr. Pathol.* **47:**225–231.

Hurd, H., and Arme, C. (1987). *Hymenolepis diminuta:* Effect of infection upon the patency of the follicular epithelium of the intermediate host *Tenebrio molitor. J. Invertebr. Pathol.* **49:**227–234.

Hurd, H., and Parry, G. (1991). Metacestode-induced depression of the production of, and response to, sex pheromone in the intermediate host *Tenebrio molitor. J. Invertebr. Pathol.* **58:**82–87.

Hurd, H., and Weaver, R. J. (1987). Evidence against the hypothesis that metacestodes of *Hymenolepis diminuta* inhibit corpora allata functioning in the intermediate host, *Tenebrio molitor. Parasitology* **95:**93–97.

Hurd, H., Mercer, J. G., and Munn, A. E. (1987). The effect of *Hymenolepis diminuta* upon ecdysteroid activity in the haemolymph of the intermediate host *Tenebrio molitor. Parasitol. Res.* **74:**198–199.

Hurd, H., Strambi, C., and Beckage, N. E. (1990). *Hymenolepis diminuta:* An investigation of juvenile hormone titre, degradation and supplementation in the intermediate host, *Tenebrio molitor. Parasitology* **100:**445–452.

Jackson, L. L., Baker, G. L., and Henry, J. E. (1968). Effect of *Malamoeba locustae* infection on the egg lipids of the grasshopper *Melanoplus bivittatus. J. Insect Physiol.* **14:**1773–1778.

Javadian, E., and Macdonald, W. W. (1974). The effect of infection with *Brugia pahangi* and *Dirofilaria repens* on the egg-production of *Aedes aegypti. Ann. Trop. Med. Parasitol.* **68:**477–481.

Johnson, D. L., and Henry, J. E. (1987). Low rates of insecticides and *Nosema locustae* (Microsporidia, Nosematidae) on baits applied to roadsides for grasshopper control. *J. Econ. Entomol.* **80:**685–689.

Junnikkala, E. (1985). Testis development in *Pieris brassicae* parasitized by *Apanteles glomeratus*. *Entomol. Exp. Appl.* **37:**283–288.
Karavaeva, N. N. (1987). Juvenile hormone activity in the spores of *Nosema bombycis*, the pathogen of nosematosis in the silkworm. *Uzb. Biol. Zh.* **3:**6–10.
Kellen, W. R., and Lindegren, J. E. (1971). Modes of transmission of *Nosema plodiae* Kellen and Lindegren, a pathogen of *Plodia interpunctella* (Hübner). *J. Stored Prod. Res.* **7:**31–34.
Keymer, A. E. (1980). The influence of *Hymenolepis diminuta* on the survival and fecundity of the intermediate host *Tribolium confusum*. *Parasitology* **81:**405–421.
Keymer, A. E. (1981). Population dynamics of *Hymenolepis diminuta* in the intermediate host. *J. Anim. Ecol.* **50:**941–950.
Khan, A. R., and Selman, B. J. (1989). *Nosema* spp. (Microspora: Microsporidia: Nosematidae) of stored-product Coleoptera and their potential as microbial agents. *Agric. Rev.* **3:**193–223.
Kuris, A. M. (1974). Trophic interactions: Similarity of parasitic castrators to parasitoids. *Q. Rev. Biol.* **49:**129–148.
Laverdure, A.-M. (1970). Action de l'ecdysone et de l'ester méthylique du farnesol sur l'ovaire nymphal de *Tenebrio molitor* (Coléoptöre) cultive in vitro. *Ann. Endocrinol.* **31:**516–524.
Lieutier, F. (1981). Influence des nématodes parasites sur l'essaimage du scolytide *Ips sexdentatus* (Boern.). action régulatrice du froid. *Acta Oecol. Oecol. Appl.* **2:**357–368.
Lieutier, F. (1982a). Les variations pondérales du tissu adipeux et des ovaries, et les variations de longueur des ovocytes, chez *Ips sexdentatus* Boern (Coleoptera: Scolytidae); rélations avec le parasitisme par les nématodes. *Ann. Parasitol. Hum. Comp.* **57:**407–418.
Lieutier, F. (1982b). Action des nématodes endoparasites sur la ponte du scolytide *Ips sexdentatus* Boerner (Insecta: Coleoptera). *Acta Oecol. Oecol. Appl.* **3:**191–204.
Lieutier, F. (1984a). Ovarian and fat body protein concentrations in *Ips sexdentatus* (Coleoptera: Scolytidae) parasitized by nematodes. *J. Invertebr. Pathol.* **43:**21–31.
Lieutier, F. (1984b). Observations sur le parasitisme *d'Ips sexdentatus* (Insecta: Scolytidae) par *Parasitorhabditis ipsophila* (Nematoda: Rhabditidae). *Ann. Parasitol. Hum. Comp.* **59:**507–520.
Listov, M. V. (1977). The effect of pathogenic protozoa on hormone balance of flour beetles (Coleoptera, Tenebrionidae). *Entomol. Obozr.* **56:**731–735.
Liu, T. P. (1986). Comparative fine structure of the corpus allatum from healthy and *Nosema* infected honeybees. *J. Apic. Res.* **25:**163–170.
Liu, T. P. (1989). Juvenile hormone III induced ultrastructural changes in the hypopharangeal glands of honeybee *Apis mellifera* L. (Hymenoptera: Apidae) without and with infection by *Nosema apis* Zander (Microsporidae; Nosematidae). *Int. J. Insect Morphol. Embryol.* **18:**73–83.
Liu, T. P. (1990). Ultrastructure of mitochondria in the corpora allata of honeybees infected by *Nosema apis* before and after treatment with anti-nosema drugs. *Tissue Cell* **22:**511–515.
MacGuidwin, A. E., Smart, G. C., Jr., Wilkinson, R. C., and Allen, G. E. (1980). Effect of the nematode *Contortylenchus brevicornis* on gallery construction and fecundity of the southern pine beetle. *J. Nematol.* **12:**278–282.
Maema, M. (1986). Experimental infection of *Tribolium confusum* (Coleoptera) by *Hymenolepis diminuta* (Cestoda): Host fecundity during infection. *Parasitology* **92:**405–412.
Malone, L. A. (1987). Longevity and fecundity of Argentine stem weevils, *Listronotus bonariensis* (Coleoptera: Curculionidae), infected with *Microsporidium itiiti* (Protozoa: Microspora). *J. Invertebr. Pathol.* **50:**113–117.
Malone, L. A., and Wigley, P. J. (1981). Quantitative studies on the pathogenicity of *Nosema carpocapsae*, a microsporidian pathogen of the codling moth, *Cydia pomonella*, in New Zealand. *J. Invertebr. Pathol.* **38:**330–334.
Massey, C. L. (1964). The nematode parasites and associates of the fir engraver beetle, *Scolytus ventralis*, LeConte, in New Mexico. *J. Insect Pathol.* **6:**133–155.
May, R. M., and Anderson, R. M. (1990). Parasite–host coevolution. *Parasitology* **100:**S89–S101.

Mercer, C. F., and Wigley, P. J. (1987). A microsporidian pathogen of the Poroporo stem borer, *Sceliodes cordalis* (Dbld.) (Lepidoptera: Pyralidae). *J. Invertebr. Pathol.* **49:**108–115.

Mercer, J. G., Munn, A. E., Munn, Arme, C., and Rees, H. H. (1987). Analysis of ecdysteroids in different developmental stages of *Hymenolepis diminuta. Mol. Biochem. Parasitol.* **25:**61–71.

Milner, R. J. (1972). *Nosema whitei,* a microsporidian pathogen of some species of *Tribolium.* III. Effect on *T. castaneum. J. Invertebr. Pathol.* **19:**248–255.

Molyneux, D. H., and Jefferies, D. (1986). Feeding behaviour of pathogen infected vectors. *Parasitology* **92:**721–736.

Nassonov, N. V. (1910). Untersuchungen zur Naturgeschichte der Strepsiteren. *Ber. Naturwiss.-Med. Ver. Inssbruck* **33:**1–205.

Nnakumusana, E. S. (1986). Effect of *Nosema algerae* Vavra and Undeen spores infection on longevity and fecundity of larval instars of *Aedes aegypti* L. and *Culex fatigans* Weid. *Indian J. Exp. Biol.* **24:**786–788.

Onstad, D. W., and Maddox, J. V. (1989). Modeling the effects of the microsporidium, *Nosema pyrausta,* on the population dynamics of the insect *Ostrinia nubilalis. J. Invertebr. Pathol.* **53:**410–421.

Papillon, M., and Cassier, P. (1978). Perturbations morphologiques et physiologiques dues a la présence du protozoaire parasite *Malameba locustae* (K. et T.) chez *Schistocerca gregaria* (Forsk.). *Acrida* **7:**101–114.

Rabindra, R. J., Balasubramanian, M., and Jayaraj, S. (1981). The effects of *Farnocystis tribolii* on the growth and development of the flour beetle *Tribolium castaneum. J. Invertebr. Pathol.* **38:**345–351.

Raina, S. K., Rai, M. M., and Khurad, A. M. (1987). Grasshopper and locust control using microsporidian insecticides. *In* "Biotechnology of Invertebrate Pathology and Cell Culture" (K. Maramorosch ed.), pp. 345–366. Academic Press, Orlando, FL.

Read, A. F. (1990). Parasites and the evolution of host sexual behaviour. *In* "Parasitism and Host Behaviour" (C. B. Barnard and J. M. Behnke, eds.), pp. 117–158. Taylor & Francis, London.

Reed-Larson, D. A., and Brown, J. J. (1990). Embryonic castration of the codling moth, *Cydia pomonella,* by an endoparasitoid, *Ascogaster quadridentata. J. Insect Physiol.* **36:**111–118.

Reynolds, D. G. (1971). Parasitization of *Culex fatigans* by *Nosema stegomyiae. J. Invertebr. Pathol.* **18:**429.

Rogers, W. P. (1973). Juvenile and moulting hormones from nematodes. *Parasitology* **67:**105–113.

Röseler, I., and Röseler, P. F. (1973). Änderungen im Muster der Haemolymphproteine von Adulten Koniginen der Hummelart *Bombus terrestris. J. Insect Physiol.* **19:**1741–1752.

Röseler, P. F., Röseler, I., and Strambi, A. (1980). The activity of corpora allata in dominant and subordinated females of the wasp *Polistes gallicus. Insect Soc.* **27:**97–107.

Roux, C. (1973). Etude comparée de la protéinémie chez les imagos de *Polistes* (Hyménoptöres, Vespidés) sains ou parasités par le Strepsiptöre *Xenos vesparum* Rossi. *C.R. Hebd. Seances Acad. Sci.* **276:**3159–3162.

Sajup, A. S., and Lewis, L. C. (1988). Histopathology of transovarial transmission of *Nosema pyrausta* in the European corn borer, *Ostrinia nubilalis. J. Invertebr. Pathol.* **52:**147–153.

Schvester, D. (1957). Contribution de l'étude de Coléoptéres scolytids. *Ann. Inst. Natl. Rech. Agron., Ser. C* **8:**1–162.

Shieh, J.-N., and Rossignol, P. A. (1992). Opposite effects of host anaemia on blood feeding rate and fecundity of mosquitos. *Parasitology* **105:**159–163.

Siegel, J. P., Maddox, J. V., and Ruesink, W. G. (1986). Lethal and sublethal effects of *Nosema pyrausta* on the European corn borer (*Ostrinia nubilalis*) in central Illinois. *J. Invertebr. Pathol.* **48:**167–173.

Smith, J. E., and Dunn, A. M. (1991). Transovarial transmission. *Parasitol. Today* **7:**146–148.

Strambi, A. (1966). Action de *Xenos vesparum* Rossi (Strepsiptére) sur la neurosécrétion des fondatrices filles de *Polistes gallicus* L. (Hyménoptère, Vespide) en diapause. *C.R. Hebd. Seances Acad. Sci.* **263:**533–535.

Strambi, A., and Girardie, A. (1973). Effect de l'implanation de corpora allata actifs de *Locusta migratoria* (Orthoptére) dans des femelles de L. (Hyménoptère) saines et parasitées par *Xenos vesparum* Rossi (Insecte, Strepsiptére). *C.R. Hebd. Seances Acad. Sci.* **276:**3319–3322.

Strambi, A., and Strambi, C. (1973). Influence du développement du parasite *Xenos vesparum* Rossi (Insecte, Strepsiptère) sur le système neuroendocrinien des femelles de *Polistes* (Hyménoptère, Vespide) au début de leur vie imaginale. *Arch. Anat. Microsc. Morphol. Exp.* **62:**39–54.

Strambi, C., Strambi, A., and Augier, R. (1982). Protein levels in the haemolymph of the wasp *Polistes gallicus* L. at the beginning of imaginal life and during overwintering. Action of the strepsipteran parasite *Xenos vesparum* Rossi. *Experientia* **38:**1189–1191.

Thompson, A. C., and Sikorowski, P. P. (1979). Effects of *Nosema heliothidis* on fatty and amino acids in larvae and pupae of the bollworm, *Heliothis zea. Comp. Biochem. Physiol. A.* **63A:**325–328.

Thong, C.H.S., and Webster, J. M. (1975a). Effects of *Contortylenchus reversus* (Nematoda: Sphaerulariidae) on haemolymph composition and oocyte development in the beetle *Dendroctonus pseudopsugae* (Coleoptera: Scolytidae). *J. Invertebr. Pathol.* **26:**91–98.

Thong, C.H.S., and Webster, J. M. (1975b). Effects of the bark beetle nematode, *Contortylenchus reversus,* on gallery construction, fecundity, and egg viability of the Douglas fir beetle, *Dendroctonus pseudotsugae* (Coleoptera: Scolytidae). *J. Invertebr. Pathol.* **26:**235–238.

Tomalak, M., Michalski, J., and Grocholski, J. (1984). The influence of nematodes on the structure and genitalia of *Tomicus piniperda* (Coleoptera: Scolytidae). *J. Invertebr. Pathol.* **43:**358–362.

Tomalak, M., Welch, H. E., and Galloway, T. (1990). Pathogenicity of allantonematidae (Nematoda) infecting bark beetles (Coleoptera: Scolytidae) in Manitoba. *Can. J. Zool.* **68:**89–100.

Weiser, J. (1974). Microsporidia in invertebrates: host–parasite relations at the organismal level. *In* "Comparative Pathobiology" (T. C. Cheng, ed.), pp. 163–201. Plenum, New York.

Windels, M. B., Chaing, H. C., and Furgala, B. (1976). Effects of *Nosema pyrausta* on pupa and adult stages of the European corn borer *Ostrinia nubilalis. J. Invertebr. Pathol.* **27:**239-242.

Wülker, W. (1975). Parasite-induced castration and intersexuality in insects. *In* "Intersexuality in the Animal Kingdom" (R. Reinboth, ed.), pp. 121–134. Springer-Verlag, New York.

Chapter 5

Behavioral Effects of Parasites and Pathogens in Insect Hosts

David R. Horton
Yakima Agricultural Laboratory
USDA-ARS
Yakima, Washington

Janice Moore
Department of Biology
Colorado State University
Fort Collins, Colorado

I. Introduction

Responses of insect hosts to parasitism include physiological, morphological, and behavioral effects, many of which affect parasite or host fitness. Changes in host biology associated with parasitism are often inferred to be under parasite control, particularly when the changes benefit the parasite (e.g., Vinson and Iwantsch, 1980). The entomological literature emphasizes the effects of parasitism on host physiology (Vinson and Iwantsch, 1980; Beckage, 1985; Jones, 1985); effects on host behavior are less well described (but see McAllister and Roitberg, 1987; Brodeur and McNeil, 1989). Nonetheless, behavioral changes caused by parasitism are common, and one may hypothesize that the changes are products of natural selection acting on the host or the parasite.

Here we summarize a diffuse literature describing parasite-induced behavioral changes, covering a variety of micro- (e.g., fungi and viruses) and macroparasitic (e.g., parasitoids and nematodes) organisms (*sensu* Anderson and May, 1979). Three broad categories of behavioral alterations include: (1)

changes in microhabitat preferences, including diurnal activity by normally nocturnal hosts; (2) changes in activity level; and (3) reproductive activity expressed under inappropriate conditions, that is, by castrated hosts. Modifications of feeding behavior in insect vectors are excluded, as this literature has been reviewed (Molyneux and Jeffries, 1986).

Behavioral modifications caused by parasitism can be interpreted to have arisen (see also Minchella, 1985): (1) through natural selection and beneficial to the parasite; (2) through natural selection and beneficial to the host; and (3) as a consequence of pathology or stress, which may or may not benefit either party. Moore (1984), Jones (1985), Moore and Gotelli (1990), and Combes (1991) emphasized the importance of deciding among these hypotheses if one is to understand the evolutionary significance of behavioral alterations. We will first consider how behavioral changes might benefit the parasite, emphasizing parasite dissemination and survival. Second, behavioral changes that may benefit the host through either kin selection ("host suicide"; Shapiro, 1976) or increased host body temperature ("behavioral fever") are discussed. Finally, research areas that should enhance our understanding of behavioral modifications will be proposed.

II. Review of Altered Behavior in Parasitized Insects

A. Changes in Microhabitat Preference

Parasitized insects often show changes in microhabitat preference (Table 1), including movement to elevated locations, movement to concealed or exposed locations, changes in temperature preferences, and altered response to light; additionally, nocturnal insects may become active during the day. Elevation-seeking behavior appears commonly in hosts infected by fungal and viral pathogens [e.g., fungal "summit disease" (Marikovsky, 1962) and viral "tree-top disease" (Steinhaus, 1949)], although macroparasites such as parasitoids and the trematode "brain-worm" (Anokhin, 1966) also induce the behavior. Insects not typically found on plants may climb to the tops of vegetation (Thaxter, 1888; Bünzli and Büttiker, 1959); subterranean insects often move to the soil surface upon being parasitized (Table 1).

Movement to exposed locations is induced by a variety of parasites, including fungi, viruses, parasitoids, trematodes, and cestodes (Table 1). Movement to concealed locations by aphids parasitized by a braconid wasp has been documented (Brodeur and McNeil, 1989, 1990). Both of these effects possibly reflect changes in light reactions of parasitized hosts (Bro-

Table 1

Change in Microhabitat Preference by Parasitized Insects

Parasite or pathogen	Host	References[a]
A. Elevation-seeking behavior		
Fungi	Lepidoptera: several species of caterpillars	24, 43, 48, 49
	Orthoptera: grasshoppers	24, 32, 38, 43
	cricket	43
	Hymenoptera: ants	7, 19, 24
	Coleoptera	28[b], 43
	Diptera	3
	Homoptera: aphids	14[b], 31
Viruses	Lepidoptera: several species of caterpillars	1, 9, 11[b], 16[b], 17, 21, 22, 29, 41, 46
Rickettsiella	Coleoptera	27[b]
Parasitoids	Lepidoptera	37, 42
	Hymenoptera: ant	12
Trematodes	Hymenoptera: ants	2, 33
	Diptera: mosquito larva	45[c]
B. Behavior seeking exposed locations		
Fungi	Diptera	23, 25
Viruses	Lepidoptera	15, 36, 47
	Hymenoptera: sawfly	39
Parasitoid	Homoptera: aphid	5
Trematode	Hymenoptera: ant	8
Cestode	Coleoptera	13[d]

(continues)

[a]References: (1) Allen (1921); (2) Anokhin (1966); (3) Baird (1957); (4) Boorstein and Ewald (1987); (5) Brodeur and McNeil (1989); (6) Brodeur and McNeil (1990); (7) Bünzli and Büttiker (1959); (8) Carney (1969); (9) Edland (1965); (10) Eilenberg (1987); (11) Fowler and Robertson (1972); (12) Gosswald (cited in Wickler, 1976); (13) Graham (1966); (14) Harper (1958); (15) Hostetter and Biever (1970); (16) Kalmakoff and Moore (1975); (17) King and Atkinson (1928); (18) Lewis (1960); (19) Loos-Frank and Zimmerman (cited in Romig *et al.*, 1980); (20) Louis *et al.* (1986); (21) Lounsbury (1913); (22) Lower (1954); (23) MacLeod *et al.*, 1973; (24) Marikovsky (1962); (25) Miller and McClanahan (1959); (26) Moore (1983a); (27) Niklas (cited in Martignoni, 1964); (28) Nirula (1957); (29) Paschke and Hamm (1961); (30) Poinar and Gyrisco (1964); (31) Rockwood (1950); (32) Roffey (1968); (33) Romig *et al.* (1980); (34) Sato *et al.* (1983); (35) Schmid-Hempel and Schmid-Hempel (1990); (36) Semel (1956); (37) Shapiro (1976); (38) Skaife (1925); (39) Smirnoff (1960); (40) Smirnoff (1965); (41) Stairs (1965); (42) Stamp (1981); (43) Thaxter (1888); (44) Watson *et al.* (1993); (45) Webber *et al.* (1987a); (46) Whitlock (1974); (47) Wygant (1941); (48) Yen (1962); (49) Yendol and Paschke (1967).

[b]Soil-dwelling insects to surface.

[c]Parasitized higher in water column.

[d]Decreased photophobia.

Table 1 (*continued*)

Parasite or pathogen	Host	References[a]
C. Behavior seeking concealed locations		
Parasitoid	Homoptera: aphid	5, 6
D. Change in reaction to light		
Parasitoid	Lepidoptera	18[e]
Acanthocephala	Dictyoptera: cockroach	26[d]
E. Diurnal behavior by nocturnal insects		
Virus	Lepidoptera	40
Parasitoid	Lepidoptera	34
Nematode	Coleoptera	30
F. Change in temperature preference		
Rickettsiella	Orthoptera: cricket	20
Microsporidium	Orthoptera: grasshopper	4
Fungus	Diptera	44
G. Change in oviposition or foraging sites		
Fungus	Diptera	10
Diptera	Hymenoptera: bumblebee	35

[e]Increased photophobia.

deur and McNeil, 1989), as noted also in other hosts (Lewis, 1960; Graham, 1966; Moore, 1983a). Diurnal activity by normally nocturnal hosts was observed in parasitized caterpillars and beetles (Table 1). Finally, parasitized crickets, grasshoppers, and flies select temperatures higher than those selected by healthy conspecifics (Table 1).

B. Changes in Level of Activity

Viruses cause gregarious sawflies to engage in solitary wandering behavior (Smirnoff, 1960, 1961, 1965); virus-infected lepidopteran larvae may likewise exhibit undirected wandering (Lower, 1954; Smith *et al.*, 1956; Tamashiro and Huang, 1963). Cockroaches parasitized by *Moniliformis moniliformis* spend more time in activity (Moore, 1983a) and move a greater distance in an arena (Wilson and Edwards, 1986) than do parasite-free insects. Foraging worker bumblebees (*Bombus*) parasitized by conopid flies spend more time away from the nest than do healthy bees (Schmid-Hempel and Muller, 1991). Queen bumblebees parasitized by the nematode *Sphaerularia bombi* fly close to the ground, occasionally alighting and digging a depression in the soil (Poinar and van der Laan, 1972; Lundberg and Svensson, 1975). Juvenile nematodes are expelled into the hole, and the process is repeated. Flight is continued long after healthy bees have initiated a brood, hence parasitized bees have been called "the eternal seekers" (Poinar and van der Laan, 1972).

Other parasitized insects show reduced activity. Although this effect may arise largely from pathology, the decrease in activity can affect parasite or host fitness (see the following). Some pathogens cause hosts to become sluggish (Henry and Oma, 1981; Johnson, 1989); others produce paralysis (Kulinčević *et al.*, 1970; Reinganum *et al.*, 1970). Ants parasitized by cestodes (Plateaux, 1972) or trematodes (Carney, 1969) move slowly compared to healthy conspecifics. Trematode-parasitized ants (*Leptothorax*) may stay in the nest, unlike healthy ants (Plateaux, 1972). Larval *Aedes aegypti* parasitized with a trematode (*Plagiorchis noblei*) show reduced activity (Webber *et al.*, 1987a).

C. Reproductive Behavior by Castrated Insects

Hosts that have been castrated by nematodes may nonetheless retain normal or slightly modified reproductive behavior. Blackflies (Simuliidae) castrated by mermithid nematodes exhibit behavior similar to that shown by healthy flies. Both healthy and parasitized females fly upstream to oviposit, but parasitized females release nematodes instead of eggs (Grunin, 1949; Hunter and Moorhouse, 1976; Colbo and Porter, 1980; Molloy, 1981). Muscid and sciarid flies show "mock oviposition" (Nappi, 1973), depositing nematodes instead of fly eggs (Poinar, 1965; Stoffolano, 1970, 1973). Parasitism by nematodes produces intersexes in some chironomids; midges retain female reproductive behaviors, but deposit nematodes instead of eggs (Rempel, 1940).

Fungi in the genus *Massospora* infect cicadas (Homoptera). In the adult insect, they are restricted to the abdominal tissues (Thaxter, 1888). Successive abdominal segments become infected and slough off, eventually producing a cicada composed primarily of head and thorax (Goldstein, 1929). *M. cicadina* consumes the eggs of its host (Goldstein, 1929); even at early stages of infection, females fail to oviposit (Lloyd *et al.*, 1982). Despite the loss of body segments, infected cicadas circulate among the mating population (Speare, 1921; Soper, 1963). Infected males continue to sing (Soper *et al.*, 1976) and attract healthy females with which they attempt to copulate (Soper, 1963). Healthy males may attempt to copulate with infected females (Soper *et al.*, 1976).

III. Potential Benefits to Parasite

A. Increased Parasite Dissemination

Several behavioral alterations should enhance parasite dissemination. For example, a fungus- or virus-loaded cadaver on the ground should be a poorer source of inoculum than one at treetop level. Viruses that cause elevation-seeking behavior in caterpillars weaken the host's epidermis; flaccid cadavers ("wilt disease") eventually break and spill infectious virus onto the foliage below (e.g., Tanada, 1963). Elevation-seeking behavior by hosts infected by fungal pathogens exposes neighbors to an infective rain of spores (MacLeod *et al.*, 1966). Fungus-infected soldier beetles (*Chauliognathus*) grip the plant with mandibles and spread elytrae and hindwings. The behavior exposes the surface of the abdomen, from which fungal spores are produced (Carner, 1980).

Dissemination may also be enhanced by increases in host activity. Pathogens that attack the host's gut often are released in the host's feces or regurgitations (e.g., Smirnoff, 1961), contaminating foliage eaten by healthy insects. These pathogens tend to kill their hosts slowly, prolonging dissemination. For example, a larval sod webworm (*Crambus trisectus*) may pass up to 100 million *Nosema* spores in its feces prior to death, and contain an additional 200 million spores at death (Maddox, 1973). Grasshoppers (*Melanoplus*) may produce upwards of 10^9 spores of *Nosema locustae* (Henry, 1972). Pathogens that cause sluggishness make the host vulnerable to cannibalism by healthy conspecifics (Henry and Oma, 1981), or to predators that in turn disseminate the pathogen (Young and Yearian, 1989).

For parasites requiring a vertebrate host, elevation-seeking behavior, changes in activity, and increased exposure all enhance transmission (Moore, 1984). For instance, ants parasitized by the "brain-worm" trematode attach themselves to the tips of grass stems where they are ingested by grazing sheep, the definitive host of the parasite (Anokhin, 1966). The increased activity and decreased photophobia shown by cockroaches parasitized by *M. moniliformis* could make cockroaches conspicuous to the parasite's definite host, *Rattus* (Moore, 1983a). Similar benefits result from reduced host activity. Trematodes cause reductions in activity of larval *Aedes aegypti*, leading to increased predation by the vole definitive host (Webber *et al.*, 1987a,b).

Behavior of castrated hosts disseminates the parasite. The nematode-depositing activities of bumblebees (Section II.B) likely spread nematodes over the wintering habitat of the bee, producing infections the following spring (Poinar and van der Laan, 1971). Upstream "oviposition" behavior by castrated blackflies helps maintain the mermithid parasite in the stream (Colbo and Porter, 1980). "Mock oviposition" by muscids spreads nematodes over areas frequented by healthy flies (Nappi, 1973). Mating attempts by *Massospora*-infected cicadas may transmit fungal spores to healthy insects (Soper *et al.*, 1976).

B. Increased Parasite Survival

Movement to new microhabitats by hosts can favor parasites by reducing predation or hyperparasitism. Stamp (1981) noted that *Euphydryas phaeton* caterpillars parasitized by a braconid wasp moved to elevated locations, lowering rates of hyperparasitism. Aphids containing parasitoids moved to concealed locations, resulting in decreased rates of hyperparasitism (Brodeur and McNeil, 1989) and, potentially, increased winter survival for the parasitoid (Brodeur and McNeil, 1990). A literature review (Fritz, 1982) concluded that predation rates often were lower on parasitized hosts than on healthy hosts, possibly because of changes in host behavior. Host castration can be viewed as a parasite survival strategy if it redirects resources used by the host for reproduction toward processes beneficial to the parasite (Baudoin, 1975).

IV. Potential Benefits to Host

A. Increased Survival of Kin

Shapiro (1976) hypothesized that behavioral changes could increase the host's inclusive fitness. Larval *Chlosyne harrissii* parasitized by a braconid wasp perch at the tops of vegetation where they are highly conspicuous (Shapiro, 1976). Shapiro speculated that if nearby caterpillars were related to the parasitized host, the host could gain in inclusive fitness by exposing itself to predators ("host suicide"). Predation would kill the parasitoid and eliminate danger to relatives from the activities of that particular wasp.

The host suicide hypothesis would seem to require several fairly restrictive conditions (Shapiro, 1976; Smith Trail, 1980; Stamp, 1981; Moore, 1984; McAllister and Roitberg, 1987): parasitized hosts should be reproductively dead; kin should be aggregated [e.g., as in social insects (see McAllister and Roitberg, 1987)]; host and parasite phenology should allow the developing parasite to eventually parasitize kin of the infected host; and behavioral changes should increase mortality of the host. A number of these conditions qualitatively appear to be met in insect host/parasite associations. Parasitized insects suffer from predation by a variety of animals (White and Dutky, 1940; Smirnoff, 1959; Hostetter and Biever, 1970; Tostowaryk, 1971). Social insects show parasitism-induced behavioral changes that may increase mortality. For example, the elevation-seeking response of parasitized ants increases their risk of predation (Section III.A). Finally, parasitic castration is common, and should minimize the cost of suicide to the host; that is, benefits need not be too great for suicide to be advantageous (Moore, 1984), since the host is evolutionarily dead (Kuris, 1974).

McAllister and Roitberg (1987) showed that pea aphids (*Acyrthosiphon pisum*) parasitized by a braconid wasp engaged in behavior that the authors interpreted as being suicidal. Parasitized aphids from hot dry habitats dropped from the host plant, resulting in desiccation, in response to the approach of a predator. In wetter areas, where the behavior was unlikely to result in desiccation, aphids were less likely to drop from the host (see also Latta, 1987; Tomlinson, 1987; McAllister and Roitberg, 1988; McAllister *et al.*, 1990).

Finally, kin may benefit from host behavior even if the behavioral change is not "suicidal." Pathogen-infected hosts that move to new microhabitats may avoid being cannibalized by healthy kin. Also, although some behavioral changes (e.g., elevation-seeking) may disseminate infective propagules over large areas, the behavior could reduce transmission to nearby hosts; that is, unless the pathogen is highly infective and requires only a few propagules to begin a new infection, widespread dissemination of propagules could reduce inoculum density to the extent that nearby kin are unlikely to contact an infective dose (C. MacVean, personal communication).

B. Host Defense: Behavioral Fever

Hosts would benefit if altered behavior assisted in combating the parasite. Many ectotherms respond to parasitism by raising body temperatures; that is, parasitized individuals select warmer microhabitats than do healthy animals ("behavioral fever"; see Kluger, 1979, 1990). This option may be available to insects, since increased temperatures are often detrimental to parasite success (Table 2; see also Heinrich, 1981). Several behavioral changes are consistent with a fever explanation. Movement to exposed locations may increase body temperatures. Marikovsky (1962) suggested that fungus-infected insects benefited from "solar irradiation" and noted that basking grasshoppers occasionally recovered from the pathogen. Diurnal activity by normally nocturnal hosts may raise body temperatures, as might movement to the surface by subterranean hosts. Increased activity, particularly flight, also raises body temperature (Heinrich, 1981).

Five studies have shown that insects challenged with toxins or pathogens elevated their body temperatures (Bronstein and Conner, 1984; Louis *et al.*, 1986; Boorstein and Ewald, 1987; McClain *et al.*, 1988; Watson *et al.*, 1993); in two cases this was shown to increase host survival (Boorstein and Ewald, 1987; Watson *et al.*, 1993; see also Louis *et al.*, 1986; Carruthers *et al.*, 1992).

V. Considerations for Future Research

With some exceptions (e.g., Stamp, 1981; Boorstein and Ewald, 1987; McAllister and Roitberg, 1987; Brodeur and McNeil, 1990), evidence tying behavioral alterations to host or parasite success has been anecdotal. Full understanding will require experimentation, specifically: (1) quantitative descriptions of behavioral changes from experiments that incorporate suitable

Table 2

Selected Examples of Effects of High Temperature on Survival of Parasitized Hosts or Mortality of Parasites

Parasite or pathogen	Host	Effects of temperature on host or parasite	Reference[a]
Viruses	Caterpillars	Host survival at 32.2°C > 26.7°C, 21.1°C, 15.6°C	2
		Host survival at 35°C > 25°C	7
		Host survival at 35°C, 30°C > 25°C	11
		Host survival at 40°C > 37°C > 33°C > 30°C	10
Fungi	Aphids	Fungal growth on agar ceased at 27–36°C	5
	Caterpillar	Host survival at 31°C > 26°C, 21°C, 16°C	3
		Host survival at 30°C > 25°C, 20°C, 15°C	8
Microsporidium	Caterpillar	Loss of parasite at culture temperatures of 35°C	12
Bacterium	Grasshopper	Host survival at 28°C > 20°C	9
Parasitoids	Aphid	Upper lethal temperature lower for parasitoid than for host	4
	Caterpillar	Upper lethal temperature lower for parasitoid than for host	1, 6

[a]References: (1) Allen and Smith (1958); (2) Boucias *et al.* (1980); (3) Boucias *et al.* (1984); (4) Force and Messenger (1964); (5) Hall and Bell (1960); (6) Kaya and Tanada (1969); (7) Kobayashi *et al.* (1981); (8) Mohamed *et al.* (1977); (9) Pospelov (1926); (10) Steinhaus and Dineen (1960); (11) Tanada and Tanabe (1965); (12) Wilson and Sohi (1977).

controls; (2) quantitative tests of whether either party benefits; and (3) tests separating behavioral changes that are present because of natural selection operating on the contemporary association from those that are present only fortuitously [i.e., as carryovers from some historical association (Moore and Gotelli, 1990) or as consequences of pathology]. The latter two approaches are important in that treatments of altered host biology may overemphasize the importance of the parasite in the interaction (see also Minchella, 1985); experimentation should reduce this temptation. We conclude by discussing these three approaches.

Although it is tempting to infer behavioral alterations from differences in the distributions of healthy and parasitized individuals, this can be misleading (Stamp, 1981). Jaynes (1954) noted that spruce budworm larvae collected from the upper tree crown were more likely to be parasitized by *Apanteles* wasps than were caterpillars lower in the tree, whereas there were no distribution differences for caterpillars parasitized by *Glypta* wasps. Hence, one may hypothesize either that *Apanteles* induced behavioral changes in the host while *Glypta* did not, or that the two parasitoids differed in foraging behavior. Doane (1970) indicated that the treetop accumulation of virus-killed gypsy moths (*Lymantria dispar*) was due to the activities of both healthy and infected caterpillars; infected caterpillars died at treetop level, whereas healthy insects moved down the tree as they aged. In sum, demonstrating true behavioral modifications requires experimentation with suitable controls.

Second, from studies reported here, it is not possible to determine the beneficiary of a given behavioral alteration; indeed, some alterations, depending upon the parasite, could be consistent with all four hypotheses presented in this paper (Table 3). Costs and benefits may accrue to both parasite and host. For example, development rates of *M. moniliformis* in cockroaches increase as temperature increases from 20 to 28°C; at temperatures higher than 28°C, parasites show developmental abnormalities (Lackie, 1972). That is, a slight increase in host temperature benefits the parasite, whereas a large increase does not. Moreover, transmission requires ingestion by a vertebrate host. Hence, even if a behavioral change were, say, a fever response, if the activity also increased host conspicuousness, the parasite might benefit through enhanced dissemination.

Table 3

Potential Benefits of Host Behavioral Changes to Parasite or Host[a]

	Parasite adaptive		Host adaptive	
Host behavioral modification	Increased dissemination	Increased survival	Increased survival of kin	Behavioral fever
Elevation-seeking	*	*	*	*
Movement to exposed locations	*	*	*	*
Movement to concealed locations	0	*	0	0
Preference for higher temperature	*	0	0	*
Altered light reaction	*	0	*	*
Increase in activity	*	0	*	*
Decrease in activity	*	0	*	0
Diurnal activity by nocturnal insects	*	0	*	*
Reproductive activity by castrated hosts or castration itself	*	*	0	0

[a]*-Modification potentially consistent with hypothesis for at least a subset of parasites that induce the behavioral change; representative examples discussed in text. 0-Modification not obviously consistent with hypothesis or not explored in text.

Thus, experimentation is critical for determining the beneficiary of a change in behavior. Given the number of possible outcomes (Table 3), this approach could become quite complex. Familiarity with the system may eliminate some hypotheses. For example, some systems may be less subject to kin selection than are others (Smith Trail,1980; Stamp, 1981; Moore, 1984; McAllister and Roitberg, 1987). In one caterpillar–parasitoid system, Stamp (1981) noted that the gregarious nature of the host might initially lead one to look for host suicide. However, she also noted that suicide was unlikely to evolve because the host was aposematic and unpalatable; results of her study were consistent with the latter hypothesis. On the basis of host and parasite biology, Moore (1984) concluded that host suicide was unlikely to evolve in arthropod–Acanthocephala associations. Conversely, in a system more amenable to kin selection, behavior consistent with host suicide was demonstrated (McAllister and Roitberg, 1987).

Lastly, we suggest that a comparative approach, optimally one that considers host and parasite phylogenies, may be necessary to fully understand the evolution of behavioral alterations (Jones, 1985; Moore and Gotelli, 1990). A behavioral alteration may show up in a parasite/host association because it evolved in an ancestral association and was retained in descendant taxa (Moore and Gotelli, 1990); benefits to descendant taxa might have changed or be absent altogether. Parasites could benefit from an evolved host behavior; that is, the parasite may elicit a preexisting host behavior rather than induce a novel behavior (Moore, 1984). For example, Wickler (1976) suggested that the behavior shown by trematode-parasitized ants, that is, attaching themselves by mandibles to the upper portion of a plant, is similar to a phylogenetically old "sleeping" behavior found in other Hymenoptera. Fever may make hosts conspicuous, such that the parasite benefits via dissemination to the definitive host. Lastly, the parasite may benefit from a host pathologic response. Cram (1931) noted that grasshoppers infected with the nematode *Tetrameres americana* were droopy (the parasite encysts in grasshopper muscle), hence might be vulnerable to the galliform final host. Although choice of encystment site may have evolved through selection favoring increased predation by the definitive host, this interpretation is preliminary in the absence of other information.

The comparative approach can take one of two directions: study of a single parasite taxon in a variety of hosts, or study of a variety of parasites in a single host taxon [for an example of the former, see Jones' (1985) discussion about the effects of *Chelonus* parasites on metamorphosis in Lepidoptera]. The present review suggests that there is a great deal of convergence in the behavior of parasitized insects. Indeed, a number of behavioral changes noted here have parallels in non-insect systems (Holmes and Bethel, 1972; Moore, 1983b, 1984; Moore and Gotelli, 1990). Elevation-seeking behavior is particularly widespread. The fact that parasites as distantly related as fungi, viruses, parasitoids, and trematodes induce this same behavior in their insect hosts might suggest that elevation-seeking is a generalized insect response to parasitism, and that any benefits to the parasite are fortuitous. In some noninsect systems host behavior (hence benefits?) differs depending upon the parasite (Bethel and Holmes, 1973; Rau, 1983, 1984; Moore and Lasswell, 1986). Comparisons of host behavior in a single insect species parasitized by taxonomically unrelated parasites would be of value.

Acknowledgments

For comments on versions of this manuscript we thank Ann Baker, John Capinera, Deb Kendall, Jim Krysan, Chuck MacVean, and Tom Unruh. Support for this project was received from the National Science Foundation (Grant BSR-8452076), the Monsanto Company, the Burroughs Wellcome Fund, the Exxon Education Foundation, and the Whitehall Foundation, as well as from NSF grant BSR-8817495 to J. Moore and N. Gotelli. Additional support for D. Horton was obtained from the Colorado Agricultural Experiment Station, USDA-ARS, the Washington Tree Fruit Research Commission, Yakima, and the Winter Pear Bureau, Portland, Oregon.

References

Allen, H. W. (1921). Notes on a bombylid parasite and a polyhedral disease of the southern grass worm, *Laphygma frugiperda*. *J. Econ. Entomol.* **14:**510–511.

Allen, W. W., and Smith, R. F. (1958). Some factors influencing the efficiency of *Apanteles medicaginis* Muesebeck (Hymenoptera: Braconidae) as a parasite of the alfalfa caterpillar, *Colias philodice eurytheme* Boisduval. *Hilgardia* **28:**1–42.

Anderson, R. M., and May, R. M. (1979). Population biology of infectious diseases. Part I. *Nature (London)* **280:**361–367.

Anokhin, I. A. (1966). Daily rhythm in ants infected with metacercariae of *Dicrocoelium lanceatum*. *Dokl. Akad. Nauk SSSR* **166:**757–759.

Baird, R. B. (1957). Notes on a laboratory infection of Diptera caused by the fungus *Empusa muscae* Cohn. *Can. Entomol.* **89:**432–435.

Baudoin, M. (1975). Host castration as a parasitic strategy. *Evolution (Lawrence, Kans.)* **29:**335–352.

Beckage, N. E. (1985). Endocrine interactions between endoparasitic insects and their hosts. *Annu. Rev. Entomol.* **30:**371–413.

Bethel, W. M., and Holmes, J. C. (1973). Altered evasive behavior and responses to light in amphipods harboring acanthocephalan cystacanths. *J. Parasitol.* **59:**945–956.

Boorstein, S. M., and Ewald, P. W. (1987). Costs and benefits of behavioral fever in *Melanoplus sanguinipes* infected by *Nosema acridophagus*. *Physiol. Zool.* **60:**586–595.

Boucias, D. G., Johnson, D. W., and Allen, G. E. (1980). Effects of host age, virus dosage, and temperature on the infectivity of a nucleopolyhedrosis virus against velvetbean caterpillar, *Anticarsia gemmatalis*, larvae. *Environ. Entomol.* **9:**59–61.

Boucias, D. G., Bradford, D. L., and Barfield, C. S. (1984). Susceptibility of the velvetbean caterpillar and soybean looper (Lepidoptera: Noctuidae) to *Nomuraea rileyi:* Effects of pathotype, dosage, temperature, and host age. *J. Econ. Entomol.* **77,** 247–253.

Brodeur, J., and McNeil, J. N. (1989). Seasonal microhabitat selection by an endoparasitoid through adaptive modification of host behavior. *Science* **244,** 226–228.

Brodeur, J., and McNeil, J. N. (1990). Overwintering microhabitat selection by an endoparasitoid (Hymenoptera: Aphidiidae): Induced phototactic and thigmokinetic responses in dying hosts. *J. Insect Behav.* **3:**751–763.

Bronstein, S. M., and Conner, W. E. (1984). Endotoxin-induced behavioural fever in the Madagascar cockroach, *Gromphadorhina portentosa. J. Insect Physiol.* **30:**327–330.

Bünzli, G. H., and Büttiker, W. W. (1959). Fungous diseases of lamellicorn larvae in southern Rhodesia. *Bull. Entomol. Res.* **50:**89–96.

Carner, G. R. (1980). *Entomophthora lampyridarum,* a fungal pathogen of the soldier beetle, *Chauliognathus pennsylvanicus. J. Invertebr. Pathol.* **36:**394–398.

Carney, W. P. (1969). Behavioral and morphological changes in carpenter ants harboring dicrocoeliid metacercariae. *Am. Midl. Nat.* **82:**605–611.

Carruthers, R. I., Larkin, T. S., and Firstencel, H. (1992). Influence of thermal ecology on the mycosis of a rangeland grasshopper. *Ecology* **73:**190–204.

Colbo, M. H., and Porter, G. N. (1980). Distribution and specificity of Mermithidae (Nematoda) infecting Simuliidae (Diptera) in Newfoundland. *Can. J. Zool.* **58:**1483–1490.

Combes, C. (1991). Ethological aspects of parasite transmission. *Am. Nat.* **138:**866–880.

Cram, E. B. (1931). Developmental stages of some nematodes of the Spiruoidea parasitic in poultry and game birds. *U.S. Dep. Agric., Tech. Bull.* **227.**

Doane, C. C. (1970). Primary pathogens and their role in the development of an epizootic in the gypsy moth. *J. Invertebr. Pathol.* **15:**21–33.

Edland, T. (1965). A granulosis of *Eupsilia transversa* Hufn. (Lepidoptera, Noctuidae) in west Norway. *Entomophaga* **10:**331–333.

Eilenberg, J. (1987). Abnormal egg-laying behaviour of female carrot flies (*Psila rosae*) induced by the fungus *Entomophthora muscae. Entomol. Exp. Appl.* **43:**61–65.

Force, D. C., and Messenger, P. S. (1964). Duration of development, generation time, and longevity of three hymenopterous parasites of *Therioaphis maculata,* reared at various constant temperatures. *Ann. Entomol. Soc. Am.* **57:**405–413.

Fowler, M., and Robertson, J. S. (1972). Iridescent virus infection in field populations of *Wiseana cervinata* (Lepidoptera: Hepialidae) and *Witlesia* sp. (Lepidoptera: Pyralidae) in New Zealand. *J. Invertebr. Pathol.* **19:**154–155.

Fritz, R. S. (1982). Selection for host modification by insect parasitoids. *Evolution* (*Lawrence, Kans.*) **36:**283–288.

Goldstein, B. (1929). A cytological study of the fungus *Massospora cicadina,* parasitic on the 17-year cicada, *Magicicada septendecim. Am. J. Bot.* **16:**394–401.

Graham, G. L. (1966). The behavior of beetles, *Tribolium confusum,* parasitized by the larval stage of a chicken tapeworm, *Raillietina cesticillus. Trans. Am. Microsc. Soc.* **85:**163.

Grunin, K. Y. (1949). A mistake in instinct resulting from parasitic castration in *Prosimulium hirtipes* Fries (Diptera, Simuliidae). *Dokl. Akad. Nauk SSSR* **66:**305–307.

Hall, I. M., and Bell, J. V. (1960). The effect of temperature on some entomophthoraceous fungi. *J. Insect Pathol.* **2:**247–253.

Harper, A. M. (1958). Notes on behaviour of *Pemphigus betae* Doane (Homoptera: Aphididae) infected with *Entomophthora aphidis* Hoffm. *Can. Entomol.* **90:**439–440.

Heinrich, B. (1981). Ecological and evolutionary perspectives. *In* "Insect Thermoregulation" (B. Heinrich, ed.), pp. 235–302. Wiley, New York.

Henry, J. E. (1972). Epizootiology of infection by *Nosema locustae* Canning (Microsporidia: Nosematidae) in grasshoppers. *Acrida* **1:**111–120.

Henry, J. E., and Oma, E. A. (1981). Pest control by *Nosema locustae,* a pathogen of grasshoppers and crickets. *In* "Microbial Control of Pests and Plant Diseases 1970–1980" (H. D. Burges, ed.), pp. 573–586. Academic Press, New York.

Holmes, J. C., and Bethel, W. M. (1972). Modification of intermediate host behaviour by parasites. *Zool. J. Linn. Soc.* **51:**123–149.

Hostetter, D. L., and Biever, K. D. (1970). The recovery of virulent nuclear-polyhedrosis virus of the cabbage looper, *Trichoplusia ni,* from the feces of birds. *J. Invertebr. Pathol.* **15:**173–176.

Hunter, D. M., and Moorhouse, D. E. (1976). Sexual mosaics and mermithid parasitism in *Austrosimulium bancrofti* (Tayl.) (Diptera, Simuliidae). *Bull. Entomol. Res.* **65:**549–553.

Jaynes, H. A. (1954). Parasitization of spruce budworm larvae at different crown heights by *Apanteles* and *Glypta. J. Econ. Entomol.* **47:**355–356.

Johnson, D. L. (1989). The effects of timing and frequency of application of *Nosema locustae* (Microspora: Microsporida) on the infection rate and activity of grasshoppers (Orthoptera: Acrididae). *J. Invertebr. Pathol.* **54:**353–362.

Jones, D. (1985). Endocrine interaction between host (Lepidoptera) and parasite (Cheloninae: Hymenoptera): Is the host or the parasite in control? *Ann. Entomol. Soc. Am.* **78:**141–148.

Kalmakoff, J., and Moore, S. G. (1975). The ecology of nucleopolyhedrosis virus in porina (*Wiseana* spp.) (Lepidoptera: Hepialidae). *N.Z. Entomol.* **6:**73–76.

Kaya, H. K., and Tanada Y. (1969). Responses to high temperature of the parasite *Apanteles militaris* and of its host, the armyworm, *Pseudaletia unipuncta. Ann. Entomol. Soc. Am.* **62:**1303–1306.

King, K. M., and Atkinson, N. J. (1928). The biological control factors of the immature stages of *Euxoa ochrogaster* Gn. (Lepidoptera, Phalaenidae) in Saskatchewan. *Ann. Entomol. Soc. Am.* **21:**167–188.

Kluger, M. J. (1979). "Fever: Its Biology, Evolution and Function." Princeton Univ. Press, Princeton, NJ.

Kluger, M. J. (1990). The febrile response. *In* "Stress Proteins in Biology and Medicine" (R. I. Morimoto, A. Tissières, and C. Geogopoulos, eds.), pp. 61–78. Cold Spring Harbor Lab. Press, Cold Spring Harbor, New York.

Kobayashi, M., Inagaki, S., and Kawase, S. (1981). Effect of high temperature on the development of nuclear polyhedrosis virus in the silkworm, *Bombyx mori. J. Invertebr. Pathol.* **38:**386–394.

Kulinčević, J. M., Stairs, G. R., and Rothenbuhler, W. C. (1970). Virus causing paralysis of adult honeybees in Ohio. *J. Invertebr. Pathol.* **16:**423–426.

Kuris, A. M. (1974). Trophic interactions: Similarity of parasitic castrators to parasitoids. *Q. Rev. Biol.* **49:**129–148.

Lackie, J. M. (1972). The effect of temperature on the development of *Moniliformis dubius* (Acanthocephala) in the intermediate host, *Periplaneta americana. Parasitology* **65:**371–377.

Latta, B. (1987). Adaptive and non-adaptive suicide in aphids. *Nature* (*London*) **330:**701.

Lewis, F. B. (1960). Factors affecting assessment of parasitization by *Apanteles fumiferanae* Vier. and *Glypta fumiferanae* (Vier.) on spruce budworm larvae. *Can. Entomol.* **92:**881–891.

Lloyd, M., White, J., and Stanton, N. (1982). Dispersal of fungus-infected periodical cicadas to new habitat. *Environ. Entomol.* **11:**852–858.

Louis, C., Jourdan, M., and Cabanac, M. (1986). Behavioral fever and therapy in a rickettsia-infected Orthoptera. *Am. J. Physiol.* **250:**R991–R995.

Lounsbury, C. P. (1913). Caterpillar wilt disease. *Agric. J. Union S. Afr.* **5:**448–452.

Lower, H. F. (1954). A granulosis virus attacking the larvae of *Persectania ewingii* Westw. (Lepidoptera: Agrotidae) in South Australia. *Aust. J. Biol. Sci.* **7:**161–167.

Lundberg, H., and Svensson, B. G. (1975). Studies on the behaviour of *Bombus* Latr. species (Hym., Apidae) parasitized by *Sphaerularia bombi* Dufour (Nematoda) in an alpine area. *Nor. J. Entomol.* **22:**129–134.

MacLeod, D. M., MacBain Cameron, J. W., and Soper, R. S. (1966). The influence of environmental conditions on epizootics caused by entomogenous fungi. *Rev. Roum. Biol., Ser. Bot.* **11:**125–134.

MacLeod, D. M., Tyrrell, D., Soper, R. S., and De Lyzer, A. J. (1973). *Entomophthora bullata* as a pathogen of *Sarcophaga aldrichi*. *J. Invertebr. Pathol.* **22:**75–79.

Maddox, J. V. (1973). The persistence of the Microsporidia in the environment. *Misc. Publ. Entomol. Soc. Am.* **9:**99–104.

Marikovsky, P. I. (1962). On some features of behavior of the ants *Formica rufa* L. infected with fungous disease. *Insectes Soc.* **9:**173–179.

Martignoni, M. E. (1964). Pathophysiology in the insect. *Annu. Rev. Entomol.* **9:**179–206.

McAllister, M. K., and Roitberg, B. D. (1987). Adaptive suicidal behaviour in pea aphids. *Nature (London)* **328:**797–799.

McAllister, M. K., and Roitberg, B. D. (1988). Assumptions about suicidal behaviour of aphids. *Nature (London)* **332:**494–495.

McAllister, M. K., Roitberg, B. D., and Laurence Weldon, K. (1990). Adaptive suicide in pea aphids: Decisions are cost sensitive. *Anim. Behav.* **40:**167–175.

McClain, E., Magnuson, P., and Warner, S. J. (1988). Behavioural fever in a Namib Desert tenebrionid beetle *Onymacris plana*. *J. Insect Physiol.* **34:**279–284.

Miller, L. A., and McClanahan, R. J. (1959). Note on occurrence of the fungus *Empusa muscae* Cohn on adults of the onion maggot, *Hylemya antiqua* (Meig.) (Diptera: Anthomyiidae). *Can. Entomol.* **91:**525–526.

Minchella, D. J. (1985). Host life-history variation in response to parasitism. *Parasitology* **90:**205–216.

Mohamed, A.K.A., Sikorowski, P. P., and Bell, J. V. (1977). Susceptibility of *Heliothis zea* larvae to *Nomuraea rileyi* at various temperatures. *J. Invertebr. Pathol.* **30:**414–417.

Molloy, D. P. (1981). Mermithid parasitism of blackflies (Diptera: Simuliidae). *J. Nematol.* **13:**250–256.

Molyneux, D. H., and Jeffries, D. (1986). Feeding behaviour of pathogen-infected vectors. *Parasitology* **92:**721–736.

Moore, J. (1983a). Altered behavior in cockroaches (*Periplaneta americana*) infected with an Archiacanthocephalan, *Moniliformis moniliformis*. *J. Parasitol.* **69:**1174–1176.

Moore, J. (1983b). Responses of an avian predator and its isopod prey to an acanthocephalan parasite. *Ecology* **64:**1000–1015.

Moore, J. (1984). Altered behavioral responses in intermediate hosts—an acanthocephalan parasite strategy. *Am. Nat.* **123:**572–577.

Moore, J., and Gotelli, N. J. (1990). A phylogenetic perspective on the evolution of altered host behaviours: A critical look at the manipulation hypothesis. *In* "Parasitism and Host Behaviour" (C. J. Barnard and J. M. Behnke, eds.), pp. 193–233. Taylor & Francis, New York.

Moore, J., and Lasswell, J. (1986). Altered behavior in isopods (*Armadillidium vulgare*) infected with the nematode *Dispharynx nasuta*. *J. Parasitol.* **72:**186–189.

Nappi, A. J. (1973). Effects of parasitization by the nematode, *Heterotylenchus autumnalis*, on mating and oviposition in the host, *Musca autumnalis*. *J. Parasitol.* **59:**963–969.

Nirula, K. K. (1957). Observations on the green muscardine fungus in populations of *Oryctes rhinoceros* L. *J. Econ. Entomol.* **50:**767–770.

Paschke, J. D., and Hamm, J. J. (1961). A nuclear polyhedrosis of *Rachiplusia ou* (Guenée). *J. Insect Pathol.* **3:**333–334.

Plateaux, L. (1972). Sur les modifications produites chez une formi par la présence d'un parasite cestode. *Ann. Sci. Nat.: Zool. Biol. Anim.* [12] **14:**203–220.

Poinar, G. O., Jr. (1965). The bionomics and parasitic development of *Tripius sciarae* (Bovien)

(Sphaerulariidae: Aphelenchoidea), a nematode parasite of sciarid flies (Sciaridae: Diptera). *Parasitology* **55:**559–569.

Poinar, G. O., Jr., and Gyrisco, G. G. (1964). Effect of light on the behavior of the alfalfa weevil, *Hypera postica*. *Ann. Entomol. Soc. Am.* **57:**213–215.

Poinar, G. O., Jr., and van der Laan, P. A. (1972). Morphology and life history of *Sphaerularia bombi*. *Nematologica* **18:**239–252.

Pospelov, V. P. (1926). The influence of temperature on the maturation and general health of *Locusta migratoria* L. *Bull. Entomol. Res.* **16:**363–367.

Rau, M. E. (1983). The open-field behaviour of mice infected with *Trichinella spiralis*. *Parasitology* **86:**311–318.

Rau, M. E. (1984). The open-field behaviour of mice infected with *Trichinella pseudospiralis*. *Parasitology* **88:**415–419.

Reinganum, C., O'Loughlin, G. T., and Hogan, T. W. (1970). A nonoccluded virus of the field crickets *Teleogryllus oceanicus* and *T. commodus* (Orthoptera: Gryllidae). *J. Invertebr. Pathol.* **16:**214–220.

Rempel, J. G. (1940). Intersexuality in Chironomidae induced by nematode parasitism. *J. Exp. Zool.* **84:**261–289.

Rockwood, L. P. (1950). Entomogenous fungi of the family Entomophthoraceae in the Pacific Northwest. *J. Econ. Entomol.* **43:**704–707.

Roffey, J. (1968). The occurrence of the fungus *Entomophthora grylli* Fresenius on locusts and grasshoppers in Thailand. *J. Invertebr. Pathol.* **11:**237–241.

Romig, T., Lucius, R., and Frank, W. (1980). Cerebral larvae in the second intermediate host of *Dicrocoelium dendriticum* (Rudolphi, 1819) and *Dicrocoelium hospes* Looss, 1907 (Trematodes, Dicrocoeliidae). *Z. Parasitenkd.* **63:**277–286.

Sato, Y., Tanaka, T., Imafuku, M., and Hidaka, T. (1983). How does diurnal *Apanteles kariyai* parasitize and egress from a nocturnal host larva? *Konchu* **51:**128–139.

Schmid-Hempel, P., and Schmid-Hempel, R. (1990). Endoparasitic larvae of conopid flies alter pollination behavior of bumblebees. *Naturwissenschaften* **77:**450–452.

Schmid-Hempel, R., and Muller, C. B. (1991). Do parasitized bumblebees forage for their colony? *Anim. Behav.* **41:**910–912.

Semel, M. (1956). Polyhedrosis wilt of cabbage looper on Long Island. *J. Econ. Entomol.* **49:**420–421.

Shapiro, A. M. (1976). Beau geste? *Am. Nat.* **110:**900–902.

Skaife, S. H. (1925). The locust fungus, *Empusa grylli,* and its effects on its host. *S. Afr. J. Sci.* **22:**298–308.

Smirnoff, W. A. (1959). Predators of *Neodiprion swainei* Midd. (Hymenoptera: Tenthredinidae) larval vectors of virus diseases. *Can. Entomol.* **91:**246–248.

Smirnoff, W. A. (1960). Observations on the migration of larvae of *Neodiprion swainei* Midd. (Hymenoptera: Tenthredinidae). *Can. Entomol.* **92:**957–958.

Smirnoff, W. A. (1961). A virus disease of *Neodiprion swainei* Middleton. *J. Insect Pathol.* **3:**29–46.

Smirnoff, W. A. (1965). Observations on the effect of virus infection on insect behavior. *J. Invertebr. Pathol.* **7:**387–388.

Smith, O. J., Hughes, K. M., Dunn, P. H., and Hall, I. M. (1956). A granulosis virus disease of the western grape skeletonizer and its transmission. *Can. Entomol.* **88:**507–515.

Smith Trail, D. R. (1980). Behavioral interactions between parasites and hosts: Host suicide and the evolution of complex life cycles. *Am. Nat.* **116:**77–91.

Soper, R. S. (1963). *Massospora levispora,* a new species of fungus pathogenic to the cicada, *Okanagana rimosa*. *Can J. Bot.* **41:**875–878.

Soper, R. S., Delyzer, A. J., and Smith, L.F.R. (1976). The genus *Massospora* entomopathogenic for cicadas. Part II. Biology of *Massospora levispora* and its host *Okanagana rimosa,* with notes on *Massospora cicadina* on the periodical cicadas. *Ann. Entomol. Soc. Am.* **69:**89–95.

Speare, A. T. (1921). *Massospora cicadina* Peck, a fungous parasite of the periodical cicada. *Mycologia* **13:**72–82.

Stairs, G. R. (1965). Artificial initiation of virus epizootics in forest test caterpillar populations. *Can. Entomol.* **97:**1059–1062.

Stamp, N. E. (1981). Behavior of parasitized aposematic caterpillars: Advantageous to the parasitoid or the host? *Am. Nat.* **118:**715–725.

Steinhaus, E. A. (1949). "Principles of Insect Pathology." McGraw-Hill, New York.

Steinhaus, E. A., and Dineen, J. P. (1960). Observations on the role of stress in a granulosis of the variegated cutworm. *J. Insect Pathol.* **2:**55–65.

Stoffolano, J. G., Jr. (1970). Nematodes associated with the genus *Musca* (Diptera: Muscidae). *Bull. Entomol. Soc. Am.* **16:**194–203.

Stoffolano, J. G., Jr. (1973). Host specificity of entomophilic nematodes—a review. *Exp. Parasitol.* **33:**263–284.

Tamashiro, M., and Huang, S. (1963). A cytoplasmic polyhedrosis of *Cactoblastis cactorum* (Berg). *J. Insect Pathol.* **5:**397–399.

Tanada, Y. (1963). Epizootiology of infectious diseases. *In* "Insect Pathology" (E. A. Steinhaus, ed.), Vol. 2, pp. 423–475. Academic Press, New York.

Tanada, Y., and Tanabe, A. M. (1965). Resistance of *Galleria mellonella* (Linnaeus) to the *Tipula* iridescent virus at high temperatures. *J. Invertebr. Pathol.* **7:**184–188.

Thaxter, R. (1888). The Entomophthoreae of the United States. *Mem. Boston Soc. Nat. Hist.* **4:**133–201.

Tomlinson, I. (1987). Adaptive and non-adaptive suicide in aphids. *Nature (London)* **330:**701.

Tostowaryk, W. (1971). Relationship between parasitism and predation of diprionid sawflies. *Ann. Entomol. Soc. Am.* **64:**1424–1427.

Vinson, S. B., and Iwantsch, G. F. (1980). Host regulation by insect parasitoids. *Q. Rev. Biol.* **55:**143–165.

Watson, D. W., Mullens, B. A., and Petersen, J. J. (1993). Behavioral fever response of *Musca domestica* (Diptera: Muscidae) to infection by *Entomophthora muscae* (Zygomycetes: Entomophthorales). *J. Invertebr. Pathol.* **61:**10–16.

Webber, R. A., Rau, M. E., and Lewis, D. J. (1987a). The effects of *Plagiorchis noblei* (Trematoda: Plagiorchiidae) metacercariae on the behavior of *Aedes aegypti* larvae. *Can. J. Zool.* **65:**1340–1342.

Webber, R. A., Rau, M. E., and Lewis, D. J. (1987b). The effects of *Plagiorchis noblei* (Trematoda: Plagiorchiidae) metacercariae on the susceptibility of *Aedes aegypti* larvae to predation by guppies (*Poecilia reticulata*) and meadow voles (*Microtus pennsylvanicus*). *Can. J. Zool.* **65:**2346–2348.

White, R. T., and Dutky, S. R. (1940). Effect of the introduction of milky diseases on populations of Japanese beetle larvae. *J. Econ. Entomol.* **33:**306–309.

Whitlock, V. H. (1974). Symptomology of two viruses infecting *Heliothis armigera. J. Invertebr. Pathol.* **23:**70–75.

Wickler, W. (1976). Evolution-oriented ethology, kin selection, and altruistic parasites. *Z. Tierpsychol.* **42:**206–214.

Wilson, G. G., and Sohi, S. S. (1977). Effect of temperature on healthy and microsporidia-infected continuous cultures of *Malacosoma dissitria* hemocytes. *Can. J. Zool.* **55:**713–717.

Wilson, K., and Edwards, J. (1986). The effects of parasitic infection on the behaviour of an intermediate host, the American cockroach, *Periplaneta americana,* infected with an acanthocephalan, *Moniliformis moniliformis. Anim. Behav.* **34:**942–944.

Wygant, N. D. (1941). An infestation of the pandora moth, *Coloradia pandora* Blake, in lodgepole pine in Colorado. *J. Econ. Entomol.* **34:**697–702.

Yen, D. F. (1962). An *Entomophthora* infection in the larva of the tiger moth, *Creatonotus gangis* (Linnaeus). *J. Insect Pathol.* **4:**88–94.

Yendol, W. G., and Paschke, J. D. (1967). Infection of a looper complex by *Entomophthora sphaerosperma*. *J. Invertebr. Pathol.* **9:**274–276.

Young, S. Y., and Yearian, W. C. (1989). Nuclear polyhedrosis virus-infected and healthy *Anticarsia gemmatalis* larvae as prey for *Nabis roseipennis* adults in the laboratory. *J. Invertebr. Pathol.* **54:**139–143.

Chapter 6

Redirection of Host Metabolism and Effects on Parasite Nutrition

S. N. Thompson
Department of Entomology
University of California
Riverside, California

I. Introduction

This chapter outlines the effects of parasitism by protelean insect parasites or parasitoids[1] on the intermediary metabolism of their insect hosts. Successful parasitic relationships involving koinobiontic parasitoids (Section II) require that the parasite be "adapted to the physiochemical conditions of the host" and "utilize host nutrient in a manner compatible with host survival" (Smyth, 1976). To fulfill these requirements, alterations in the host's physiol-

[1]The term *parasite* is used here to describe the activity and behavior of the parasitic larval stages of "protelean insect parasites" (Askew, 1971). Parasitoid is used in reference to the free-living adult stage.

ogy often are necessary (Vinson, 1984). This "redirection" of the host is a multidimensional phenomenon usually accompanied by observable changes in growth, development, and behavior (Thompson, 1990). The host remains a distinct and viable entity, but its fate is permanently altered (Thompson, 1983a). The process of redirection is critical for the success of the parasite and has important consequences in the evolution of these parasite–host relationships.

The physiology of the parasitized host reflects changes in the normal function and activity of the host's tissues and organ systems. The observed effects are manifested through changes in metabolism. Indeed, metabolism is responsible for maintaining physiological function and although not macroscopically evident, physiological integration is achieved through metabolic, that is, biochemical, regulation. Even the simplest physiological adjustment involves complex networks of biochemical reactions and their associated regulatory mechanisms integrated through the host's nervous and endocrine systems. The molecular and biochemical bases for the redirection of host physiology are the subject of intense study. Changes in intermediary metabolism that may be adaptive in relation to nourishment of the developing parasite larvae are the subject of this discussion.

II. Insect Parasitism and Host Viability

Protelean insect parasites comprise the largest group of parasitic animals. However, because the fate of the insect host is, in most cases, death, insect parasites were historically considered to be primitive—" their feeding behavior intermediate between the parasitic and predaceous ends of the behavioral continuum" (Knutson and Berg, 1966). It is now appreciated that the parasitic stages of insect parasites exhibit a great diversity of "feeding behavior." Askew and Shaw (1986) distinguished idiobiontic parasitoids, which paralyze and kill their host quickly, and koinobiontic species, which develop in a host that remains alive and continues to feed. In the latter case, the host and parasite often exhibit an array of physiological, behavioral, and developmental interactions.

Establishing the viability of the parasite–host complex is important for assessing the pathological consequences of parasitism to the host. Viability is inextricably linked with energy. The orderly processing of energy is the function of metabolism and is achieved through the regulation of biochemical pathways. In this regard, the adenylate nucleotide pool plays a ubiquitous and pivotal role. Adenylates influence the activity of many regulatory enzymes and their ratios are measures of the phosphorylation potential as predicted by the theory of metabolic regulation through fractional amplification (see Newsholme and Start, 1973). Because many regulatory enzymes are affected by adenylate ratios rather than by the absolute concentrations of individual nucleotides, the energy charge ratio [EC= ATP + $\frac{1}{2}$ ADP)/ATP + ADP + AMP)] has proven to be of considerable importance in metabolic regulation (Atkinson, 1968). Chapman *et al.* (1971) demonstrated EC to be an overall indicator of viability *in vivo*. EC is a qualitative measure of the potential energy of the organism and therefore provides a parameter for evaluation the overall metabolic stress induced by parasitism.

Studies by Thompson and Yamada (1984) on the adenylate nucleotide pool of the insect host *Trichoplusia ni* (Hübner) parasitized by the solitary ichneumonid *Hyposter exiguae* (Viereck) clearly demonstrated that the energy status of the insect host is maintained following parasitization despite extensive alterations in physiology and behavior. The concentrations and ratios of adenylate nucleotides in the fat body of fourth-instar *T. ni* were determined 3 days postoviposition. The level of AMP tended to increase and ATP decreased, but the EC was only slightly altered. Subsequent *in vivo* ^{13}P-NMR studies with *Manduca sexta* (Linnaeus) parasitized by the gregarious braconid *Cotesia congregata* (Masson) [=*Apanteles congregatus* (Say)] demonstrated that the overall energy status of this insect host also was minimally affected by parasitism (Thompson *et al.*, 1988).

III. Altered Metabolite Levels in Host Tissues

A. Carbohydrates

Dahlman and Vinson (1975) were the first to observe elevated hemolymph trehalose levels in parasitized insects. The mean trehalose concentration in *Heliothis virescens* (Fabricius) parasitized late in the second instar by the bra conid *Microplitis croceipes* (Cresson) increased from approximately 5 to 40 m*M* by 6 days postoviposition, and thereafter declined. In contrast, the trehalose level in hemolymph of control insects increased from 4 to 16 m*M* during the same period. The fat body glycogen level of parasitized *H. virescens* also increased following parasitization, from a mean of 81 μg/mg fresh weight to approximately 100 μg/mg 5 days postoviposition. Thereafter, glycogen level decreased and, at the time of parasite emergence, the fat body of parasitized larvae was almost glycogen depleted (Dahlman and Vinson, 1980). Glycogen also increased in fat body of control larvae, but to a lesser extent. The authors suggested that glycogen mobilization may have contributed to the elevation in hemolymph trehalose observed in parasitized insects. Elevated hemolymph trehalose in *H. virescens* was later reported by Dahlman and Vinson

(1976) during parasitization by the ichneumonid parasitoid *Campoletis sonorensis* (Cameron).

Trichoplusia ni larvae parasitized late in the third stadium by *H. exiguae* displayed sharply elevated tissue carbohydrate levels during their development (Thompson, 1982a). Hemolymph trehalose concentration increased from 12 to 15 m*M* at the time of oviposition to a mean of approximately 35 m*M* during the fifth stadium, 2 to 4 days after parasitization. The trehalose level in control larvae remained almost constant. The glycogen level in the fat body of parasitized larvae also increased during the fourth stadium from a mean of approximately 50 μg/mg fresh weight to 140 μg/mg by 2 days postoviposition. Thereafter, glycogen declined. In contrast, the glycogen level in fat body of control larvae declined throughout the fourth stadium from 50 μg/mg to approximately 10 μg/mg, and increased during the fifth stadium to approximately 120 μg/mg.

Altered host tissue carbohydrate levels have also been reported during other insect parasite–host relationships. Führer (1972) reported an increased level of glycogen in fat body, midgut, and hypodermis of *Pieris brassicae* Linneaus parasitized by *Cotesia* (=*Apanteles*) *glomeratus* (Linneaus). On the basis of histological studies, Hawlitzky and Boulay (1986) suggested that tissues of *Anagasta kuehniella* Zeller parasitized by *Phanerotoma flavitestacea* Fisch. had higher levels of glycogen than were observed in control larvae. Thompson *et al.* (1990) observed elevated glycogen in histological sections of *M. sexta* fat body following parasitization by *C. congregata*. The hemolymph trehalose level of *M. sexta*, however, was reduced by parasitism, verifying earlier reports by Dahlman (1969, 1975) on this parasite–host relationship.

B. Proteins and Amino Acids

Junnikkala (1966) was among the first to report alterations in nitrogen metabolites following parasitization by an insect parasitoid. Tissue amino acids and proteins were examined in *P. brassicae* larvae parasitized when one day old by the gregarious parasitoid *C. glomerata*. Following parasitization, the total concentration of ninhydrin-positive compounds in the host's hemolymph fluctuated and was only occasionally higher than the levels in hemolymph of unparasitized controls. During the fifth stadium, however, the level of hemolymph protein was significantly lower in parasitized larvae. The predominant protein, designated "peptide 1" and probably a storage protein, in hemolymph of unparasitized last instar larvae was absent from hemolymph of parasitized insects.

The hemolymph amino acid compositions of parasitized and unparasitized *P. brassicae* were qualitatively and quantitatively similar until the end of the fifth stadium. At that time the levels of several amino acids, including proline, serine, glutamate/glutamine, and glycine, were significantly higher in hemolymph of parasitized larvae. In contrast, tyrosine was present in hemolymph from parasitized larvae at much lower levels than in controls. The changes in nitrogen metabolites appeared to reflect the altered developmental pattern of the parasitized larvae. The latter failed to metamorphose to the pupal stage. The author pointed out, however, that, during its second stadium, the parasite feeds on the host's fat body, which is almost completely absent by the time the parasite molts to the third instar and emerges from the host to pupate. Some of the changes in nitrogen metabolites may, therefore, have been due to tissue damage.

Several other studies have demonstrated qualitative and quantitative changes in hemolymph protein composition in parasitized insect hosts as largely evidenced by electrophoretic patterns (see Vinson and Iwantsch, 1980; Beckage *et al.*, 1987, 1989; Jones, 1989). Because the nature and functions of these proteins are largely unknown, and their potential roles in intermediary metabolism unclear, proteins will not be discussed further here.

Barras *et al.* (1969) examined hemolymph amino acid levels in the host *H. virescens* during parasitism by the solitary parasite *Cardiochiles nigriceps* Viereck. Although no qualitative differences were reported, the hemolymph amino acid concentration of fourth-instar host larvae parasitized in the second stadium was approximately half that of hemolymph from unparasitized controls. Parasite larvae were late first or early second instars at that stage of host development. Because these parasite stages develop within the host hemocoel and feed on hemolymph, the authors concluded that the decreased amino acids observed in the host were the result of parasite utilization. Thompson (1986a) also observed decreased hemolymph amino acid levels in the host *T. ni* parasitized by *H. exiguae*. In contrast to these results, Vinson and Iwantsch (cf. Vinson, 1990b) observed an increase in amino acid levels in the hemolymph of *H. virescens* parasitized by *M. croceipes* or *C. sonorensis*.

C. Lipids

Several studies suggested that qualitative and quantitative changes in lipid metabolites may accompany parasitism. Barras *et al.* (1970) reported that the relative levels of various lipid classes in the hemolymph and solid tissues of *H. virescens* were slightly altered following parasitization by *Cardiochiles nigriceps* Viereck, although triglyceride was the major lipid of tissues of both parasitized and unparasitized individuals. Differences in the relative levels of fatty acids in various lipid fractions were also evident between control and parasitized insects. The triglycerides of parasitized *H. virescens*, for example, contained significantly lower relative amounts of unsaturated fatty acids.

Thompson (1982b) reported that the concentration of total lipid in whole fifth-instar larvae of *T. ni* was near constant following parasitization by *H. exiguae,* whereas the concentration in controls increased throughout the fifth stadium. Moreover, the relative level of triglyceride was constant in parasitized larvae and the increased lipid in controls was largely due to triglyceride deposition.

Inhibition of lipophorin-mediated lipid transport in hemolymph of *Philosamia cynthia* (Drury) during parasitism by the dipterous parasite *Blepharipa sericariae* Cornalia [=*B. zebina* (Walker)] was demonstrated by Hayakawa (1986). A small peptide secreted by the parasite larva was demonstrated to be responsible. Later, Hayakawa (1987) reported that the factor inhibited lipophorin uptake of diacylglyceride from locust fat body. Recently, Horwood and Hales (1991) reported that the lipid content of the orthopteran host *Chortoicetes terminifera* (Walker) was significantly elevated during parasitism by the dipterous parasite *Trichopsidea oestracea* (Westwood) and suggested that this effect may have resulted from an inhibition of lipid transport.

IV. Metabolic Alteration in Insect Hosts

The foregoing studies demonstrating that host tissue metabolite levels are altered during parasitism strongly suggest that host metabolism is directly affected by parasitism. Very few studies, however, have been conducted to demonstrate the nature of those biochemical lesions. At the present time our knowledge is limited to a few studies of the intermediary metabolism of carbohydrates.

A. *Trichoplusia ni* Parasitized by *Hyposoter exiguae*

Following the demonstration by Thompson (1982a) that elevated hemolymph trehalose concentration and increased fat body glycogen deposition occurred in *T. ni* following parasitization by *H. exiguae,* Thompson and Cohen (1984) reported that excretion of uric acid by parasitized larvae was significantly greater than that of controls. The authors suggested that the uric acid production resulted from amino acid deamination and that utilization of glucogenic amino acids for *de novo* carbohydrate synthesis may have been responsible for the elevated carbohydrate reserves in parasitized hosts.

Thompson and Binder (1984) demonstrated elevated gluconeogenic enzyme activity in *T. ni* larvae 3 days postoviposition. The maximal enzyme velocity of both phosphofructokinase (PFKase-catalyzing glycolytic conversion of glucose to pyruvate) and fructose bisphosphatase (FBPase-catalyzing gluconeogenesis) was significantly higher in fat body isolated from parasitized larvae. The activity of FBPase, however, was elevated to a greater extent, resulting in a potential severalfold increase in gluconeogenic flux.

Subsequently, Thompson (1986a) reported that the total hemolymph glucogenic amino acid concentration in parasitized larvae was decreased approximately 35% three days after parasitization.

The regulation of FBPase in fat body of fourth-instar *T. ni* was examined by Thompson (1985). Adenosine monophosphate (AMP) and fructose-2,6-bisphosphate (F2,6BPate), major effectors of FBPase in liver (Pilkis *et al.*, 1981), were also modulators of this enzyme in *T. ni*. Their effects in *T. ni* were similar to those reported in rat liver (Van Schaftingen and Hers, 1981). Both effectors were inhibitory and the inhibition caused by each was enhanced in the presence of the other. Because previous results demonstrated that fat body AMP level was elevated following parasitization (Thompson and Yamada, 1984), these findings regarding AMP modulation initially appeared inconsistent with the conclusion that increased carbohydrate reserves in parasitized *T. ni* resulted from enhanced gluconeogenesis. In fat body preparations from parasitized larvae, however, both AMP and F2,6BPate were less inhibitory. The findings demonstrated that gluconeogenesis may increase despite minor increases in AMP level, although the basis for the decreased sensitivity of FBPase to AMP was unexplained.

Nutritional studies further indicated that the regulation of *de novo* carbohydrate synthesis in *T. ni* may be altered by parasitization (Thompson, 1986b). Hemolymph trehalose level of insects reared on artificial media with increased levels of sucrose was constant. In parasitized larvae, however, the hemolymph trehalose level was markedly elevated by increasing the dietary carbohydrate level. Thus, it was concluded that the regulation of carbohydrate biosynthesis in response to dietary sugar may be abnormally regulated in parasitized *T. ni*.

B. *Manduca sexta* Parasitized by *Cotesia congregata*

Several *in vivo* NMR studies have been conducted to discern the metabolic basis for elevated glycogen deposition in *M. sexta* parasitized by *C. congregata*. Metabolism of [^{13}C]2-pyruvate by fifth-instar larvae was examined by Thompson *et al.* (1988). The labeled substrate was metabolized much slower by parasitized larvae and the products into which the label was incorporated differed markedly between parasitized and unparasitized insects. In control larvae, pyruvate was incorporated primarily into fat following oxidation to acetyl-CoA. The difference in fat content between control and parasitized larvae was striking and evident from their respective natural abundance ^{13}C-NMR spectra. In parasitized larvae ^{13}C from pyruvate was incorporated into amino acids after decarboxylation to oxaloacetate and metabolism through the Krebs cycle. Significant gluconeogenesis from pyruvate, however, was not observed in either control or parasitized larvae.

Subsequent NMR studies reported by Thompson *et al.* (1990) demonstrated that incorporation of [^{13}C]1-glucose into storage metabolites differed between parasitized and unparasitized *M. sexta*. Control insects incorporated label from glucose into fat and [^{13}C]1-trehalose. [^{13}C]1-Trehalose and [^{13}C]1-glycogen were the major products of labeled glucose in parasitized larvae. In contrast to control larvae, parasitized insects did not incorporate label into fat. Differences in the metabolism of preformed (dietary) glucose, therefore, partially explained the differences in storage metabolite levels observed between parasitized and unparasitized *M. sexta*.

V. Basis for Metabolic Alterations: Hormones, Viruslike Particles, and Parasite Mediation

Redirection of host physiology is often initiated in response to parasitoid-derived factors injected into the host with the parasite egg during oviposition. Thus, effects of parasitism are often noted even before the parasite egg hatches. Several entities are responsible (Stoltz, 1986). Although these various factors may interact, often producing synergistic effects (see, e.g., Tanaka and Vinson, 1991), perhaps the most notable are the polydnaviruses (Polydnaviridae—Stoltz *et al.*, 1984; Whitfield, 1990), viruslike particles (see Federici, 1991) originating from the calyx region of the reproductive tract of most ichneumonid and braconid parasitoids (Stoltz *et al.*, 1981). Numerous studies have demonstrated that polydnaviruses play a critical role in suppression of the host's defense response (Schmidt and Schuchmann-Feddersen, 1989; Vinson, 1990a), as well as many other physiological effects.

Some effects of parasitism on host metabolite levels also appear to be initiated by polydnavirus and/or activity of other parasitoid-derived factors initially occurring independent of the parasite itself. Changes in carbohydrate reserves in *H. virescens* and *T. ni* in response to parasitism by *M. croceipes* (Dahlman and Vinson, 1975) and *H. exiguae* (Thompson, 1982a), respectively, were observed immediately following parasitization before parasite hatching and development. Dahlman and Vinson (1977) demonstrated that elevation of hemolymph trehalose in *H. virescens* could be induced by injection of calyx fluid of *M. croceipes*.

Many physiological effects of parasitism/polydnavirus activity are hormonally mediated (Beckage, 1985; Lawrence, 1986). For example, alterations of host growth and developmental patterns are closely related to titers of ecdysone and juvenile hormone (Dover *et al.*, 1988a), and previous investigation suggested that the prothoracic glands may be among the host tissues that polydnavirus invades (Dover *et al.*, 1988b). Little is known of the effects of developmental hormones on intermediary metabolism (see Keeley, 1978) and these hormones have not been directly implicated in the metabolic changes observed in parasitized insects. Insects, however, have additional hormones that are known to play a fundamental role in metabolic regulation.

Hormones with insulin- and glucagonlike activities have been isolated from tissues of numerous insects (Steele, 1983). In *M. sexta* larvae, for example, adipokinetic hormone originating from the corpora cardiaca stimulates glycogen phosphorylase activity (Ziegler, 1990) in a manner similar to that known for the vertebrate hormone glucagon. In this regard, studies by Thompson (1985) on gluconeogenic activity in *T. ni* are of interest because the decreased inhibitory response of fructose bisphosphatase to regulation by AMP and F2,6BPate in parasitized larvae is consistent with the effect of glucagon on this metabolic pathway in liver (Van Schaftingen *et al.*, 1980). Although previous investigations have failed to demonstrate any effect of factors derived from the corpora cardiaca on gluconeogenesis (J. E. Steele, personal communication), Steele *et al.* (1988) reported that glycogen phosphorylase stimulation alone is not sufficient to explain trehalose synthesis induced by corpus cardiacum hormones.

The results of nutritional investigations indicating that dietary regulation of gluconeogenesis was altered in parasitized *T. ni* (Thompson, 1986b; Section IV.A) may further suggest that a glucagonlike regulation of *de novo* carbohydrate synthesis is involved in elevating hemolymph trehalose level in parasitized larvae. Rocha *et al.* (1973) observed markedly elevated glucagon levels in septic human subjects. Subsequently, Long *et al.* (1976) reported that the rate of gluconeogenesis from amino acids was elevated during sepsis, and that the administration of sugar failed to mediate this response. The authors suggested that the results were likely caused by a hormonal imbalance. Although investigations with vertebrate animals cannot be directly compared to studies with insects, the potential role of hormones in inducing the metabolic changes observed in parasitized insects needs to be investigated.

Almost nothing is known concerning the role of the developing parasite larvae in mediating effects of parasitism on host physiology. Such effects might be expected as the parasite larvae mature and grow, and be particularly evident during host associations of gregarious species in which the biomass of parasites is considerable relative to that of the host. Indeed, Beckage and Riddiford (1983) demonstrated that heavily parasitized *M. sexta* have a larger host mass than lightly parasitized larvae, independent of the contribution by the parasites. The results suggest that the nutritional physiology of the host is carefully synchronized with parasite development and modified in relation to parasite burden to satisfy the nutritional requirements of the developing parasites.

Previous investigations suggest that parasite respiration may induce metabolic changes in the host. Fisher (1963) demonstrated that the O_2 level in hemolymph of *Ephestia kuhniella* Hübner Gravenhorst decreased dramatically during parasitism by *Venturia* (=*Nemeritis*) *canescens* Gravenhorst. Based on the rate of oxygen consumption of the developing parasite, it was concluded that the decreased oxygen level in the host's hemolymph was due to parasite respiration. The oxygen consumption of newly hatched parasite larvae was approximately equal to the total oxygen content of the host's hemolymph at any given time. Because the tracheal system provides oxygen directly to insect tissues, the potential effects of hemolymph oxygen depletion on host physiology and metabolism are difficult to evaluate. However, oxygen consumption by parasitized *E. kuhniella* larvae was decreased during parasitism. Dahlman and Herald (1971) demonstrated that the cyclic pattern of O_2 consumption during development of *M. sexta* was dampened in parasitized larvae. Moreover, parasitized larvae displayed significantly lower O_2 consumption than their unparasitized counterparts during the latter part of each stadium. Decreased hemolymph O_2 and respiration of *M. sexta* parasitized by *C. congregata* would be consistent with the decreased rate of glucose utilization in parasitized hosts described in the foregoing (Section IV).

VI. Nutritional Consequences of Metabolic Redirection and Physiological Alteration

Dahlman and Vinson (1975) suggested that alterations in the chemical composition of host tissues may play a nutritional role and influence the insect parasite–host relationship. These authors proposed that the elevated trehalose concentration in *H. virescens* induced following parasitization by *M. croceipes* provided nutrient directly to the developing parasite. This parasite species completes its entire larval development feeding on host hemolymph. The host's fat body and other tissues are not consumed. Results and conclusions of these experiments with *M. croceipes* were contrasted with those of studies involving another braconid, *C. nigriceps*. Hemolymph trehalose in *H. virescens* was not elevated after parasitization by this species. *Cardiochiles nigriceps*, however, feeds actively on host tissues throughout its development, and the authors suggested that the parasite larvae obtain sufficient carbohydrate as well as lipid nutrient by consumption of fat body.

The nature of the changes in storage carbohydrate levels in *T. ni* parasitized by *H. exiguae* (Section III) also suggests the potential for direct nutritional benefit. The parasitic larvae of this species feed on hemolymph during their early stadia and subsequently on fat body and other solid host tissues.

Thus, elevated hemolymph trehalose levels may provide an important source of nutrient for the early instars of the parasite, whereas high levels of glycogen stored in the fat body are consumed in the third stadium.

Nutritional studies employing *in vitro* culture of insect parasites on defined artificial media would provide critical answers regarding the potential nutritional effects of altered host tissue metabolite levels on parasite growth and development. Unfortunately, such techniques are currently unavailable for kionobiontic species. *In vitro* culture methods and artificial diets are, however, available for rearing a few idiobiontic parasites, and the results of nutritional investigations suggest that the increase in hemolymph carbohydrate titer observed in hosts parasitized by koinobiontic parasites may be sufficient to provide nutritional benefit to the developing larvae. Thompson (1983b) demonstrated that increasing the level of dietary glucose in an artificial medium from 0.5 to 1.5%, an increase approximately equal to that observed in both *T. ni* following parasitization by *H. exiguae* (Thompson, 1982a; Section III) and *H. virescens* during parasitism by *M. croceipes* (Dahlman and Vinson, 1975; Section III), resulted in an increase in mean relative growth rate of larvae of the solitary chalcid parasite *Brachymeria lasus* (Walker) from 200 to 250 mg/g/day. A similar increase in dietary glucose in a defined artificial diet for the ichneumonid parasite *Exeristes roborator* (Fabricius) increased the mean growth rate from approximately 5 to 12 mg/g/day (Thompson, 1982c).

Despite evidence suggesting nutritional benefit for the altered host tissue carbohydrate levels observed in parasitized hosts, the conclusion appears inconsistent with the results of studies on some insect parasite–host relationships. The solitary parasite *C. sonorensis,* like *H. exiguae,* feeds on the host's hemolymph during its early stadia and directly on solid host tissues during the last two stadia. In this case, hemolymph trehalose in the host *H. virescens* was elevated only slightly during the early stages of parasite development, and more dramatically later when it was unlikely that the hemolymph would supply significant nutrient over that obtained by consumption of the other host tissue (Dahlman and Vinson, 1976; Section III). Conclusions regarding the significance of glycogen stores during some parasite–host relationships may be similarly criticized. In *M. sexta* parasitized by *C. congregata,* for example, the fat body glycogen level was increased but the hemolymph trehalose level declined. *Cotesia congregata,* however, feeds only on the host hemolymph and therefore the glycogen stores are unlikely to be of direct nutritional benefit for this species.

When evaluating the potential nutritional significance of host metabolite levels for developing parasites, care must be taken to consider the overall effects of parasitism on total nutrient availability. For example, in many cases, including *H. virescens* and *T. ni* parasitized by *M. croceipes* and *H. exiguae,* respectively, host growth is severely retarded following parasitization. The total nutrient available for the developing parasite(s) may, therefore, be less, despite the finding that the concentration of individual metabolites may have increased. During parasite–host relationships in which the parasite completes its development on host hemolymph alone, for example, *M. croceipes,* elevated levels of hemolymph nutrients may, nevertheless, be important for optimal parasite growth and development. However, in cases where the entire host contents are consumed at the completion of parasite development, as is the case with *H. exiguae,* elevated host hemolymph metabolite levels likely provide nutritional benefit only for the early parasitic stages.

In addition to alterations in the chemical composition of host tissues brought about by changes in intermediary metabolism, other parasitoid-induced changes in host behavior and physiology may play an interactive role in influencing parasite nutrition. Slow growth, delayed or precocious development, and/or interrupted host metamorphosis may act to ensure an adequate time period for parasite nourishment and development, as well as maximal exploitation of an often limited food supply. Decreased food consumption, a common behavior during parasitism, may play a role because "nutrient-deprived" insects develop slower (Thompson, 1982a). Moreover, previous studies indicate that modified nutritional physiology resulting from decreased food consumption may also have significant consequences. Thompson (1983c) demonstrated that decreased food consumption displayed by *T. ni* parasitized by *H. exiguae* was accompanied by a net increase in assimilation, greater than that observed in unparasitized pair-fed larvae. This may be important for providing adequate nutrient to the host and thus for maintaining hemolymph nutrient levels for developing parasites. Subsequent investigations demonstrated that dietary carbohydrate level had little effect on the hemolymph trehalose level in unparasitized *T. ni,* but markedly increased the concentration in parasitized larvae (Thompson, 1986b; Section IV.A).

VII. "Host Regulation" and Its Role in Redirecting Host Metabolism and Physiology

Vinson and Iwantsch (1980) described the overall effects of parasitism on the insect host as reflecting parasitoid regulation of host physiology. "Host regulation" is responsible for modifying the host's physiology to ensure parasite survival (Vinson, 1984). Jones (1989) subsequently used the term "redirection." Recently, interest in host regulation has focused on inhibition of the host's defense following parasitization. Clearly, "immunosuppression" is critical for successful parasitism and parasitoids have evolved a variety of means by which this is accomplished (Vinson, 1990a), in many cases being initiated by polydnavirus and/or other parasitoid-derived factors (Stoltz, 1986). Thus, viewing the insect parasitoid–host relationship as one in which the parasitoid is "in control" (Jones, 1985a,b) as a result of its ability to regulate the physiology of the host is appropriate. Indeed, this paradigm formed the basis of Dawkins' (1982) definition of a parasitized host as an "extended phenotype" of the parasite.

In contrast to the inhibition of host defense, there is a paucity of experimental evidence demonstrating any benefit to most other physiological and metabolic alterations observed in parasitized insects. Although it has been demonstrated that many alterations are brought about by parasitoid-derived factors, inferring parasitoid regulation of host biology, this by itself does not demonstrate any adaptive significance of the effects observed. In most cases the "host regulation" paradigm necessitates a subjective assessment of physiological alteration in relation to parasite survival. To understand physiological alteration within the overall context of "host regulation," some researchers have emphasized the necessity to explain the basis of alterations that fail to meet a demonstrated or hypothesized criteria of benefit. Thus, Jones *et al.* (1986) distinguished beneficial effects from "indirect stress effects"; the latter being "unintentional" (Jones *et al.*, 1981). A rational extension was the subsequent characterization of parasites as "host regulators" or "host conformers" (Lawrence, 1986). In the latter case, parasitism fails to result in observable developmental changes.

Despite limited experimental evidence, it is reasonable to assume that many alterations in host physiology are of significance for successful parasitism—for example, to ensure a suitable environment and nutritional milieu for the developing parasite (Section VI). Vinson (1984) stated: "The problem of nutritional suitability may not be lack of specific nutrients or accessory growth factors in the host, but the quality of and the ability to obtain certain nutrients present in host tissues at the proper time and to compete with host tissues for available nutrients . . . the parasitoid may actively affect these last two conditions." Viewing the insect parasite–host interaction as a product of "host regulation," however, considers altered host physiology in terms of specific parasitoid survival strategies and fails to explain how such strategies have evolved. Moreover, it provides little insight into how the relationship continues to evolve. Considering the relationship within the universal paradigm of adaptation provides another perspective and new insight into the nature of these parasitic associations.

The benefit of physiological alterations to parasite survival arises by natural selection through evolutionary processes. This evolution was described by Barnard (1990) as an "arms race" during which there occurs constant action and reaction by both symbionts in response to the activities of their partners.

Sprent (1969) discussed immunological interaction within this framework. His "adaptation tolerance" hypothesis provides a basis for understanding the dynamic nature of parasite–host coevolution. Dawkins and Krebs (1979) outlined the limitations and costs of such adaptive interaction. Little is understood concerning such interactions during the parasitic phase of the insect parasitoid–host relationship because the parasite does not reproduce in the host. Moreover, the host dies following parasite development. Thus, Vinson and Iwantsch (1980) stated that the host plays no role in the evolution of the association. Although physiological interaction may not represent a strong coevolutionary force, it is a critical component of parasitoid evolution. If, for example, the host's metabolic response during parasitism has significant nutritional consequences, then the degree to which these responses are expressed directly impacts parasite survival and, in turn, fitness and fecundity of the adult parasitoid. A less suitable host produces a less fit parasitoid, and conversely a more suitable host produces a more fit parasitoid.

During the establishment of the parasite–host relationship, alterations in host physiology reflect responses to various physiological and metabolic stresses induced by parasitism. Those responses provide a basis upon which natural selection acts, thereby enabling the parasite to exploit the modified physiological state of the host. Characteristics of parasites or factors injected into the host by the parasitoid that induce changes having beneficial effects for parasite survival are selected for. When the physiological and metabolic alterations that occur during parasitism are examined from an evolutionary perspective, and the symbionts as interacting partners, the division of pathological versus beneficial changes takes on a new and more profound significance. Effects that might otherwise be considered "dull pathological byproducts of infection" (Dawkins, 1982) are no longer ignored, but are recognized as factors playing a potentially significant role in the evolution of the parasite–host relationship. All effects of parasitism have the potential to influence parasite fitness by contributing to the variability upon which selection can act.

The evolution is evident upon considering the interactive nature of parasitic effects in influencing parasite nutrition described earlier (Section VI). Decreased food consumption, a common behavior during parasitism, might be considered an indirect stress effect (see Lawrence, 1988). Although perhaps not regulated by the parasite, decreased food consumption is nevertheless induced as a result of parasitism, and may have significant consequences for the developing parasite by delaying development. Every effect of parasitism that is deemed beneficial for the parasite need not be regulated, and every effect considered regulated does not necessarily provide benefit to the parasite. Because of the interactive nature of parasitic effects in influencing the outcome of the parasite–host relationship, the utility of "host regulation" is best appreciated by considering altered host physiology as being of overall benefit to the developing parasite, as implied by Vinson (1975, 1990b).

VIII. Enantiostasis and Metabolic Regulation

The difficulty in understanding and evaluating the significance of the physiological and metabolic alterations occurring in parasitized insects is confounded by the enantiostatic character of many processes in insects. Maintenance of physiological function in the face of changing environmental conditions has long been recognized as fundamental for survival of any organism. Homeostasis, the maintenance of a near-constant internal state, is characteristic of many animals. Less specialized animals, however, particularly invertebrates, typically function over a wide range of internal conditions and are not homeostatic. The hemolymph sugar level of insects, for example, is maintained only within a very broad concentration range. In *M. sexta* the hemolymph sugar level fluctuates greatly, and although hormones are involved in regulating hemolymph sugar levels in insects (Section V), hemolymph sugar may decrease to near undetectable levels when larvae are starved (Dahlman, 1969, 1973; Siegert, 1987).

Prosser (1955; see also Prosser, 1986) distinguished physiological adjustment or conformity from physiological regulation. Magnum and Towle (1977) referred to the condition of conserved physiological function in nonhomeostatic systems as enantiostasis. Enantiostasis is often characterized by balancing adjustments in one physiological process with counterchanges in others. Thus, caution must be exercised when evaluating changes in the internal milieu of parasitized insects because many alterations may simply reflect opposing adjustments in seemingly unrelated processes. The elevated trehalose level in *T. ni* parasitized by *H. exiguae* (Section III.A), for example, may represent a compensatory adaptation to the simultaneous decrease in the level of amino acids. In this manner the system is maintained in a functional steady state. Osmotic pressure may reflect the specific function preserved.

Although researchers have speculated on the potential of elevated hemolymph trehalose to provide nutrient for developing parasites, amino acids may, in fact, be more important. A decreased amino acid level observed in some hosts may result from rapid absorption by the parasite(s). Amino acids often occur in insect blood at levels exceeding that of trehalose. In unparasitized fifth-instar *T. ni*, the level of glutamine alone is approximately threefold that of trehalose. Moreover, *in vitro* nutritional studies on the idiobiontic parasites described in Section VII indicate that amino acids may totally spare carbohydrate as an energy source for larval development (Thompson, 1976).

IX. Conclusion

Redirection of host physiology characterizes many insect parasite–host relationships. Although our understanding of the metabolic basis for these changes is still rudimentary, redirection often includes alterations in tissue levels of lipids, carbohydrates, and nitrogen metabolites. Metabolic studies suggest that some changes reflect expected adjustments brought about by the host's normal enantiostatic regulatory mechanisms in response to the stress of parasitism. In other cases, however, the normal regulatory mechanisms appear to be altered or inoperative. In those instances it might be concluded that regulation of host metabolism by the parasite has superseded the host's normal regulatory functions. In either case the changes observed may have nutritional consequences for the developing parasite larvae.

The various manners in which we view insect parasitism reflect more than "semantic" differences (Fisher, 1986). They influence in a fundamental way our understanding of insect parasitism (see Goff, 1982)—our interpretation of experimental data and our approach to further research. The concept of "host regulation" has and continues to provide a valuable framework within which the nature of the insect parasite–host relationship can be understood. It is important to recognize the dynamic nature of the insect parasite–host interaction and the evolutionary basis for the development of this complex symbiosis.

References

Askew, R. R. (1971). "Parasitic Insects." American Elsevier, New York.

Askew, R. R., and Shaw, M. R. (1986). Parasitoid communities: Their size, structure and development. *In* "Insect Parasitoids" (J. Waage and D. Greathead, eds.), pp. 225–264. Academic Press, London.

Atkinson, D. E. (1968). The energy charge of the adenylate pool as a regulatory parameter. Interaction with feedback modifiers. *Biochemistry* **7:**4030–4034.

Barnard C. J. (1990). Parasitic relationships. *In* "Parasitism and Host Behaviour" (C. J. Barnard and J. M. Behnke, eds.), pp. 1–33. Taylor & Francis, New York.

Barras, D. J., Wiygul, G., and Vinson, S. B. (1969). Amino acids in the haemolymph of the tobacco budworm, *Heliothis virescens* Fab. as affected by its habitual parasite *Cardiochiles nigriceps* Viereck. *Comp. Biochem. Physiol.* **31:**707–714.

Barras, D. J., Joiner, R. L., and Vinson, S. B. (1970). Neutral lipid composition of the tobacco budworm, *Heliothis virescens* (Fab.) as affected by its habitual parasite, *Cardiochiles nigriceps* Viereck. *Comp. Biochem. Physiol.* **36:**775–783.

Beckage, N. E. (1985). Endocrine interactions between endoparasitic insects and their hosts. *Annu. Rev. Entomol.* **30:**271–413.

Beckage, N. E., and Riddiford, L. M. (1983). Growth and development of the endoparasitic wasp *Apanteles congregatus:* Dependence on host nutritional status and parasite load. *Physiol. Entomol.* **8:**231–241.

Beckage, N. E., Templeton, T. J., Nielsen, B. D., Cook, D. I., and Stoltz, D. B. (1987). Parasitism-induced hemolymph polypeptides in *Manduca sexta* (L.) larvae parasitized by the braconid wasp *Cotesia congregata* (Say). *Insect Biochem.* **17:**439–455.

Beckage, N. E., Nesbit, D. J., Nielsen, B. D., Spence, K. D., and Barman, M.A.E. (1989). Alteration of hemolymph polypeptides in *Manduca sexta* larvae parasitized by *Cotesia congregata:* A two-dimensional electrophoretic analysis and comparison with major bacteria-induced proteins. *Arch. Insect Biochem. Physiol.* **10:**29–45.

Chapman, A. G., Fall, L., and Atkinson, D. E. (1971). Adenylate energy change in *Escherichia coli* during growth and starvation. *J. Bacteriol.* **108:**1072–1086.

Dahlman, D. L. (1969). Trehalose levels in parasitized and nonparasitized tobacco hornworm, *Manduca sexta,* larvae. *Ann. Entomol. Soc. Am.* **63:**615–617.

Dahlman, D. L. (1973). Starvation of the tobacco hornworm, *Manduca sexta.* 1. Changes in hemolymph characteristics of 5th-stage larvae. *Ann. Entomol. Soc. Am.* **66:**1023–1029.

Dahlman, D. L. (1975). Trehalose and glucose levels in hemolymph of diet-reared, tobacco leaf-reared and parasitized tobacco hornworm larvae. *Comp. Biochem. Physiol. A* **50A:**165–167.

Dahlman, D. L., and Herald, F. (1971). Effects of the parasite, *Apanteles congregatus,* on respiration of tobacco hornworm, *Manduca sexta,* larvae. *Comp. Biochem. Physiol. A* **40A:**871–880.

Dahlman, D. L., and Vinson, S. B. (1975). Trehalose and glucose levels in the hemolymph of *Heliothis virescens* parasitized by *Microplitis croceipes* or *Cardiochiles nigriceps. Comp. Biochem. Physiol. B* **52B:**465–468.

Dahlman, D. L., and Vinson, S. B. (1976). Trehalose level in the hemolymph of *Heliothis virescens* parasitized by *Campoletis sonorensis. Ann. Entomol. Soc. Am.* **69:**523–524.

Dahlman, D. L., and Vinson, S. B. (1977). Effect of calyx fluid from an insect parasitoid on host hemolymph dry weight and trehalose content. *J. Invertebr. Pathol.* **29:**227–229.

Dahlman, D. L., and Vinson, S. B. (1980). Glycogen content in *Heliothis virescens* parasitized by *Microplitis croceipes. Comp. Biochem. Physiol. A* **66A:**625–630.

Dawkins, R. (1982). Host phenotypes of parasite genes. *In* "The Extended Phenotype: The Gene as the Unit of Selection," pp. 209–227. Freeman, Oxford.

Dawkins, R., and Krebs, J. R. (1979). Arms races between and within species. *Proc. R. Soc. London, Ser. B* **205:**489–511.

Dover, B. A., Davies, D. H., and Vinson, S. B. (1988a). Dose-dependent influence of *Campoletis sonorensis* polydnavirus on the development and ecdysteroid titers of last-instar *Heliothis virescens* larvae. *Arch. Insect Biochem. Physiol.* **8:**113–126.

Dover, B. A., Davies, D. H., and Vinson S. B. (1988b). Degeneration of last-instar *Heliothis virescens* prothoracic glands by *Campoletis sonorensis* polydnavirus. *J. Invertebr. Pathol.* **51:**80–91.

Federici, B. F. (1991). Viewing polydnaviruses as genetic secretions of endoparasitic Hymenoptera. *Redia* **74:**463–466.

Fisher, R. C. (1963). Oxygen requirements and the physiological suppression of supernumerary insect parasitoids. *J. Exp. Biol.* **40:**531–540.

Fisher, R. C. (1986). George Salt and the development of experimental insect parasitology. *J. Insect Physiol.* **32:**249–253.

Führer, E. (1972). Abnorme glycogenspeicherung in larven von *Pieris brassicae* L. als folge des parasitismus von *Apanteles glomeratus* L. *Z. Angew. Entomol.* **70:**370–374.

Goff, L. J. (1982). Symbiosis and parasitism: Another viewpoint. *BioScience* **34:**255–256.

Hawlitzky, N., and Boulay, C. (1986). Effects of the egg-larval parasite, *Phanerotoma flavitestacea*

Fisch. (Hymenoptera, Braconidae) on the dry weight and chemical composition of its host *Anagasta kuehniella* Zell. (Lepidoptera, Pyralidae). *J. Insect Physiol.* **32:**269–274.
Hayakawa, Y. (1986). Inhibition of lipid transport in insects by a factor secreted by the parasite, *Blepharipa sericariae. FEBS Lett.* **195:**122–124.
Hayakawa, Y. (1987). Inhibition of lipid transport in insects by a parasitic factor. *Comp. Biochem. Physiol. B* **87B:**279–283.
Horwood, M. A., and Hales, D. F. (1991). Fat body changes in a locust, *Chortoicetes terminifera* (Walker) (Orthoptera: Acrididae), parasitized by a nemestrinid fly. *Arch. Insect Biochem. Physiol.* **17:**53–63.
Jones, D. (1985a). Endocrine interaction between host (Lepidoptera) and parasite (Cheloninae: Hymenoptera): Is the host or the parasite in control? *Ann. Entomol. Soc. Am.* **7:**141–148.
Jones, D. (1985b). Parasite regulation of host insect metamorphosis: A new form of regulation in pseudoparasitized larvae of *Trichoplusia ni. J. Comp. Physiol. B* **155B:**583–590.
Jones, D. (1989). Protein expression during parasite redirection of host (*Trichoplusia ni*) biochemistry. *Insect Biochem.* **19:**445–455.
Jones, D., Jones, G., and Hammock, B. D. (1981). Developmental and behavioral responses of larval *Trichoplusia ni* to parasitization by an imported braconid parasite *Chelonus* sp. *Physiol. Entomol.* **6:**387–394.
Jones, D., Jones, G., Rudnicka, M., Click, A., Reck-Malleczewen, V., and Iwaya, M. (1986). Pseudoparasitism of host *Trichoplusia ni* by *Chelonus* spp. as a new model system for parasite regulation of host physiology. *J. Insect Physiol.* **32:**315–328.
Junnikkala, E. (1966). Effect of braconid parasitization on the nitrogen metabolism of *Pieris brassicae* L.: A study of the nitrogenous compounds in the hemolymph of larvae at different ages reared in controlled laboratory conditions. *Ann. Acad. Sci. Fenn., Ser. A4* **100:**1–83.
Keeley, L. L. (1978). Endocrine regulation of fat body development and function. *Annu. Rev. Entomol.* **23:**329–352.
Knutson, L. V., and Berg, C. O. (1966). Parasitoid development in snail-killing sciomyzid flies. *Trans. Am. Microsc. Soc.* **85:**164–165.
Lawrence, P. O. (1986). Host–parasite hormonal interactions: An overview. *J. Insect Physiol.* **32:**295–298.
Lawrence, P. O. (1988). Hormonal interactions between parasitoids and hosts: Adaptations to stress. *In* "Endocrinological Frontiers in Physiological Insect Ecology" (F. Sehnal, D. Denlinger, and A. Zabza, eds.), pp. 423–435. Wroclaw Tech. Univ. Press, Wroclaw, Poland.
Long, C. L., Kinney, J. M., and Geiger, J. W. (1976). Nonsuppressibility of gluconeogenesis by glucose in septic patients. *Metab., Clin. Exp.* **25:**193–201.
Magnum, C. P., and Towle, D. W. (1977). Physiological adaptation to unstable environments. *Am. Sci.* **65:**67–75.
Newsholme, E. A., and Start, C. (1973). "Regulation in Metabolism." Wiley, New York.
Pilkis, S. J., El-Maghrabi, M. R., Pilkis, J., and Claus, T. (1981). Inhibition of fructose-1,6-bisphosphatase by fructose-2,6-bisphosphate. *J. Biol. Chem.* **256:**3619–3622.
Prosser, C. L. (1955). Physiological adaptation in animals. *Biol. Rev. Cambridge Philos. Soc.* **30:**229–262.
Prosser, C. L. (1986). "Adaptational Biology: Molecules to Organisms." Wiley, New York.
Rocha, D. M., Santeanio, F., Faloona, G. R., and Unger, R. H. (1973). Abnormal pancreatic α-cell function in bacterial infections. *N. Engl. J. Med.* **288:**700.
Schmidt, O., and Schuchmann-Feddersen, I. (1989). Role of virus-like particles in parasitoid–host interaction of insects. *Subcell. Biochem.* **15:**91–119.
Siegert, K. J. (1987). Carbohydrate metabolism in starved fifth instar larvae of *Manduca sexta. Arch. Insect Biochem. Physiol.* **4:**151–160.
Smyth, J. D. (1976). "Introduction to Animal Parasitology." Hodder & Stoughton, London.

Sprent, J.F.A. (1969). Evolutionary aspects of immunity in zooparasitic infections. *In* "Immunity to Parasitic Animals" (G. J. Jackson, R. Herman, and I. Singer, eds.), pp. 3–62. North-Holland Publ., Amsterdam.

Steele, J. E. (1983). Control of carbohydrate metabolism in insects. *In* "Endocrinology of Insects" (R.G.H. Downer, and H. Laufer, eds.), pp. 429–439. Liss, New York.

Steele, J. E., McDougall, G. E., and Shadwick, R. (1988). Trehalose efflux from cockroach fat body *in vitro:* Paradoxical effects of corpus cardiacum and methylxanthines. *Insect Biochem.* **18:**585–590.

Stoltz, D. B. (1986). Interactions between parasitoid-derived products and host insects: An overview. *J. Insect Physiol.* **32:**347–350.

Stoltz, D. B., Krell, P. J., and Vinson, S. B. (1981). Polydisperse viral DNA's in ichneumonid ovaries: A survey. *Can. J. Microbiol.* **27:**123–130.

Stoltz, D. B., Krell, P., Summers, M. D., and Vinson, S. B. (1984). Polydnaviridae—a proposed family of insect viruses with segmented, double-stranded, circular DNA genomes. *Intervirology* **21:**1–4.

Tanaka, T., and Vinson, S. B. (1991). Interaction of venoms with the calyx fluids of three parasitoids, *Cardiochiles nigriceps, Microplitis croceipes* (Hymenoptera: Braconidae), and *Campoletis sonorensis* (Hymenoptera: Ichneumonidae) in effecting a delay in the pupation of *Heliothis virescens* (Lepidoptera: Noctuidae). *Ann. Entomol. Soc. Am.* **84:**87–92.

Thompson, S. N. (1976). Effects of dietary amino acid level and nutritional balance on larval survival and development of the hymenopterous parasitoid *Exeristes roborator. Ann. Entomol. Soc. Am.* **69:**835–838.

Thompson, S. N. (1982a). Effects of parasitization by the insect parasite *Hyposoter exiguae* on the growth, development and physiology of its host *Trichoplusia ni. Parasitology* **84:**491–510.

Thompson, S. N. (1982b). Effects of the insect parasite, *Hyposoter exiguae,* on the total body glycogen and lipid levels of its host, *Trichoplusia ni. Comp. Biochem. Physiol. B* **72B:**233–237.

Thompson, S. N. (1982c). *Exeristes roborator:* Quantitative determination of *in vitro* larval growth rates in synthetic media with different glucose concentrations. *Exp. Parasitol.* **54:**229–234.

Thompson, S. N. (1983a). Biochemical and physiological effects of metazoan endoparasites on their host species. *Comp. Biochem. Physiol. B* **74B:**183–211.

Thompson, S. N. (1983b). *Brachymeria lasus:* Effects of nutrient level on *in vitro* larval growth of a chalcid insect parasite. *Exp. Parasitol.* **55:**312–319.

Thompson, S. N. (1983c). The nutritional physiology of *Trichoplusia ni* parasitized by the insect parasite. *Hyposoter exiguae,* and the effects of parallel-feeding. *Parasitology* **87:**15–28.

Thompson, S. N. (1985). Effects of AMP and fructose 2,6 bisphosphate on fructose 1,6 bisphosphatase activity in *Trichoplusia ni* parasitized by the insect parasite, *Hyposoter exiguae. J. Parasitol.* **71:**117–119.

Thompson, S. N. (1986a). Effect of the insect parasite *Hyposoter exiguae* (Viereck) on the carbohydrate metabolism of its host *Trichoplusia ni* (Hubner). *J. Insect Physiol.* **32:**287–293.

Thompson, S. N. (1986b). Effects of dietary carbohydrate on the nutritional physiology and blood sugar level of *Trichoplusia ni* parasitized by the insect parasite, *Hyposoter exiguae. Parasitology* **92:**25–30.

Thompson, S. N. (1990). Physiological alterations during parasitism and their effects on host behavior. *In* "Parasitism and Host Behavior" (C. J. Barnard and J. M. Behnke, eds.), pp. 64–94. Taylor & Francis, New York.

Thompson, S. N., and Binder, B. F. (1984). Altered carbohydrate levels and gluconeogenic enzyme activity in *Trichoplusia ni* parasitized by the insect parasite, *Hyposoter exiguae. J. Parasitol.* **70:**644–651.

Thompson, S. N., and Cohen, A. C. (1984). Nitrogen elimination in *Trichoplusia ni* parasitized by the insect parasite, *Hyposoter exiguae. Comp. Biochem. Physiol. A* **78A:**785–786.

Thompson, S. N., and Yamada, K. A. (1984). Energy status in the fat body of *Trichoplusia ni* parasitized by the insect parasite *Hyposoter exiguae*. *J. Invertebr. Pathol.* **44:**46–51.

Thompson, S. N., Lee, R.W.-K., and Beckage, N. E. (1988). Preliminary studies in the metabolism of [^{13}C] labelled metabolites in *Manduca sexta* parasitized by the insect parasite, *Cotesia congregata*. *Colloq. IRNA* **48:**73–79.

Thompson, S. N., Lee, R.W.-K, and Beckage, N. E. (1990). Metabolism of parasitized *Manduca sexta* examined by nuclear magnetic resonance. *Arch. Insect Biochem. Physiol.* **13:**127–143.

Van Schaftingen, E., and Hers, H.-G. (1981). Inhibition of fructose-1,6-bisphosphatase by fructose-2,6-bisphosphate. *Proc. Natl. Acad. Sci. U.S.A.* **78:**2861–2863.

Van Schaftingen, E., Hue, L., and Hers, H.-G. (1980). Control of fructose-6-phosphate/fructose-1,6-bisphosphate cycle in isolated hepatocytes by glucose and glucagon. *Biochem. J.* **192:**887–901.

Vinson, S. B. (1975). Biochemical coevolution between parasitoids and their hosts. *In* "Evolutionary Strategies of Parasitic Insects and Mites" (P. W. Price, ed.), pp. 14–48. Plenum, New York.

Vinson, S. B. (1984). Parasitoid–host relationship. *In* "Chemical Ecology of Insects" (W. J. Bell and R. T. Carde, eds.), pp. 205–233. Chapman & Hall, London.

Vinson, S. B. (1990a). How parasites deal with the immune system of their host: An overview. *Arch. Insect Biochem. Physiol.* **13:**3–27.

Vinson S. B. (1990b). Physiological interactions between the host genus *Heliothis* and its guild of parasitoids. *Arch. Insect Biochem. Physiol.* **13:**63–81.

Vinson, S. B., and Iwantsch, G. F. (1980). Host regulation by insect parasitoids. *Q. Rev. Biol.* **55:**143–165.

Whitfield, J. B. (1990). Parasitoids, polydnaviruses and endosymbiosis. *Parasitol. Today* **6:**381–384.

Ziegler, R. (1990). Biological effects of synthetic AKH in *Manduca sexta* and estimates of the amount of AKH in corpora cardiaca. *Arch. Insect Biochem. Physiol.* **15:**111–116.

Chapter 7

Teratocytes: Developmental and Biochemical Characteristics

Douglas L. Dahlman
Department of Entomology
University of Kentucky
Lexington, Kentucky

S. Bradleigh Vinson
Department of Entomology
Texas A&M University
College Station, Texas

I. Introduction

Scientists have reported unusually large cells dispersed throughout the hemolymph of certain parasitized insects. These cells were assigned various names that reflected a proposed function and/or embryological source, but they remained a curiosity until the 1960s, when increased knowledge about insect immunology, physiology, biochemistry, virology, and molecular biology, as well as advances in instrumentation techniques, made it possible to look beyond descriptive biology. The function of these cells, most commonly called teratocytes, was first reviewed by Salt (1968). Two important reviews on their embryological origin (the serosal membrane of the parasitoid embryo) subsequently appeared (Ivanova-Kasas, 1972; Tremblay and Caltagirone, 1973). Hawlitzky and Lauge (1979) discussed the hypothetical roles of the serosal membrane in a single egg-larval parasitoid and Vinson and Iwantsch's 1980 review on host regulation by insect parasitoids included a major section on teratocytes. Interest in the role of teratocytes in the regulation of host physiology, biochemistry, endocrinology, immunology, etc., has accelerated during the last 10 years. Dahlman (1990, 1991) provided overviews of teratocyte functions and reviews by Beckage (1990) and Coudron (1991) included sections on teratocytes.

Prior to 1980, teratocytes were generally thought to rapidly absorb nutrients that were consumed later by the developing parasitoid. Clearly the view today is that teratocytes do more than simply supply an additional source of nutrients to the developing parasitoid larva. Important interactions between teratocytes and other components of the parasitoid system (venom, accessory gland products, and, in some cases, polydnavirus) are being investigated.

II. Origins

Developing embryos of certain parasitic Hymenoptera are enclosed within an external membrane that, for different species, may have different origins and fates. The polar bodies are considered to be the most likely ultimate source of these embryonic membranes. According to Tremblay and Caltagirone (1973) the polar bodies are the daughter cells of the primary and secondary oocytes formed during egg maturation. The specific terms assigned to the membranes that develop from these cells vary with the insect taxon. In the Braconidae, which contain the majority of reported instances of teratocytes, the polar bodies give rise to a serosal membrane that in turn dissociates to yield individual cells after the egg hatches (Fig. 1).

Tremblay and Caltagirone (1973) stated that the serosal membrane enveloping the embryo falls into one of three types: (1) a true serosa, (2) a deutoserosa (pseudoserosa), or (3) a trophamnion. The serosa is derived from the first cleavage nucleus. Sources of the true serosa in braconids may be from either serosal delamination or proliferation of polar bodies (Tremblay and Calvert, 1971; Tremblay and Caltagirone, 1973). A serosal membrane originating by delamination also has been described for an ichneumonid, although no evidence of teratocytes has been reported for this group. The deutoserosal membrane, in contrast to a true serosa, is extensively derived from the embryo proper but also contains cells from poorly defined extraembryonic tissues. The trophamnionic membrane has no corresponding structure in the embryonic development of insects in general and, in the opinion of Tremblay and Caltagirone (1973), the term is not embryologically appropriate but it has been used so frequently "that it seems advisable to preserve it."

Undoubtedly various combinations of these membranes will be found upon closer examination of the embryological development of various species of

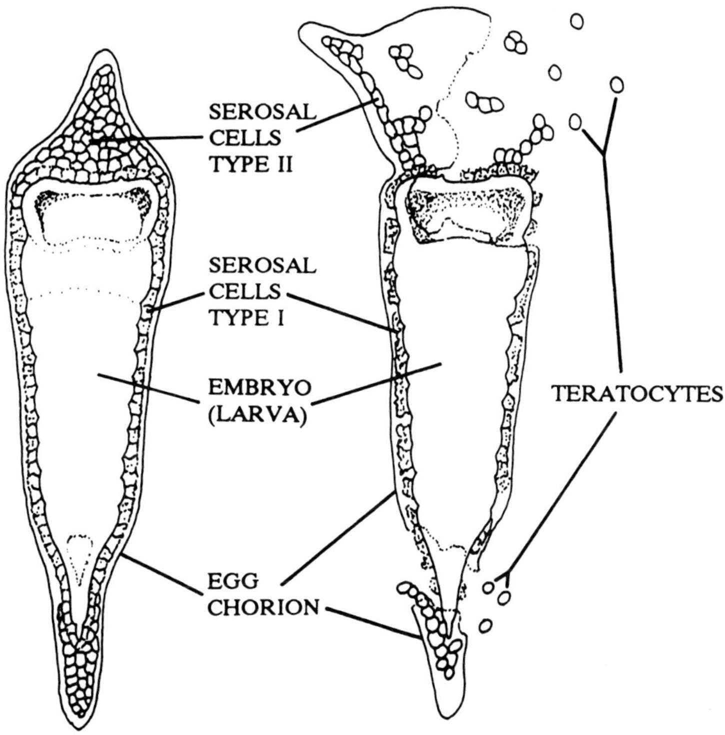

Figure 1 Well-developed (left) and hatching larvae (right) of *Cardiochiles nigriceps* showing the location of the two types of serosal cells that enclose the embryo. Only Type II cells form teratocytes when liberated from the embryo. The Type I serosal cells degrade (F. Pennacchio, S. B. Vinson, and E. Tremblay, Texas A&M University, unpublished).

Hymenoptera. In fact, at least one intermediate example is known where the trophamniotic membrane is derived from both polar nuclei and cleavage nuclei (Tremblay and Calvert, 1971). Membranes of the trophamniotic type also are found in endoparasitic Hymenoptera that develop polyembryonically (Ivanova-Kasas, 1972).

III. Distribution within Taxonomic Groups

If the strict definition of teratocytes is followed (i.e., individual cells released into the host hemocoel as the result of the dissociation of the serosal membrane of the embryo), these cells are restricted to a few families of the Hymenoptera (Braconidae, Platygasteridae, and Scelionidae). A single teratocyte per parasitoid was reported for one trichogrammatid species (Voegele *et al.*, 1974), but Strand (1986) convincingly argued that this single cell is a polar body and should not be considered a teratocyte. However, in a broader sense, nondissociated tissues of the same origin (i.e., polar bodies) have been described from several species of the superfamily Chalcidoidea. Speculation on prerequisites for continued activity of postemergence extraembryonic tissue would be that the parasitoid egg must be deposited within the host so that the tissues have access to host body fluids for sustenance. Thus, for example, examination of homopteran hosts parasitized by Pipunculidae (Diptera) might provide additional insight into the importance of these extraembryonic tissues. Nevertheless, other factors must be necessary for success because teratocytes have not been reported from many other families of parasitic Hymenoptera.

IV. Developmental Characteristics

A. Morphology

1. Abundance, Persistence, and Growth

Some of the synonyms used for teratocytes reflect observations on the size, function, and possible origins of these specialized extraembryonic cells. They have been called giant cells, trophic cells, trophamnion cells, and trophoserosa cells, but the term teratocytes, originally suggested by Hollande (1920), is currently most commonly used. Although a continuum between no dissociation and total dissociation of the extraembryonic membrane probably can be found in various species, this discussion will be primarily restricted to those membranes that dissociate.

Newly oviposited eggs of endoparasitic Hymenoptera typically are small with relatively little yolk. During embryogenesis, egg volume increases several hundredfold with concomitant growth of the serosal membrane and stretching of the chorion. Differentiation among the cells forming the serosal membrane probably occurs during embryogenesis. In *Cardiochiles nigriceps* the serosal membrane cells have two fates. Those cells next to the epithelial cells of the embryo begin to degenerate just prior to larval hatching, although they may remain for a while even after hatching. Smith (1952) described a similar phenomenon. However, anterior and posterior to the embryo the serosa is several layers thick (Fig. 1) and when these cells separate from the embryo, instead of degenerating, they are released and develop into the teratocytes (F. Pennacchio, S. B. Vinson, and E. Tremblay, unpublished). The situation in other species has not been described in detail, but the serosal membrane is thought to dissociate into individual teratocytes. The number of teratocytes reported from individual parasitoid species ranges from as few as 8 to as many as 800 (Vinson and Iwantsch, 1980). Once dissociated, they do not undergo further cell division, although the ploidy level may increase manyfold (Strand and Wong, 1991). Although precise details are species specific, the volume of individual teratocytes usually increases manyfold during their residence within the host. Initial diameters range from 5 to 86 μm with a 5- to 10-fold increase in diameter being quite common (Vinson and Iwantsch, 1980). One example of over a 3000-fold increase in size has been recorded in a naturally parasitized host (Smith, 1952), while even larger sizes have been obtained when the teratocytes were transferred to nonparasitized hosts (Vinson, 1970; Strand and Wong, 1991). Teratocytes with diameters as great as 1 mm have been found in *Heliothis virescens* pupae that were injected as fifth-instar larvae with *C. nigriceps* teratocytes (Pennacchio *et al.*, 1993b).

In most reported cases, the initial number of teratocytes was significantly reduced by the time the parasitoid completed its larval development (see number with time for *Perilitus coccinellae*, Fig. 2). The cause for this loss is related to the stage of host parasitized, the physiology of the parasitoid–host interaction, and the behavior of the parasitoid. Some parasitoid larvae feed

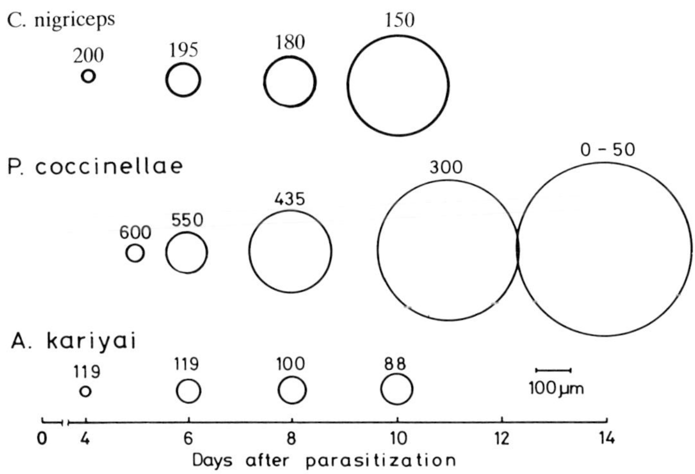

Figure 2 The diameters and numbers of teratocytes for three species of braconids are depicted. The teratocytes of *Perilitus coccinellae* reach over 450 μm in diameter whereas those of *Apanteles kariyai* rarely exceed 100 μm. Data for *Cardiochiles nigriceps* provided by S. B. Vinson (Texas A&M University, unpublished), for *P. coccinellae* provided by K. Kadono-Okuda (Tsukuba University, unpublished), and for *A. kariyai* provided by T. Okuda (National Institute of Sericulture and Entomological Science, Tsukuba, Japan, unpublished).

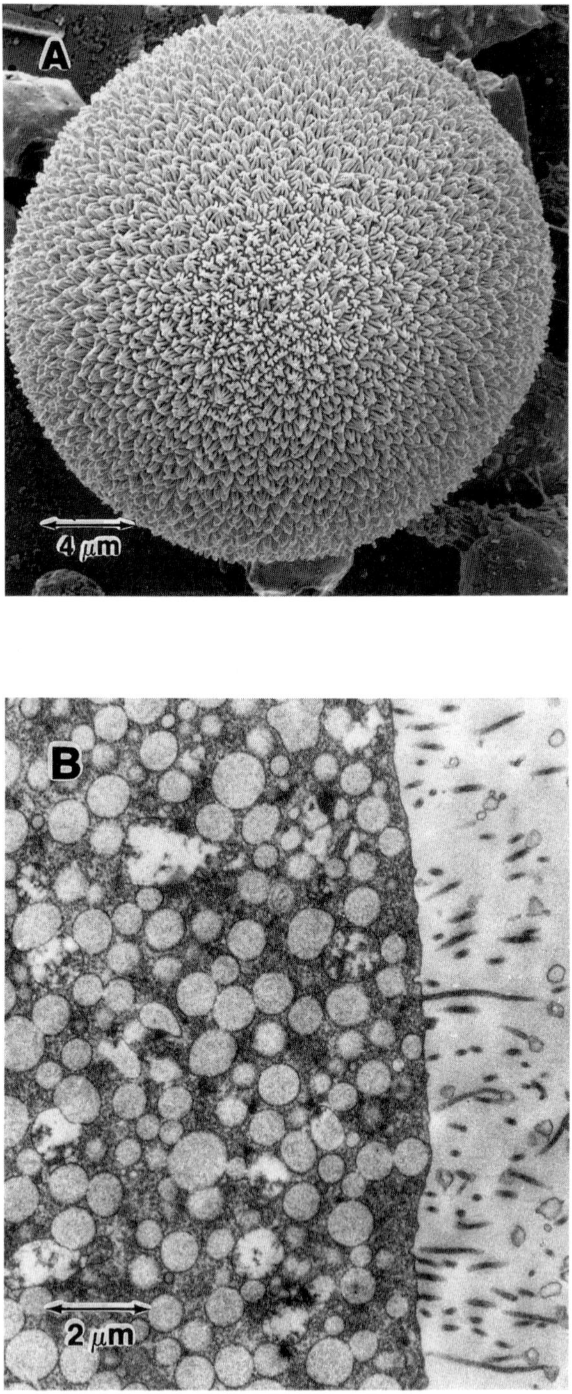
A
4 μm
B
2 μm

almost exclusively on the enlarged teratocyte cells (Schlinger and Hall, 1960). There is clear evidence of teratocytes present in the digestive tract of some parasitoid larvae (Sluss and Leutenegger, 1968; Arakawa and Kitano, 1989; Strand and Wong, 1991). In other cases it is presumed that the cells are physically destroyed as the result of the growth of the parasitoid within the limited confines of the host (Loan and Holdaway, 1961). In those parasitoid species that do not completely consume their host, it is more common that at least some of the teratocytes persist alive in the host after the parasitoid has exited (Tawfik, 1961; Kitano, 1965). On the other hand, the initial complement of teratocytes released upon hatch of a *Microplitis croceipes* egg was still found in the body cavity of its host after the parasitoid larva exited (Vinson and Lewis, 1973). The available data suggest that the evolution of teratocytes has taken two pathways. In some parasitoid–host relationships the teratocyte numbers decrease prior to the general consumption of host tissues by the developing parasitoid, whereas in others the teratocytes either do not significantly decrease in number or they are only consumed when the tissues of the host are consumed. For example, in the midpoint of their development in the host, the number of teratocytes from *Apanteles kariyai* eggs has not decreased at all (Fig. 2). In contrast, 6% of the *C. nigriceps* teratocytes are lost by the midpoint of its development and about 20% of the teratocytes from *P. coccinellae* are gone by 7 days.

2. SEM and TEM Ultrastructure

Early reports of teratocytes in parasitized hosts included line drawings, photographs of whole cells, sections dyed with various cytological stains, etc. (Vinson, 1970). More recent reports have included excellent transmission electron microscopy (TEM) photomicrographs showing ultrastructure (Sluss and Leutenegger, 1968; Tremblay and Iaccarino, 1971; Vinson and Scott, 1974; Strand *et al.*, 1985, 1986; Tanaka and Wago, 1990; Pennacchio *et al.*, 1993a; de Buron *et al.*, 1993). Information from scanning electron microscopy (SEM) studies also has appeared (Dahlman, 1991). Both TEM and SEM studies show that teratocytes are covered with a dense mat of microvilli (Fig. 3A). This expanded surface area is considered to enhance both the potential absorptive and secretory functions of the teratocytes. Ultrastructural studies show large extensively ramified nuclei, extensive reticular and swollen endoplasmic reticulum (Fig. 3B), sequestering mitochondria, orga-

---

Figure 3 (A) Scanning electron micrograph of a fixed teratocyte of *Microplitis croceipes* (5 days after hatching) showing extensive microvillae. (B) Thin section of a fixed teratocyte of *Cardiochiles nigriceps* showing the extensive swollen vacuoles surrounded by ribosomes and rough endoplasmic reticulum.

nized Golgi, myelinlike structures, putative autophagic vacuoles, and cellular outpocketings. Many of these features are also shared by nondissociated serosal membranes (Lawrence, 1990), trophamnions (Koscielski *et al.*, 1978), and gut epithelial cells.

B. Comparisons among Egg, Egg-Larval, Larval, Larval-Pupal, and Adult Parasitoids

All stages of the life cycle of insects are susceptible to parasitization by one or more species of parasitic insects. With the exception of parasitoids that only attack the pupal stage, teratocytes have been observed in parasitized hosts of all other life stages.

Three species of scelionid egg parasitoids have been reported to have teratocytes (Gerling and Orion, 1973; Strand *et al.*, 1985; Volkoff and Colazza, 1992). Teratocytes have been documented in several species of *Chelonus* (Braconidae) (Bühler *et al.*, 1985; Jones, 1987), which are egg/larval parasitoids. The single report of teratocytes in a platygasterid also is an egg/larval parasitoid (Hill and Emery, 1937).

The larval stage of hosts is by far the most common in which teratocytes have been reported. However, it also is possible that this simply reflects the fact that most of the examples thus far studied have been larval parasitoids. Orders hosting parasitoids with teratocytes include Lepidoptera, Hemiptera, and Homoptera. A few of the species parasitizing lepidopterous larvae are gregarious (Kitano, 1965; Beckage, 1990) and particularly large numbers of teratocytes are present in the hemocoel. Nymphs (larvae) of aphids (Spencer, 1926) and mirids (Cohen and DeBolt, 1984) are hosts for several species of braconids that yield teratocytes, and adults of certain families of Coleoptera (Jackson, 1928; Smith, 1952; Sluss, 1968), as well as adult aphids, are hosts for braconids that release teratocytes. *Diachasmimorpha* (=*Biosteres*) *longicaudata*, the larval/pupal braconid parasitoid of the tephritid fruit fly, *Anastrepha suspensa*, releases an intact serosal membrane that may play a role similar to that of the teratocytes (Lawrence, 1990). It is interesting to note that we have not found any records of teratocytes associated with pupal parasitoids. Perhaps the internal histolysis of tissues associated with reallocation of structural components obviates the need for teratocytes that may serve a lytic/dissociation function in other types of hosts.

C. Comparisons between Single and Multicell Tissues

As defined, teratocytes are individual cells released from an extraembryonic membrane subsequent to the hatch of the parasitoid egg. As such, they are restricted to a few families of Hymenoptera. However, from a functional point of view, extraembryonic tissue from the parasitoid that persists in the host posteclosion may be similar whether or not the cells of the membrane have dissociated. Given that the teratocytes arise from polar bodies, it can be argued that tissues from the same source might have similar functions. The role of polar bodies in endoparasitism has been discussed (Tremblay and Caltagirone, 1973), and Ivanova-Kasas (1972) addressed a somewhat more specialized aspect of this subject in a review on insect polyembryony. Clearly a variety of extraembryonic tissues may persist within the hosts and these tissues more than likely express a diversity of biological activities that contribute to the success of the parasitoid.

V. Physiological and Biochemical Characteristics

Since experimental studies on the functions of teratocytes have only recently been initiated, much remains to be explained with respect to specific roles played by these cells in the parasitoid–host interaction. This is especially true with egg parasitoids but the information from even the most frequently studied group (lepidopterous larval parasitoids) is very limited. Historically, the primary roles of teratocytes have been reported to be trophic, immunosuppressive, and secretory.

A. Trophic Functions

The tremendous increase in volume of teratocytes followed by their disappearance toward the completion of the parasitoid's larval growth leads to the conclusion that teratocytes serve as a source of specific nutrients, particularly for the later instars of the parasitoid. In fact, complete cells and fragments have been found in the digestive tracts of some parasitoids (Sluss, 1968; Arakawa and Kitano, 1989; Strand and Wong, 1991). The dense mat of microvilli observed on teratocytes (Fig. 3A) could be considered an adaptation for absorption of nutrients from the host's hemolymph. Although it is possible that teratocytes may synthesize specific nutrients required for successful parasitoid development, it is also possible that teratocytes release factors that either alter host nutrients or influence the physiological environment in which the parasitoid develops. For example, the teratocytes of *Telenomus heliothidis* secrete histolytic enzymes that digest the content of the egg host (Strand *et al.*, 1986).

In at least two species, the success of *in vitro* rearing of parasitoid larvae was enhanced by including teratocytes in the rearing media (Strand *et al.*, 1988; Greany *et al.*, 1989), perhaps as the result of the production of histolytic

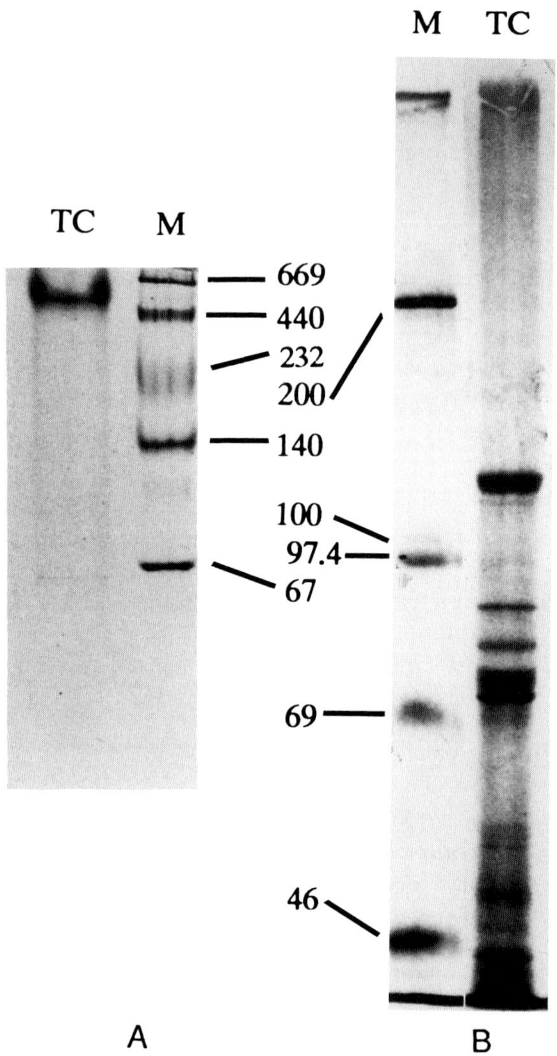

Figure 4 (A) Native PAGE (2.5 to 20% gradient) of an 11-day teratocyte homogenate of *Perilitus coccinellae* stained with Coomassie Brilliant Blue (T. Okuda, National Institute of Sericulture and Entomological Science, Tsukuba, Japan, unpublished). (B) Autoradiograph of ^{35}S-labeled proteins from an SDS-PAGE of a 9-day teratocyte homogenate of *Cardiochiles nigriceps* (A. Mourad and S. B. Vinson, Texas A&M University, unpublished). TC = teratocytes, M = marker.

enzymes that hydrolyze complex molecules found in the rearing media. In one report where teratocytes did not improve *in vitro* growth, the rearing media may have been deficient in one or more critical components because it supported only marginal growth without teratocytes; therefore, it may not have been possible to demonstrate any supplemental effects of teratocytes (Rotundo *et al.,* 1988). In contrast, *in vitro* studies with *C. nigriceps* by Pennacchio *et al.* (1992b) demonstrated both growth and molting from first to second instar without active teratocytes.

It is unlikely that teratocytes are eaten by *M. croceipes* larvae because the original number remain in the host after the parasitoid has exited (Vinson and Lewis, 1973). A significant proportion of the teratocytes from *Cotesia* (=*Apanteles*) *glomerata* also remain at the termination of parasitoid larval development (Hashimoto and Kitano, 1971). It has been suggested that the teratocytes might disintegrate in the host's hemolymph and thus release or replenish nutrient supplies that are then available to the parasitoid. However, the best evidence for true trophic function comes from T. Okuda (personal communication). He found that teratocytes from *P. coccinellae* contain one major protein band (Fig. 4A), whereas the reduction in teratocyte numbers from 600 to less than 50 (see Fig. 2, K. Kodono-Okuda, personal communication) suggests that teratocytes from *P. coccinellae* are actually consumed by the parasitoid. As mentioned earlier, teratocytes could also play a trophic role by digesting host hemolymph proteins, etc.

B. Immunosuppressive Functions

In 1968 George Salt speculated that teratocytes aid in resistance to host defense mechanisms. He argued that teratocyte growth is achieved by absorbing nutrients from host hemolymph with the concomitant depletion of hemolymph material. As a consequence, hematopoiesis is retarded and potential hemocytic reactions of the host to the young parasitoid are depressed. He further argued that the presence of large numbers of teratocytes after the parasitoid left its host "was a sheer waste of the teratocytes if their function was to nourish; but was understandable if they had already performed their function of protecting the young larvae from a defense reaction." Salt (1971) reported several interesting observations in which eggs and larvae of the ichneumonid wasp *Nemeritis canescens,* which does not produce teratocytes, were implanted into unnatural hosts. He observed that the eggs and larvae were rapidly encapsulated, except in several hosts that had previously, but unknown to Salt, been parasitized by a braconid. In each case he observed teratocytes and considered this to be evidence that teratocytes are a generic way to prevent host defense reactions. However, these results may today be interpreted to be due to the presence of the polydnaviruses (Vinson and Stoltz, 1986).

Kitano (1969, 1974) and Vinson (1972) confirmed Salt's observations by demonstrating that parasitoid larvae were not encapsulated when teratocytes were introduced into the host's hemocoel, even in the absence of either host venom or calyx fluid (polydnavirus). However, such results must be interpreted with caution as polydnavirus DNA sequences also occur in teratocytes (Dahlman, 1991). In other species, the teratocytes contribute to, but are not entirely responsible for, inhibition of encapsulation (Tanaka and Wago, 1990). Vinson (1972) found that encapsulation was less effective when the teratocytes had been in the host for several days before the introduction of the parasitoid larva. Generally, younger teratocytes were more effective than older teratocytes in inhibiting encapsulation.

Teratocytes may also contribute to host immunosuppression by inhibiting hemolymph phenoloxidase (PO) activity. Though depressed PO activity in parasitized larvae was reported by Sroka and Vinson (1978), the independent contribution of teratocytes to this process has only recently been reported. Kitano *et al.* (1990) found that PO inhibition was greater in younger parasitoids of *C. glomerata* (i.e., young teratocytes). However, PO inhibition may not have any relation to prevention of encapsulation because when proPO was activated with injections of zymosan, as evidenced by significant melanization formation, the parasitoid larvae were not encapsulated. In another species, only older teratocytes of *A. kariyai* had an inhibitory effect on host PO activity (Tanaka and Wago, 1990). In this case, the PO inhibitory factor was released into the medium when teratocytes were incubated in Grace's medium for 48 hr.

C. Secretory Functions

Evidence acquired from ultrastructural studies (Fig. 3A) and experimental approaches using *in vitro* techniques can be used to demonstrate the secretory nature of teratocytes. The ultrastructural features of teratocytes indicate that they are capable of the synthesis and secretion of protein and perhaps other substances. Many of these same features are shared by trophamnion membranes (Koscielski *et al.*, 1978) and nondissociated serosal membranes (Lawrence, 1990). The specific aspects of the secretory role will vary according to the physiological requirements dictated by the parasitoid.

Although only one species of egg parasitoid (*Telenomus heliothidis*, a scelionid) has been studied in detail (Strand *et al.*, 1985, 1986; Strand, 1986), a major role of teratocytes in this system is to release enzymes that digest the host tissues and provide nutrients for the developing parasitoid. These teratocytes produce large numbers of electron-dense vesicular bodies that are presumably released into the host egg. Detectable levels of acid phosphatase, esterase, and leucine amino peptidase were found in teratocyte extracts. In addition, *in vitro* media to which teratocytes were added "cleared up," suggesting a hydrolytic role of the teratocytes or a teratocyte secretory product. To our knowledge, this is the only species' teratocytes that have been studied for secretory enzymes, but it seems likely that teratocytes of other egg parasitoids also work in a similar manner. It is unlikely that teratocytes associated with parasitoids using larval and adult hosts will secrete enzymes that indiscriminately digest host tissue because the integrity of most hosts is maintained by the parasitoid until the parasitoid completely devours the host's internal organs just before the completion of larval development. However, it is possible that teratocytes release enzymes that attack only specific tissues, for example, collagenese, which would attack the collagen sheath surrounding the fat body. Such action would permit selective release of fat body cells, which would not kill the host, but would contribute extra nutrients to the parasitized larva. This may in fact account for the often-noted friability of fat body and the opalescence of parasitized host blood. It would also explain enhanced *in vitro* teratocyte growth in the presence of fat body (Greany *et al.*, 1989) and decreased fat body mass in parasitized larvae (Dahlman and Vinson, 1980; Zhang *et al.*, 1992).

Brewer *et al.* (1973), using native polyacrylamide gel electrophoresis, found several proteins in teratocytes of *C. nigriceps* that were absent in nonparasitized hemolymph, but they did not compare the protein profile to hemolymph of naturally parasitized larvae. It seems likely that teratocytes are a source of these proteins that appear toward the end of the parasitoid's life within its host (Brewer *et al.*, 1973; Beckage, 1990; Lawrence, 1990). In their 1980 review, Vinson and Iwantsch mentioned that teratocytes rapidly absorbed ^{14}C-labeled amino acids, synthesized proteins, and secreted some of them back into the medium (Fig. 4B). Beckage (1990), using similar techniques, reported that teratocytes of *Cotesia congregata* synthesized proteins *in vitro* that differed from those associated with the parasitoid. Dahlman (1991) made similar observations with teratocytes from *M. croceipes*, a parasitoid of *H. virescens* and *Helicoverpa zea*.

In addition to release of enzymes, only two specific functions have been suggested for the proteinaceous teratocyte secretory products. Both Kitano *et al.* (1990) and Tanaka and Wago (1990) believe that teratocytes are responsible for inhibiting PO activity in the hemolymph. It is thought that PO inhibition may interfere with the host's ability to defend itself against the parasitoid larva. A second specific activity of teratocytes is the inhibition of hemolymph juvenile hormone esterase (JHE). It has been known for several years that the JHE titers in *H. virescens* parasitized by *M. croceipes* were reduced (Dahlman *et al.*, 1990). Zhang and Dahlman (1989) and Zhang *et al.* (1992) conclusively demonstrated that injection of teratocytes derived from *in vivo* sources completely inhibited production of JHE when at least one embryo equivalent of *M. croceipes* teratocytes was used. Dahlman (1991) found that culture media in which *in vitro M. croceipes* teratocytes had been held (but not control media) caused larval mortality and delayed growth when injected into young fifth-instar *H. virescens*. It also is important to note that Hayakawa (1990) purified and reported on the structure of (Hayakawa, 1991) a 4.5-kD peptide from the hemolymph of *Pseudaletia separata* parasitized by *A. kariyai* that repressed JHE activity when injected into nonparasitized hosts. It is reasonable to postulate that it is a teratocyte product.

Secretion of nonprotein products by teratocytes is also possible. Führer and Elsufty (1979) found a fungistatic substance in the teratocytes of *C. glomerata*. Joiner *et al.* (1973) reported juvenile hormone (JH) activity in extracts from *C. nigriceps* teratocytes. Extracts from older teratocytes had greater JH activity in the *Galleria* bioassay. They were not sure, and it has not been demonstrated in any other system whether the teratocytes synthesized JH or whether they simply served as a depot for JH possibly acquired from the host. Although *Locusta migratoria* is not a parasitoid, the fact that its serosal cells synthesize JH (Hartmann *et al.*, 1987) gives credibility to the report of Joiner *et al.* (1973). Grossniklaus-Bürgin and Lanzrein (1990) suggested that teratocytes from a *Chelonus* species released JH (no evidence provided), whereas Beckage and Riddiford (1982) found only slight JH activity in the cellular fraction (which would include the teratocytes) from hemolymph of *Manduca sexta* parasitized by *C. congregata*.

D. General Effects on Host Growth and Development

Experiments that investigated the possible role of teratocytes in the absence of confounding factors such as the parasitoid larva, poison gland products, and/or calyx fluid (polydnavirus) were first conducted by Vinson (1970). He showed that younger teratocytes from *C. nigriceps* delayed *H. virescens* larval development more than older teratocytes. Most of the treated larvae eventually pupated or developed abnormalities resulting from incomplete larval–pupal ecdyses. Zhang and Dahlman (1989), in their studies with *M. croceipes*, observed both age- and dose-dependent effects of teratocytes. These effects were expressed as delayed larval growth and mortality and incomplete larval–pupal ecdysis. Wani *et al.* (1990) found that older teratocytes of *A. kariyai* prolonged the larval stage of its host, *P. separata*. Recently Pennacchio *et al.* (1992a) conducted a detailed examination of the *C. nigriceps–H. virescens* parasite–host complex. They showed that higher doses of teratocytes (equivalent to 2, 4, or 8 embryo equivalents) had a greater impact on the prevention of larval–pupal ecdysis. The age of the host at the time of treatment also was important, with older larvae being less responsive. It also was interesting to note that even the apparently "normal" pupae obtained from teratocyte-treated larvae failed to develop into adults. Zhang and Dahlman (1989) found that even a 0.25 embryo equivalent of teratocytes prevented pupation of 80% of the treated mature *H. virescens* larvae. However, at least some of the pupae that did form developed into adults. They also reported that younger hosts were more sensitive and that younger teratocytes were more potent.

Zhang and Dahlman (1989) observed that injected teratocytes of *M. croceipes* caused many of the same developmental aberrations observed in truly parasitized hosts. Dahlman *et al.* (1990) reported reduced levels of ecdysteroid in parasitized *H. virescens* at the time of the normal surge associated with larval–pupal ecdysis. Zhang *et al.* (1992) found a similar reduction in ecdysteroid titer in teratocyte-treated larvae. In contrast, in the very same host but with a different braconid parasitoid (*C. nigriceps*), Pennacchio *et al.* (1991, 1992a) reported a threefold elevation of ecdysteroid titer in larvae receiving two embryo equivalents of teratocytes as new fifth instars. They attempted to explain this difference by suggesting increased production of ecdysteroids as the result of interference with normal feedback mechanisms that control the rate of ecdysteroid synthesis. They also recognized that inactive ecdysteroid degradation products might respond in the RIA used to quantify ecdysteroid titer, thus giving falsely high results.

It is curious that the teratocytes from two larval endoparasitoids using the same host species would produce such different results in ecdysteroid titer. However, when the host is truly parasitized by either species, ecdysteroid titers are reduced (Dover *et al.*, 1987; Dahlman *et al.*, 1990; Pennacchio *et al.*, 1991, 1993b). It must be kept in mind that the mode of feeding by these two parasitoids is quite different. *Cardiochiles nigriceps* feeds on hemolymph through the first few days following hatch but then begins to feed on selected host tissue until the time it emerges from its host and consumes the entire carcass (Vinson and Barras, 1970). Pennacchio *et al.* (1992, 1993b) proposed that the teratocytes from *C. nigriceps* regulate protein metabolism, possibly by inactivation of 20-hydroxyecdysone. This results in increased titers of hemolymph proteins at a time correlated with rapid parasitoid larval growth. Thus, the parasitoid more efficiently exploits its natural food source and reaches the critical size necessary for molting and exiting from its host. In contrast, *M. croceipes* apparently derives its nutrients from its host's hemolymph although the host's fat body fails to properly develop (Dahlman and Vinson, 1980). The host may remain alive (although inactive and developing no further than the cell formation stage; Webb and Dahlman, 1985) for more than a week after the parasitoid emerges (Vinson and Lewis, 1973). Undoubtedly various feedback mechanisms must be involved with both of these parasitoid–host interactions with respect to their different means of interfering with the hormonal titers of their hosts.

VI. Interactions of Teratocytes, Venoms, and Polydnavirus

A substantial number of studies have been conducted on the importance of interactions between the polydnaviruses and venoms of parasitic wasps (Tanaka, 1987; Dover *et al.*, 1987, 1988; Stoltz *et al.*, 1988; Strand and Dover, 1991; Tanaka and Vinson, 1991). However, relatively little of this research has included teratocytes as a third component in the tests. Wani *et al.* (1990) showed that young teratocytes from *A. kariyai* did not inhibit growth and development of the host. However, when hosts were first injected with venom and calyx fluid and subsequently injected with young teratocytes at the approximate time teratocytes would normally be released in a parasitized host, both larval–larval and larval–pupal development were impaired. Strand and Wong (1991) found that an embryo equivalent dose of *Microplitis demolitor* teratocytes had an inconsistent effect on the development of its host, *Pseudoplusia includens*, but when the injection of teratocytes was preceded with an injection of calyx fluid or polydnavirus plus venom, the developmental alterations were very similar to those observed during normal parasitism. In fact, *D. molitor* teratocytes were encapsulated when injected into nonparasitized *P. includens* larvae but remained unencapsulated when injected into hosts that had previously been injected with a mixture of polydnavirus and venom (Strand and Noda, 1991). Pennacchio *et al.* (1992a) suggest that inhibition of ecdysteroid production by venom and polydnavirus explains why ecdysteroid titers are low in parasitized hosts as compared to the high titers seen when hosts are only treated with teratocytes. Jones (1987) concluded that polydnavirus and venom (but not teratocytes) were necessary for the production of a factor that caused precocious metamorphosis/suppressed prepupal development in *Trichoplusia ni* parasitized by a *Chelonus* sp. Tanaka (1987) found that about 50% of the *Microplitis mediator* eggs with completely developed serosal membranes (nondissociated teratocytes) were protected from encapsulation when injected with calyx fluid, whereas none were protected without the calyx fluid.

VII. Conclusion and Summary

Teratocytes are unusual extraembryonic cells associated with certain groups of endoparasitic Hymenoptera. They play an important, but incompletely defined, role in the total interaction between the larval stage of the parasitoid and its host. It is not clear why they are restricted to only certain families, although it is likely that related tissues perform similar roles in other groups of parasitoids. Specific functions vary between parasitoid species and the developmental stage of the host. Trophic, immunosuppressive, and secretory functions have been well documented. Titers of the principle hormones in the host may be altered by teratocytes. Interactions at several levels of complexity are possible, not only between the teratocytes and the host but also between other parasitoid-derived factors such as venom and nonpathogenic polydnaviruses resident within both the parasitoid and the extraembryonic tissues. It is likely that the success of the parasitoid is dependent upon each type of interaction. Much remains to be learned and the coming decade will yield considerable information on this complex and interesting aspect of insect parasitology.

Acknowledgments

We gratefully acknowledge the support of USDA Competitive Research Grant 89-37250-4705 (D.L.D.) and a grant from R. J. Reynolds (D.L.D.). We are grateful for the critical review and constructive suggestions provided by Dr. P. D. Greany. This is paper 91-07-147 of the Kentucky Agriculture Experiment Station, Lexington, KY 40546-0091.

References

Arakawa, T., and Kitano, H. (1989). A possible reason for the decrease of the number of teratocytes in the body cavities of *Peiris rapae crucivora* Boisduval (Lepidoptera:Pieridae) parasitized by *Apanteles glomeratus* L. *Appl. Entomol. Zool.* **24:**229–231.

Beckage, N. E. (1990). Parasitic effects on host development. *In* "New Directions in Biological Control: Alterations for Suppressing Agricultural Pests and Diseases" (R. Baker and P. Dunn, eds.), pp. 497–515. Liss, New York.

Beckage, N. E., and Riddiford, L. M. (1982). Effects of parasitism by *Apanteles congregatus* on the endocrine physiology of the tobacco hornworm *Manduca sexta*. *Gen. Comp. Endocrinol.* **47:**308–322.

Brewer, F. D., Glick, B., and Vinson, S. B. (1973). A comparative study of selected tissues from the parasitoid *Cardiochiles nigriceps*, its susceptible host *Heliothis virescens* and a resistant host *Heliothis zea* by polyacrylamide gel electrophoresis. *Comp. Biochem, Physiol. B* **46B:**567–574.

Bühler, A., Hanzlik, T. N., and Hammock, B. D. (1985). Effects of parasitization of *Trichoplusia ni* by *Chelonus* sp. *Physiol. Entomol.* **10:**383–394.

Cohen, A. C., and DeBolt, J. W. (1984). Fatty acid and amino acid composition of teratocytes from *Lygus hesperus* (Miridae:Hemiptera) parasitized by two species of parasites, *Leiophron uniformis* (Braconidae:Hymenoptera) and *Peristenus stygicus* (Braconidae:Hymenoptera). *Comp. Biochem. Physiol. B* **79B:**335–337.

Coudron, T. (1991). Host-regulating factors associated with parasitic Hymenoptera. In "Naturally Occurring Pest Bioregulators" (P. A. Hédin ed.), pp. 41–65. Am. Chem. Soc., Washington, D.C.

Dahlman, D. L. (1990). Evaluation of teratocyte functions: An overview. *Arch. Insect Biochem. Physiol.* **13:**159–166.

Dahlman, D. L. (1991). Teratocytes and host/parasitoid interactions. *Biol. Control* **1:**118–126.

Dahlman, D. L., and Vinson, S. B. (1980). Glycogen content in *Heliothis virescens* parasitized by *Microplitis croceipes*. *Comp. Biochem. Physiol. A* **66A:**625–630.

Dahlman, D. L., Coar, D. L., Koller, C. N., and Neary, T. J. (1990). Contributing factors to reduced ecdysteroid titers in *Heliothis virescens* parasitized by *Microplitis croceipes*. *Arch. Insect Biochem. Physiol.* **13:**29–39.

de Buron, I., Nesbit, D. J., Tan, F. F., and Beckage, N. E. (1993). Development of *Cotesia congregata* teratocytes in host larvae of the tobacco hornworm, *Manduca sexta:* Correlation with morphological and biochemical changes in the cells. *Dev. Biol.* (in press).

Dover, B. A., Davies, D. H., Strand, M. R., Gray, R. S., and Vinson, S. B. (1987). Ecdysteroid-titre reduction and developmental arrest of last-instar *Heliothis virescens* larvae by calyx fluid from the parasitoid *Campoletis sonorensis*. *J. Insect Physiol.* **33:**333–338.

Dover, B. A., Davies, D. H., and Vinson, S. B. (1988). Degeneration of last instar *Heliothis virescens* prothoracic glands by *Campoletis sonorensis* polydnavirus. *J. Invertebr. Pathol.* **51:**80–91.

Führer, E., and Elsufty, R. (1979). Producktion fungistatischer metabolite durch teratocyten von *Apanteles glomeratus* L. (Hym., Braconidae). *Z. Parasitenkd.* **59:**21–25.

Gerling, D., and Orion, T. (1973). The giant cells produced by *Telenomus remus*. *J. Invertebr. Pathol.* **21:**164–171.

Greany, P. D., Ferkovich, S. M., and Clark, W. R. (1989). Progress towards development of an artificial diet and *in vitro* rearing system for *Microplitis croceipes*. *Southwest. Entomol., Suppl.* **12:**89–94.

Grossniklaus-Bürgin, C., and Lanzrein, B. (1990). Endocrine interrelationship between the parasitoid *Chelonus* sp. and its host *Trichoplusia ni*. *Arch. Insect Biochem. Physiol.* **14:**201–216.

Hartmann, R., Jendrsczok, C., and Peter, M. G. (1987). The occurrence of a juvenile hormone binding protein and in vitro synthesis of juvenile hormone by the serosa of *Locusta migratoria* embryos. *Roux's Arch. Dev. Biol.* **196:**347–355.

Hashimoto, K., and Kitano, H. (1971). Studies on the number of "giant cells" in the body cavity of *Peris rapae crucivora* attacked by *Apanteles glomeratus* L. *Zool. Mag.* **80:**323–329.

Hawlitzky, N., and Lauge, G. (1979). Rôles presumes de la membrane embryonnaire d'un parasite ovo-larvaire de Pyralidae *Phanerotoma flavitestacea* [Hym.:Braconidae]. *Entomophaga* **24:**247–253.

Hayakawa, Y. (1990). Juvenile hormone esterase activity repressive factor in the plasma of parasitized insect larvae. *J. Biol. Chem.* **265:**10813–10816.

Hayakawa, Y. (1991). Structure of a growth-blocking peptide present in parasitized insect hemolymph. *J. Biol. Chem.* **226:**7982–7984.

Hill, C. C., and Emery, W. T. (1937). The biology of *Platygaster herrickii,* a parasite of the Hessian fly. *J. Agric. Res.* **55:**199–213.

Hollande, A.-C. (1920). Oenocytoides et teratocytes du sang des chenilles. *C. R. Hebd. Seances Acad. Sci.* **179:**1341–1344.

Ivanova-Kasas, O. M. (1972). Polyembryony in insects. *In* "Developmental Systems: Insects" (S. J. Counce and C. H. Waddington, eds.), Vol. 1, pp. 243–271. Academic Press, New York.

Jackson, D. J. (1928). The biology of *Dinocampus (Perilitus) rutilus* Nees, a braconid parasite of *Sitona lecanium* L. Part I. *Proc. Zool. Soc. London* pp. 597–630.

Joiner, R. L., Vinson, S. B., and Benskin, J. B. (1973). Teratocytes as a source of juvenile hormone activity in a parasitoid–host relationship. *Nature (London) New Biol.* **246:**120–121.

Jones, D. (1987). Material from adult female *Chelonus* sp. directs expression of altered developmental programme of host Lepidoptera. *J. Insect Physiol.* **33:**129–134.

Kitano, H. (1965). Studies on the origin of giant cells in the body fluid of *Pieris rapae crucivora* attacked by *Apanteles glomeratus* L. II. Determination of their origin and speculation on their biological significance. *Zool. Mag.* **74:**192–197.

Kitano, H. (1969). Experimental studies on the parasitism of *Apanteles glomeratus* L. with special reference to its encapsulation-inhibiting capacity. *Bull. Tokyo Gakugei Univ., Ser. 4* **21:**95–136.

Kitano, H. (1974). Effects of the parasitization of a braconid, *Apanteles,* on the blood of its host, *Pieris. J. Insect Physiol.* **20:**315–327.

Kitano, H., Wago, H., and Arakawa, T. (1990). Possible role of teratocytes of the gregarious parasitoid, *Cotesia* (=*Apanteles*) *glomerata* in the suppression of phenoloxidase activity in the larval host, *Pieris rapae crucivora. Arch. Insect Biochem. Physiol.* **13:**177–185.

Koscielski, B., Koscielska, M. K., and Szroeder, J. (1978). Ultrastructure of the polygerm of *Ageniaspis fuscicollis* Dalm. (Chalcidoidea, Hymenoptera). *Zoomorphologie* **89:**279–288.

Lawrence, P. O. (1990). Serosal cells of *Biosteres longicaudatus* (Hymenoptera:Braconidae): ultrastructure and release of polypeptides. *Arch. Insect Biochem. Physiol.* **13:**199–216.

Loan, C. C., and Holdaway, F. G. (1961). *Pygostolus falcatus* (Nees), a parasite of *Sitona* species. *Bull. Entomol. Res.* **52:**473–488.

Pennacchio, F., Vinson, S. B., and Tremblay, E. (1991). Effects of *Cardiochiles nigriceps* Viereck (Hymenoptera, Braconidae) teratocytes on physiology of its host *Heliothis virescens* (F.) (Lepidoptera, Noctuidae). Insect Parasitoids, 4th European Workshop, Perugia 305 April, 1991. Redia, **74:**433–438.

Pennacchio, F., Vinson, S. B., and Tremblay, E. (1992a). Host regulation effects on *Heliothis virescens* (F.) larvae induced by teratocytes of *Cardiochiles nigriceps* Viereck (Lepidoptera, Noctuidae—Hymenoptera, Braconidae). *Arch. Insect Biochem. Physiol.* **19:**177–192.

Pennacchio, F., Vinson, S. B., and Tremblay, E. (1992b). Preliminary results on *in vitro* rearing of the endoparasitoid *Cardiochiles nigriceps* from egg to second instar. *Entomol. Exp. Appl.* **64:**209–216.

Pennacchio, F., Vinson, S. B., and Tremblay, E. (1993a). Embryonic membranes in *Cardiochiles nigriceps* Viereck: Development and ultrastructure. *Int. J. Insect Morphol. Embryol.* (in press).

Pennacchio, F., Vinson, S. B., and Tremblay, E. (1993b). Growth and development of *Cardiochiles nigriceps* Viereck (Hymenoptera, Braconidae) larvae and their synchronization with the protein and lipid titers of the haemolymph of their host, *Heliothis virescens* (F.) (Lepidoptera, Noctuidae). *J. Insect Physiol.* (in press).

Rotundo, G., Cavalloro, R., and Tremblay, E. (1988). In vitro rearing of *Lysiphlebus fabarum* [Hym.:Braconidae]. *Entomophaga* **33:**261–267.

Salt, G. (1968). The resistance of insect parasitoids to the defense reactions of their hosts. *Biol. Rev. Cambridge Philos. Soc.* **43:**200–232.

Salt, G. (1971). Teratocytes as a means of resistance to cellular defense reactions. *Nature* (*London*) **232:**639.

Schlinger, E. I., and Hall, J. C. (1960). The biology, behavior, and morphology of *Praon palitans* Muesebeck, an internal parasite of the spotted alfalfa aphid, *Therioaphis maculata* (Buckton) (Hymenoptera:Braconidae, Aphidiinae). *Ann. Entomol. Soc. Am.* **53:**144–160.

Sluss, R. (1968). Behavioral and anatomical responses of the convergent lady beetle to parasitism by *Perilitus coccinellae* (Schrank). *J. Invertebr. Pathol.* **10:**9–27.

Sluss, R. R., and Leutenegger, R. (1968). The fine structure of the trophic cells of *Perilitus coccinellae* (Hymenoptera:Braconidae). *J. Ultrastruct. Res.* **25:**441–451.

Smith, O. J. (1952). Biology and behavior of *Microctonus vittatae* Muesebeck. *Univ. Calif., Berkeley, Publ. Entomol.* **9:**315–344.

Spencer, H. (1926). Biology of the parasites and hyperparasites of aphids. *Ann. Entomol. Soc. Am.* **19:**119–157.

Sroka, P., and Vinson, S. B. (1978). Phenoloxidase activity in the haemolymph of parasitized and unparasitized *Heliothis virescens. Insect Biochem.* **8:**399–402.

Stoltz, D. B. Guzo, D., Belland, E. R., Lucarotti, C. J., and Mackinnon, E. A. (1988). Venom promotes uncoating in vitro and persistence in vivo of DNA from a braconid polydnavirus. *J. Gen. Virol.* **69:**903–907.

Strand, M. R. (1986). The physiological interactions of parasitoids with their hosts and their influence on reproductive strategies. *In* "Insect Parasitoids" (J. Waage and D. Greathouse, eds.), pp. 97–136. Academic Press, New York.

Strand, M. R., and Dover, B. A. (1991). Developmental disruption of *Pseudoplusia includens* and *Heliothis virescens* larvae by the calyx fluid and venom of *Microplitis demolitor. Arch. Insect Biochem. Physiol.* **18:**131–145.

Strand, M. R., and Noda, T. (1991). Alterations in the haemocytes of *Pseudoplusia includens* after parasitism by *Microplitis demolitor. J. Insect Physiol.* **37:**839–850.

Strand, M. R., and Wong, E. A. (1991). The growth and role of *Microplitis demolitor* teratocytes in parasitism of *Pseudoplusia includens. J. Insect Physiol.* **37:**503–515.

Strand, M. R., Quarles, J. M., Meola, S. M., and Vinson, S. B. (1985). Cultivation of teratocytes of the egg parasitoid *Telenomus heliothidis* (Hymenoptera:Scelionidae). *In Vitro Cell. Dev. Biol.* **21:**361–367.

Strand, M. R., Meola, S. M., and Vinson, S. B. (1986). Correlating pathological symptoms in *Heliothis virescens* eggs with development of the parasitoid *Telenomus heliothidis. J. Insect Physiol.* **32:**389–402.

Strand, M. R., Vinson, S. B., Nettles, W. C., Jr., and Xie, Z. N. (1988). *In vitro* culture of the egg parasitoid *Telenomus heliothidis:* The role of teratocytes and medium consumption in development. *Entomol. Exp. Appl.* **46:**71–78.

Tanaka, T. (1987). Effect of the venom of the endoparasitoid, *Apanteles kariyai* Watanabe, on the cellular defence reaction of the host, *Pseudaletia separata* Walker. *J. Insect Physiol.* **33:**413–420.

Tanaka, T., and Vinson, S. B. (1991). Interaction of venoms with the calyx fluids of three parasitoids, *Cardiochiles nigriceps, Microplitis croceipes* (Hymenoptera:Braconidae), and *Campoletis sonorensis* (Hymenoptera:Ichneumonidae) in effecting a delay in the pupation of *Heliothis virescens* (Lepidoptera:Noctuidae). *Ann. Entomol. Soc. Am.* **84:**87–92.

Tanaka, T., and Wago, H. (1990). Ultrastructural and functional maturation of teratocytes of *Apanteles kariyai. Arch. Insect Biochem. Physiol.* **13:**187–197.

Tawfik, M.F.S. (1961). Teratocytes in the larva of *Prodenia litura* (Fabricius) parasitized by *Microplitis demolitor* Wilkinson. *J. Insect Pathol.* **3:**221–224.

Tremblay, E., and Caltagirone, L. E. (1973). Fate of polar bodies in insects. *Annu. Rev. Entomol.* **18:**421–444.

Tremblay, E., and Calvert, D. (1971). Embryosystematics in the Aphidiines (Hymenoptera:Braconidae). *Boll. Lab. Entomol. Agrar. "Filippo Silvestri"* **29:**223–249.

Tremblay, E., and Iaccarino, F. M. (1971). Notizie sull'ultrastruttura dei trsofociti di *Aphidius matricariae* Hal. (Hymenoptera: Braconidae). *Boll. Lab. Entomol. Agrar. "Filippo Silvestri"* **29:**305–314.

Vinson, S. B. (1970). Development and possible functions of teratocytes in the host–parasite association. *J. Invertebr. Pathol.* **16:**93–101.

Vinson, S. B. (1972). Factors involved in successful attack on *Heliothis virescens* by the parasitoid *Cardiochiles nigriceps. J. Invertebr. Pathol.* **20:**118–123.

Vinson, S. B., and Barras, D. J. (1970). Effects of the parasitoid, *Cardiochiles nigriceps,* on the growth, development, and tissues of *Heliothis virescens. J. Insect Physiol.* **16:**1329–1338.

Vinson, S. B., and Iwantsch, G. F. (1980). Host regulation by insect parasitoids. *Q. Rev. Biol.* **55:**143–165.

Vinson, S. B., and Lewis, W. J. (1973). Teratocytes: Growth and numbers in the hemocoel of *Heliothis virescens* attacked by *Microplitis croceipes. J. Invertebr. Pathol.* **22:**351–355.

Vinson, S. B., and Scott, J. R. (1974). Ultrastructure of teratocytes of *Cardiochiles nigriceps* Viereck (Hymenoptera:Braconidae). *Int. J. Insect Morphol. Embryol.* **3:**293–304.

Vinson, S. B., and Stoltz, D. B. (1986). Cross-protection experiments with two parasitoid viruses. *Ann. Entomol. Soc. Am.* **79:**216–218.

Voegele, J., Brun, P., and Daumal, J. (1974). Les Trichogrammes. I. Modalités de la prise de possession et de l'élimination de l'hôte chez le parasite embryonnaire *Trichogramma brasiliensis. Ann. Soc. Entomol. Fr.* **10:**757–762.

Volkoff, N., and Colazza, S. (1992). Growth patterns of teratocytes in the immature stages of *Trissolcus basalis* (Woll.) (Hymenoptera:Scelionidae), an egg parasitoid of *Nezara viridula* (L.) (Heteroptera:Pentatomidae). *Int. J. Morphol. & Embryol.* **21:**323–336.

Wani, M., Yagi, S., and Tanaka, T. (1990). Synergistic effect of venom, calyx and teratocytes of *Apanteles kariyai* on the inhibition of larval pupal ecydsis of the host, *Pseudaletia separata. Entomol. Exp. Appl.* **57:**101–104.

Webb, B. A., and Dahlman, D. L. (1985). Developmental pathology of *Heliothis virescens* larvae parasitized by *Microplitis croceipes:* Parasite-mediated host developmental arrest. *Arch. Insect Biochem. Physiol.* **2:**131–143.

Zhang, D., and Dahlman, D. L. (1989). *Microplitis croceipes* teratocytes cause developmental arrest of *Heliothis virescens* larvae. *Arch. Insect Biochem. Physiol.* **12:**51–61.

Zhang, D., Dahlman, D. L., and Gelman, D. B. (1992). Juvenile hormone esterase activity and ecdysteroid titer in *Heliothis virescens* larvae injected with *Microplitis croceipes* teratocytes. *Arch. Insect Biochem. Physiol.* **20:**231–242.

Chapter 8

The Polydnavirus Life Cycle

Donald B. Stoltz
Department of Microbiology and Immunology
Dalhousie University
Halifax, Nova Scotia, Canada

I. Introduction

The polydnaviruses comprise a group of remarkable agents whose life cycles are inextricably linked with those of certain parasitic Hymenoptera (parasitoids). The nature of that linkage, and in particular its biological consequences, is the subject of this article.

These viruses are assigned to the family Polydnaviridae primarily on the basis of both genome structure and host range, although other factors such as morphology and site of replication are equally valid as diagnostic indicators for the family (Stoltz *et al.*, 1984). When extracted from purified virions, the polydnavirus genome is seen to consist of a population of double-stranded circular DNAs; within this population, there exist a number of different classes of molecules that differ in terms of both molecular weight and genetic content (Krell and Stoltz, 1980; Krell *et al.*, 1982; Theilmann and Summers, 1988; Guzo, 1988). Linear DNA sequences homologous to encapsidated viral DNAs are covalently linked to parasitoid chromosomal DNA (Fleming and Summers, 1986; Xu and Stoltz, 1991; see references in Fleming and Krell, Volume 1, Chapter 9); this appears to be a stable relationship, now thought to be required for the transmission of polydnaviruses within parasitoid populations (Stoltz, 1990).

The polydnavirus family consists of two large groups of viruses, which

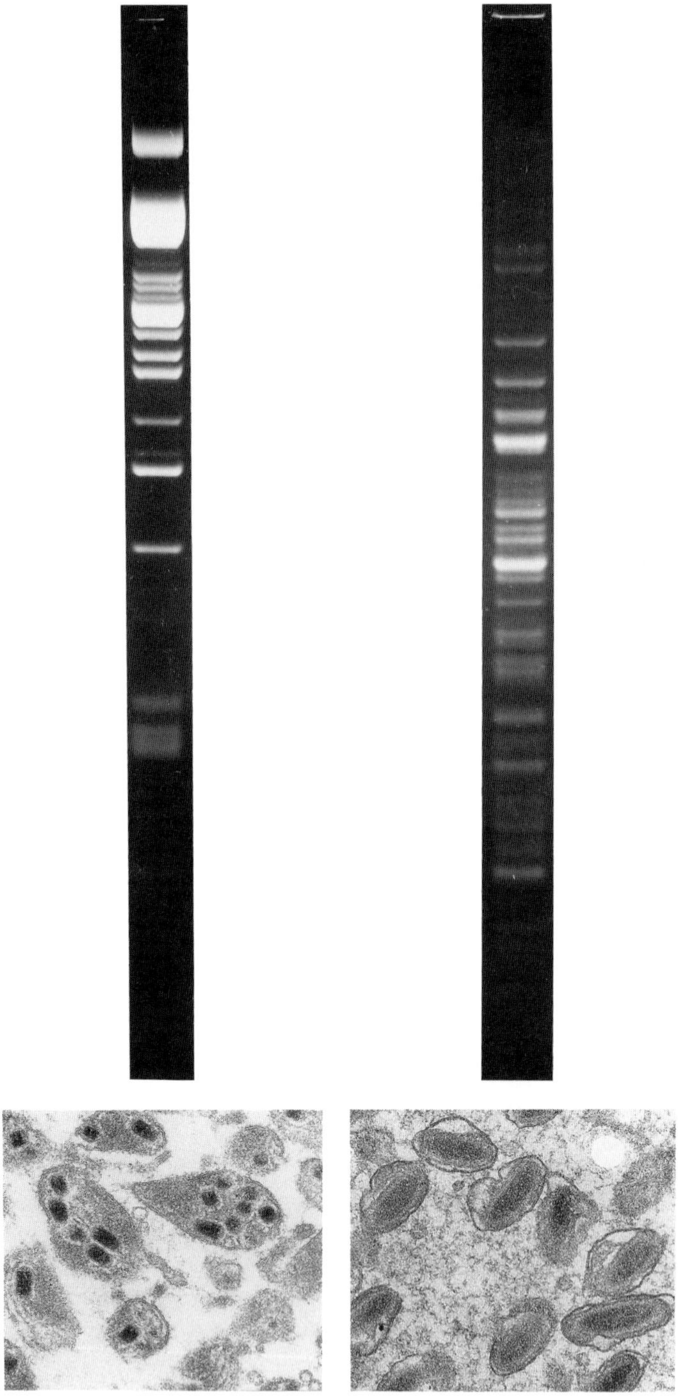

have been defined on the basis of consistent differences in both morphology and host range. Unofficially, these have been referred to as the braco- and ichnovirus groups. Bracovirus particles consist of cylindrical nucleocapsids enveloped by a single unit membrane; these have been observed only in certain species of braconid parasitoids. Ichnovirus particles consist of fusiform nucleocapsids surrounded by two unit membrane envelopes; these are restricted to the Ichneumonidae. Typical examples of polydnaviruses and (encapsidated) polydnavirus genomes are illustrated in Fig. 1.

Viral replication is restricted to the parasitoid ovary. When mature, virions enter the lumen of the reproductive tract, where they comprise the particulate fraction of a so-called "calyx fluid" (Norton *et al.*, 1975; Stoltz and Vinson, 1979a); this material, together with one or more parasitoid eggs, is injected into host animals during oviposition (Stoltz and Vinson, 1979a). Replication in parasitized animals, however, has never been observed (Stoltz and Vinson, 1979a; Theilmann and Summers, 1986; Dover *et al.*, 1989), and is therefore assumed not to occur.

II. Life Cycle

Typically, a virus life cycle consists of transmission and replication within a susceptible host; strictly speaking, both of these events are in the case of polydnaviruses restricted to certain parasitoid species. The polydnavirus life cycle, however, must be considered in a much broader context, since transmission and replication within parasitoid populations would not be possible in the absence of an additional element, namely, those insects that serve as hosts for parasitoids. As we now know, polydnaviruses are invariably present in host animals where they act to ensure successful parasitism (see Section II.B). In short, the obligatory linkage between parasitoid and host requires the presence of polydnavirus in both. I therefore argue that the polydnavirus life cycle, unlike that of other viruses, in fact comprises *two* arms. These—transmission/replication, and what may be referred to as a genetic colonization of the host (Stoltz *et al.*, 1986)—appear, respectively, to be mediated by two distinct forms of virus-specific DNA. Specifically, transmission and replication would appear to be directed by a permanently integrated viral genome

---

Figure 1 Typical examples of polydnavirus particles and genomes extracted therefrom. Below left is an electron micrograph of a bracovirus (from *Protapanteles paleacritae*) and, to the right, an ichnovirus (from *Hyposoter exiguae*). Above these, agarose gel electrophoretic profiles of DNAs extracted from virus particles are seen after ethidium bromide staining. Left, bracovirus DNA from *Cardiochiles nigriceps;* right, ichnovirus from *H. rivalis.*

(i.e., a provirus), whereas a genetic colonization of the host is effected by circular molecules delivered in encapsidated form to host cells and tissues during oviposition.

In what follows, I will deal briefly with transmission, since this subject has been recently treated in some depth (Stoltz, 1990), and even more briefly with replication, since almost nothing is known about how this occurs (Fleming and Krell, Volume 1, Chapter 9). Considerably more attention will be given to the subject of virus-directed activity in the host, since this aspect of polydnavirus biology has not recently been reviewed.

A. Transmission and Replication

It has been clear for some time that polydnavirus transmission occurs with 100% efficiency, making vertical transfer through germ line tissue likely. Nevertheless, it was initially considered possible that developing parasitoid larvae could become infected *per os,* by consuming viral DNA delivered to the host during oviposition. Moreover, it is known that such DNA can persist throughout the natural course of parasitization (Theilmann and Summers, 1986; Stoltz *et al.*, 1986). The availability of isofemale lines (i.e., derived from a single female) carrying marker polydnavirus DNA circles permitted the design of experiments aimed at determining whether *per os* transmission did in fact occur. In one such experiment, washed eggs from a *Cotesia melanoscela* line were co-injected along with bracovirus from a line that carried an extra genome segment; none of the adult female wasps resulting from this experiment carried that genome segment, suggesting that larvae either did not consume such DNA, or else could not become infected by that route. In the case of the ichneumonid parasitoid *Hyposoter fugitivus,* washed eggs were able to develop to maturity in *Malacosoma americana;* adult females still carried the normal polydnavirus associated with that species, suggesting that *per os* infection would probably be irrelevant to polydnavirus transmission, even if it *did* occur (Stoltz *et al.*, 1986).

Isofemale lines were also employed in genetic crossing experiments, the results of which clearly demonstrated that both braco- and ichnovirus genome segments segregate in Mendelian fashion. It was concluded that transmission of the polydnavirus genome most likely requires a physical linkage to the parasitoid genome; a chromosomal model for polydnavirus transmission has been discussed by Stoltz (1990). Molecular evidence for chromosomal integration of ichnovirus genome segments had previously been developed by Fleming and Summers (1986), and more recent observations now strongly suggest that chromosomal integration is almost certainly a general feature of the polydnavirus family (Xu and Stoltz, 1991; Fleming and Krell, Volume 1, Chapter 9). In short, both genetic and molecular approaches to the question of polydnavirus transmission have independently led to the same conclusion: the polydnavirus genome is transmitted in the form of an endogenous provirus.

Polydnavirus replication is restricted to the parasitoid ovary, and replication begins in the pupal stage (Norton and Vinson, 1983; Theilmann and Summers, 1986) and continues throughout the life of the adult female. It has been suggested that replication of viral DNA might be triggered by hormonal changes associated with metamorphosis and/or morphogenesis of the ovary (Norton and Vinson, 1983). In keeping with this supposition, Webb and Summers (1992) have recently demonstrated that viral DNA replication in explanted ovaries can be induced by the addition of ecdysone. The availability of *in vitro* systems such as these should very quickly lead to an understanding of the molecular basis of viral replication.

The question of which form of viral DNA serves as template for the synthesis of the circular molecules that are destined for encapsidation remains as yet unanswered. In this regard, it should first be noted that both linear (integrated) and circular (extrachromosomal) forms of viral DNA are thought to be present in most if not all parasitoid tissues (Fleming and Summers, 1986; Stoltz *et al.*, 1986), even in the absence of viral replication. Thus, it might reasonably be assumed that at least a few extrachromosomal viral DNAs are present in ovarian primordia, and therefore are potentially available to serve later on as replication templates. However, Theilmann and Summers (1986) were unable to detect the circular form of CsV genome segment B during the early pupal stages of *Campoletis sonorensis;* the cognate (corresponding) linear chromosomal sequences, on the other hand, were observed. This study suggests that extrachromosomal polydnavirus DNAs are not present in the parasitoid ovary prior to viral DNA replication, in turn suggesting that linear templates may be utilized. Genetic experimentation has provided additional evidence to the effect that extrachromosomal molecules play no role in determining the composition of the polydnavirus genome (Stoltz, 1990). Therefore, such molecules are either in some manner eliminated from the ovary prior to the onset of viral DNA replication (or were never present in that particular organ) or else do not serve as DNA replication templates.

In any case, given that polydnavirus genomes are chromosomally transmitted (Stoltz, 1990), it seems likely that linear chromosomal DNA sequences must—at least initially—represent a point of departure in the amplification/replication of polydnavirus DNA; how this might occur remains to be determined, but several possibilities are plausible. For example, amplification of viral DNA could in theory occur *in situ* (i.e., at chromosomal loci), circular viral DNAs subsequently being generated via homologous recombination; alternatively, excised, presumably circular, *copies* of chromosomal loci might serve as templates for DNA replication. Finally, such loci might be excised in their entirety prior to amplification, leaving no linear chromosomal copy behind. As a working hypothesis, I presently favor the first mechanism, since the results of genetic crossing experiments tend to militate against an involvement of circular, extrachromosomal DNA templates in polydnavirus replication (*unless* potential circular replication templates are removed from the ovary prior to the onset of viral DNA replication). Excision of circular DNAs must in any case presumably occur at some point in the replication cycle, and recent sequencing data have been interpreted by Fleming and Summers (1990) as suggesting a possible role for the CsV 540-bp repeat element in that process. It should be noted that if polydnavirus genome segments are generated from linear chromosomal templates, then amplification of any such templates need not be extensive; instead, excision could in some way be coupled to DNA replication.

The so-called "onionskin" model for *in situ* amplification (see Osheim and Miller, 1983; Stark *et al.*, 1989) may not be entirely appropriate here, and would in any case be incompatible with data reported by Theilmann and Summers (1986) to the effect that the amount of chromosomal polydnavirus-specific DNA remains constant during pupation, when viral replication first begins.

Each of the replication scenarios listed here allows certain predictions to be made. An *in situ* amplification model, for example, would not necessarily require that sequences responding to replication signals be located on extrachromosomal DNAs; they could instead be located within flanking chromosomal sequences. On the other hand, the demonstrated persistence of polydnavirus DNA in the parasitized host (see the following) might well require something equivalent to an autonomously replicating sequence.

I have argued here that the linear, integrated form of viral DNA is necessary for both virus transmission and DNA replication (whether or not this occurs from linear or circular templates). What then is the role of the circular form of viral DNA? Although its function, if any, in the parasitoid itself is presently unknown, it seems clear that circular viral genome segments are transcriptionally active in the parasitized host, there being as yet no evidence for integration of polydnavirus DNA in such animals. This aspect of the polydnavirus life cycle is considered in the next section.

B. Genetic Colonization of the Host

The results of electron microscopic studies carried out in the mid to late 1970s clearly indicated that polydnavirus particles (then commonly referred to as viruslike/calyx fluid particles) were destined for export into host animal tissues (Stoltz and Vinson, 1977, 1979b). At the same time, it became increasingly clear that at least some of the obvious changes resulting from natural parasitism could also be observed following manual injection of calyx fluid (Vinson, 1972, 1977; Norton and Vinson, 1977; Dahlman and Vinson, 1977; Vinson *et al.*, 1979); we now know that the biological activity of crude calyx fluid can, in many cases, be duplicated using gradient-purified polydnavirus particles. Studies carried out in two different laboratories have clearly demonstrated that biological activity can be abrogated by covalent cross-linking of encapsidated viral genomes *in situ* (Cook *et al.*, 1984; Guzo and Stoltz, 1985, 1987; Stoltz and Guzo, 1986; Beckage *et al.*, 1987, 1990). This suggests that biological activity is for the most part not due to some cytotoxic structural components carried into the host by invading virus particles. Instead, activity appears to be dependent upon the presence of a viable viral genome. Studies clearly indicate that circular viral DNAs persist throughout the natural course of parasitization (Stoltz *et al.*, 1986; Theilmann and Summers, 1986) and are transcriptionally active (Fleming *et al.*, 1983; Blissard *et al.*, 1986, 1987, 1989; Theilmann and Summers, 1988; Stoltz *et al.*, 1988; Webb and Summers, 1990) in one or more host tissues. Thus, there is at least indirect evidence to suggest that transcription—whether encoded or induced by polydnavirus—is a probable requirement for successful parasitism (Guzo and Stoltz, 1987).

In short, the behavior of polydnavirus DNA in host animals has all the earmarks of a genetic colonization. The term *genetic colonization* has a particular meaning in biology, since it carries the implication that the process is directed toward the accomplishment of some goal that is of benefit to the colonizing (parasitic) entity. A cursory review of the biological activities now attributed to polydnaviruses (Table 1) will suggest several arenas in which polydnaviruses could have crucial significance for successful parasitism. The most obvious of these, of course, is immunosuppression.

1. Immunosuppression

Insect larvae are clearly quite capable of mounting effective immune responses against large foreign objects. Parasitoid eggs and larvae, then, must either avoid such a response, or suppress it, or have it suppressed by material(s) injected by the female parasitoid during oviposition. In the context of polydnavirus biology, the experimental approach has been largely aimed at demonstrating virus-induced immunosuppression, particularly during the early stages of parasitization. Suppression of cellular immunity (encapsulation and nodulation) using purified ichnovirus was first described by Edson *et al.* (1981), and subsequently by Stoltz and Guzo (1986); in each instance, immunosuppression was in part attributed to an inhibition of plasmatocyte adhesive and spreading capacity (Stoltz and Guzo, 1986; Davies *et al.*, 1987), although depletion in the numbers of circulating plasmatocytes was also noted in one case (CsV in *Heliothis virescens*). As mentioned earlier, changes in

Table 1
Physiological Changes in Host Animals Attributed to the Presence of Polydnavirus

Activity	References[a]
Suppression of cellular immune response	Vinson (1977); Edson *et al.* (1981) (V); Guzo and Stoltz (1985, 1987); Stoltz and Guzo (1986) (V); Vinson and Stoltz (1986) (V); Tanaka (1987a); Strand and Noda (1991) (V).
Inhibition of weight gain	Vinson (1972); Vinson *et al.* (1979) (V).
Changes in hemocyte count or behavior	Stoltz and Guzo (1986); Guzo and Stoltz (1987); Tanaka (1987b); Davies *et al.* (1987) (V); Wago and Tanaka (1989); Strand and Noda (1991) (V); D. B. Stoltz and N. E. Beckage (V; unpublished data); Stoltz, this report.
Appearance of new hemolymph polypeptides	Cook *et al.* (1984) (V); Beckage *et al.* (1987).
Inhibition of phenoloxidase activity	Stoltz and Cook (1983) (V); Beckage *et al.* (1990) (V); Strand and Noda (1991) (V).
Inhibition of protein storage in fat body	Tanaka (1986)
Reduction in hemolymph viscosity	Davies *et al.* (1987) (V); Strand and Noda (1991); D. B. Stoltz (unpublished data).
Change in hemolymph trehalose concentration	Dahlman and Vinson (1977)
Degeneration of hemopoietic tissue	Guzo and Stoltz (1987)
Pigmentation changes	Beckage *et al.* (1990) (V).
Degeneration of the prothoracic gland	Dover *et al.* (1988a) (V).
Prolongation or arrest of development	Tanaka *et al.* (1987); Dover *et al.* (1988b) (V); Tanaka and Vinson (1991a); K. W. Schleifer and N. E. Beckage (unpublished data)
Perturbation of hormone levels	Dover *et al.* (1987, 1988b) (V); Tanaka *et al.* (1987); Tanaka and Vinson (1991b)

[a]V = gradient-purified virus used (otherwise, calyx fluid). Note that venom is required for full activity of some braconid polydnaviruses.

plasmatocyte behavior are likely not due to a direct cytotoxic effect caused by virus: CsV added to hemocyte preparations, for example, had no effect on plasmatocyte adhesion and spreading, whereas plasma from parasitized *H. virescens* larvae did (Davies *et al.*, 1987).

Considerably less information is available on the effects of purified *bracoviruses*; however, an inhibition of plasmatocyte spreading has now been demonstrated in two different systems (Strand and Noda, 1991; D. B. Stoltz and N. E. Beckage, unpublished). Certainly, calyx fluid—comprising primarily polydnavirus particles—seems to be required for successful parasitism in all of the braconid parasitoid/host systems that have been studied to date (Vinson, 1974, 1977; Guzo and Stoltz, 1985, 1987; Kitano, 1986; Tanaka, 1987a; N. E. Beckage, (personal communication). As with ichnovirus/host systems, immunosuppression is associated with obvious hematological changes (Guzo and Stoltz, 1987; Tanaka, 1987b; Wago and Tanaka, 1989; D. B. Stoltz and N. E. Beckage, unpublished), including destruction of hemopoietic tissue (Guzo and Stoltz, 1987); in one case, a direct effect of calyx fluid/venom on host hemocytes has been observed (Wago and Tanaka, 1989). A requirement for venom (for successful parasitism) complicates the picture in most (Kitano, 1982, 1986; Guzo and Stoltz, 1985, 1987; Tanaka, 1987a) but not all (Vinson, 1977; Beckage *et al.*, 1987, 1990) bracovirus/host systems (see the following).

Immunosuppression induced by parasitoids might logically be expected to reduce resistance to microbial infection in host larvae, and this has now been documented (Stoltz and Guzo, 1986; Guzo and Stoltz, 1987). In the long term, then, a generalized immunosuppression would be lethal to parasitoid as well as host unless either (a) some degree of cellular immunity returned or (b) other defense mechanisms (e.g., antibacterial proteins) were either capable of making up for that loss or else were in fact augmented. As yet, very few studies have been directed toward answering this question (but see Ross and Dunn, 1989). In the case of the ichnovirus HfV (from *Hyposoter fugitivus*), which suppresses cellular immunity in *Malacosoma disstria,* a degree of cellular immunity returns in 1 to 2 days after hatching of the parasitoid egg and the resurgent encapsulation response appears to be selective, since parasitoid larvae remain unaffected. Similar examples of this type of interaction may be found elsewhere in the literature (e.g., Vinson, 1972). It is of course possible that the surfaces of parasitoid larvae are in some instances not recognized as foreign, in which case a complete recovery of cellular immune responses could in theory be tolerated by the parasitoid.

Particularly in the case of bracovirus/host systems, venom gland secretions—in addition to polydnavirus—may be required for successful parasitism. There is some question, however, as to whether the role of venom in such cases should best be described as primary or accessory. Most studies tend to suggest an accessory function. First, venom alone has no obvious biological activity, at least in the systems examined to date; conversely, calyx fluid or polydnaviruses are on their own fully active in a variety of situations. This is especially true in the case of the ichnoviruses (Edson *et al.*, 1981; Stoltz and Guzo, 1986), but holds as well for at least two well-studied bracoviruses (Vinson, 1977; Beckage *et al.*, 1987, 1990). In the case of the bracovirus CmV (from *Cotesia melanoscela*), venom appears to act at an early stage of virus penetration (Stoltz *et al.*, 1988), promoting the uncoating of viral DNA and subsequent genetic colonization. Though necessary, this function is probably best interpreted as providing support for the more important tasks mediated by the activity of polydnavirus DNA.

Obviously, very little is as yet known concerning the biological role of parasitoid venom gland secretions, and it is therefore likely that as more information is developed, we will find that these products are necessary in ways as yet unimagined. This argument is supported by the recent finding that the CsV genome appears to have acquired a parasitoid venom gland gene (Webb and Summers, 1990); this gene would of course be carried into parasitized animals by CsV particles injected during oviposition. There would in this case be two potential sources for venom: the parasitoid venom gland itself, and gene products specified by the encapsidated form of viral DNA.

It is still not clear that melanization plays a pivotal role in insect defense responses. Studies with *Drosophila* mutants suggest that the melanization and hardening of hemocytic capsules may be secondary to the initial non-self recognition event (Rizki and Rizki, 1990). Some polydnaviruses are known to inhibit phenoloxidase activity (Stoltz and Cook, 1983; Beckage *et al.*, 1990), whereas others apparently do not (Davies *et al.*, 1987). Interesting, calyx fluid from *Apanteles kariyai promotes* melanization (of fat body cells; Tanaka, 1986)! At this juncture, it would seem reasonable to suspect that an inhibition of phenoloxidase activity may be required for successful parasitism in some biological systems, but not others. As with phenoloxidase inhibition, a number of other polydnavirus-induced changes are of intrinsic interest in their own regard, and may (or may not) causally relate to immunosuppression. These include the appearance of major new polypeptides (Cook *et al.*, 1984; Beckage *et al.*, 1987, 1990) and a reduction in hemolymph viscosity (Davies *et al.*, 1987; Strand and Noda, 1991; D. B. Stoltz, unpublished data).

Immunosuppression provides perhaps the most obvious and readily understood linkage between the two forms of polydnavirus DNA. Survival of the *linear*, chromosomally integrated genome—assumed to be carried within parasitoid eggs and larvae—depends upon the presence in host tissues of transcriptionally active, *circular* viral DNA. However, as described in the following, immunosuppression is by no means the only necessary function carried out by the latter.

2. Developmental Regulation

It is of course axiomatic that the life cycles of endoparasitoids must be well adapted to those of the hosts within which they reside. In addition to a potentially lethal immune response, the endoparasitic life stages must presumably have to contend with other aspects of host physiology that may be equally threatening; of these, the onset of metamorphosis could represent a serious threat to successful parasitism. For example, the physiology of a pupal host will be expected to differ greatly from that of a larval host; a parasitoid larva developing within a host larva would presumably have to make major adjustments in response to host pupation (perhaps by pupating itself). It is also worth noting that the pupal cuticle is, in general, likely to be much tougher than the larval cuticle, and hence more difficult to penetrate by the emerging parasitoid.

Although many parasitoids have evolved in such a way that their eggs and larvae can accommodate exposure to two different host physiologies (e.g., egg/larval parasitoids), others have developed strategies that allow them to complete development within a single type of host milieu (larval parasitoids). It has been argued that this requires that the parasitoid have the capacity to modulate, or regulate, host physiology (Vinson and Iwantsch, 1980). Since the transition from larva to pupa is under hormonal control, we may reasonably expect that many parasitoids have developed the capacity to affect the endocrine physiology of the host. The literature is in fact replete with examples of host/parasitoid systems in which the metamorphosis of host larvae is either delayed, accelerated, or arrested; this literature has recently been reviewed by Beckage (1985) and Lawrence (1986) and will not be treated further here. I will instead confine myself to a consideration of what is known concerning the role of *polydnaviruses* in the regulation of host endocrine systems. This discussion will be necessarily brief, since relatively few studies have as yet targeted the polydnaviruses as potential mediators of developmental modulation.

Polydnaviruses could in theory affect host development in a number of ways. For example, prolongation of the larval stage could be caused by an increase in juvenile hormone titer or a decrease in ecdysone levels; the former could be caused by a decrease in juvenile hormone (JH) esterase activity (e.g., Hayakawa, 1990), and the latter by an inhibition of prothoracicotropic hormone (PTTH) release or by a direct effect on the prothoracic gland. There are other possibilities, most of which have in fact been associated with endoparasitism in one system or another. Evidence for the participation of calyx fluids in hormonal perturbation was reported for both a bracovirus and an ichnovirus in 1987 (Dover *et al.*, 1987; Tanaka *et al.*, 1987). Taking these in turn, calyx fluid and venom from *A. kariyai* were shown to inhibit an increase in ecdysteroid levels required for pupation, thus prolonging the final instar in *Pseudaletia separata* host larvae. Since injection of PTTH induced pupation, it was assumed that the prothoracic glands of parasitized larvae were intact; it was also assumed that the prothoracic gland of calyx fluid/venom-injected larvae remained intact too, although PTTH injections were not specifically carried out using these larvae. Injections of either calyx fluid or venom alone had no effect. It should be noted that in this system, calyx fluid/venom merely delay pupation, whereas naturally parasitized larvae cannot pupate. Thus, other factors, possibly including teratocytes (Zhang and Dahlman, 1989), must be involved in preventing pupation. More recently, calyx fluid in combination with venom from the braconid parasitoid *Cardiochiles nigriceps* was shown to depress prothoracic gland function in *Heliothis virescens* (Tanaka and Vinson, 1991a), and inhibition of ecdysteroid release in response to PTTH was observed following *in vitro* incubation of prothoracic glands with calyx and venom fluids, suggesting a *direct* effect on this tissue. How this might occur remains to be elucidated; penetration of the prothoracic gland by virus particles has not been observed. It should be noted that the effect of venom in this system is apparently host (i.e., parasitoid) specific: a combination of *Microplitis croceipes* venom with *C. nicriceps* calyx fluid had no effect on host pupation (Tanaka and Vinson, 1991b). Calyx fluid from the ichneumonid wasp *Campoletis sonorensis* has also been shown to induce a decrease in ecdysteroid levels and developmental arrest in last-instar *H. virescens* larvae (Dover *et al.*, 1987). In this case, only the particulate fraction of calyx fluid was active, suggesting that CsV virions were involved; as with other ichneumonid/host systems, there was no apparent requirement for wasp venom. Ligation experiments suggested that the target tissue resided in the thorax, which in turn suggested that calyx fluid might be exerting some effect on the prothoracic gland itself.

Subsequent work using purified CsV has now clearly established that this polydnavirus can induce developmental arrest in *H. virescens* larvae (Dover *et al.*, 1988a,b). Furthermore, it appears that this is caused by a virus-induced degeneration of the prothoracic gland (Dover *et al.*, 1988a), which may in turn account for an observed reduction in ecdysteroid titer (Dover *et al.*, 1988b). It should be noticed that the effect is dose-dependent: injection of lower doses of CsV resulted in the degeneration of some prothoracic gland cells, but not all, and developmental arrest was replaced by a prolongation of the last larval instar.

The effect of polydnavirus on the prothoracic gland is probably not direct, since there is no evidence to suggest that virus particles can invade this tissue, which in turn raises the very important question of how some cells are able to apparently escape being damaged. To my knowledge, only one other attempt has been made to associate purified polydnavirus with an endocrine function: in this case, a bracovirus (CcV) from *Cotesia congregata* has been shown to delay the development of *Manduca sexta* larvae (N. E. Beckage, personal communication), suggesting that CcV activity can apparently be expressed in the absence of wasp venom.

It can be predicted that a number of additional polydnaviruses will in the near future be examined in terms of their potential for influencing host endocrine systems, and it can confidently be suggested that many, if not most, will be found to do so. It should also be kept in mind that other virus-induced changes in host physiology, perhaps including immunosuppression, may well have a hormonal basis. It is, for example, well known that striking changes in hemocyte morphology and function can occur during molting or metamorphosis (e.g., Kurata *et al.*, 1989), presumably in response to changes in the levels of one or more hormones.

3. Changes in Nutrient Levels and/or Accessibility

Changes in nutrient levels of various tissues have on several occasions been attributed to the activity of calyx fluid or polydnaviruses. For example, calyx fluid from the braconid *Microplitis croceipes* is known to elevate hemolymph trehalose levels in *H. virescens* larvae; venom is not required (Dahlman and Vinson, 1977). It is of interest to note that trehalose is readily absorbed by *M. croceipes* larvae (Edson and Vinson, 1977). Particularly in the case of hemolymph feeders, it may be important to divert nutrients away from tissues in which they might otherwise be stored, and hence be unavailable to the developing parasitoid. *Pseudaletia separata* larvae parasitized by *A. kariyai* may be a case in point: the fat body tissue of parasitized larvae does not accumulate the protein storage granules characteristically observed in normal larvae. Inhibition of protein deposition in this tissue can also be demonstrated following injection of calyx fluid and venom (Tanaka, 1986). Another possible example of nutrient diversion may be presented by the phenomenon of parasitized-induced castration of host larvae (Reed-Larsen and Brown, 1990); conceivably, parasitoid larvae may have to compete with host gonads for nutrients. It has been suggested—but as yet not demonstrated—that polydnaviruses could be involved in this process of premature gonad atrophy (Junnikkala, 1985).

The scope of biological activities that can be attributed to polydnaviruses will likely be found, ultimately, to be much broader than is indicated by the list provided in Table 1. However, viewed more closely, some of these seemingly disparate phenomena could well comprise a cascade of related events, precipitated by polydnavirus acting upon some primary target. These unusual viruses should be seen as extremely useful tools that, used judiciously, should tell us much about normal host physiology, and how it may be manipulated, if need be, to our advantage. In the following section, I provide a brief overview of how some aspects of polydnavirus research may be explored in the not-too-distant future.

III. Future Directions

In recent years, polydnavirus research has made significant progress on a number of important issues, and the establishment of a firm base of knowledge upon which to build further; development of this exciting field seems likely to advance rapidly. In addition, with the demonstration that purified virus has biological activity in a variety of contexts, it would seem likely that a number of laboratories having no interest in polydnaviruses per se may nonetheless come to perceive them as valuable tools for studying insect physiology. Clearly, given the complexity of polydnavirus/parasitoid/host systems, a multifaceted approach will continue to be productive and should be encouraged wherever possible. In any case, we are now at a stage where it has become at least possible to formulate, and to some extent address, a number of fundamental questions. A few of these are considered next.

1. Is the Polydnavirus Genome a Genetic Mosaic?

Chromosomal transmission of polydnavirus genomes naturally raises the question of whether these entities are indeed viruses (they could just as readily be referred to as, say, nuclear secretions). This in turn raises the question of whether it would have been easier for a parasitoid to evolve a new "virus" or, alternatively, co-opt the services of one already in existence. Such questions are by no means idle ones, and indeed the polydnavirus phenomenon requires that some thought be given as to what the word "virus" really means. In this regard, it is of interest to note that definitions for "virus" are both scarce and, perhaps fortunately, somewhat nebulous; I therefore offer the following:

> **virus:** infectious entity whose genome, consisting of either DNA or RNA, is minimally packaged within a protein coat designed to facilitate entry into a susceptible host cell (we could add that lacking ribosomes, all viruses must necessarily replicate within cellular hosts).

The word "infection" is used here in the broadest possible sense, so as to include such phenomena as oncogenesis, genetic colonization, and so on; the term "susceptible" does not necessarily imply that a productive replication cycle will ensue. In fact, it matters little whether one thinks of polydnaviruses as being viruses or nuclear secretions, since either designation would appear to be legitimate. Given that, is the question being addressed in this section really itself legitimate? Put another way, is it in fact possible to distinguish between cellular and viral genes in the polydnavirus context? (The reader should keep in mind that all viral genes must originally have been cellular.)

To the question of whether there are classical viral genes within the polydnavirus repertoire, the answer must be: yes—provided only that our definition for "virus" is reasonable. For example, polydnavirus particles demonstrably possess protein coats that are structurally and functionally analogous to viral capsids (Stoltz and Vinson, 1979a). There is indeed no reason not to refer to them as capsids, whether or not the genes encoding them are thought of as cellular or viral in terms of origin. However, moving beyond a consideration of those genes involved in particle morphogenesis, there is at present no easy way to define other patently viral genes in the polydnavirus context. This, however, should not preclude reasoned speculation, and to that end I will specifically suggest that the polydnavirus genome will eventually be shown to include not only viral and parasitoid genes, but traces of host genes as well (i.e., from the parasitized host); presumably, all three of these entities have coevolved over millions of years. A number of hypotheses relevant to this issue may be subject to scientific inquiry, the results of which could be of considerable interest, regardless of whether or not the hypotheses themselves prove valid:

(a) If the polydnaviruses are derived from classical viruses, then strategies employed by the latter to modulate host physiology (e.g., the baculovirus ecdysteroid glucosyl transferase; O'Reilly and Miller, 1989) may well be represented within polydnavirus genomes. Such genes, if present, will probably turn out to have been assimilated from the host organism. It is of interest in this regard to note that a polydnaviruslike particle found in *Venturia canescens* exhibits immunological cross-reactivity with one or more host (*Ephestia kuhniella*) proteins, including hemolin, a known antibacterial protein (Berg *et al.*, 1988; Schmidt and Theopold, 1991).

(b) Similarly, genes employed by parasitoids to regulate host physiology may now have acquired polydnavirus excision/encapsidation signals. Such genes could in theory represent duplicate copies of coexisting parasitoid genes. An interesting example of this may already have been discovered; specifically, Webb and Summers (1990) propose that a parasitoid venom gland gene is encapsidated within CsV particles. Presumably, continuous expression of this gene in host larvae would be of benefit to the developing parasitoid.

(c) Conversely, genes no longer required in (parasitized) animals, in which polydnaviruses do not replicate, may have lost the capacity to excise from the parasitoid genome; prime candidates here would be genes requiring expression only in the parasitoid ovary (e.g., those encoding virion structural proteins). Still, it is of interest to note that some CsV genes not expressed in host larvae are nevertheless encapsidated (Theilmann and Summers, 1988; Blissard *et al.*, 1989). The significance of such observations, while undoubted, remains to be clarified.

(d) If the polydnaviruses are phylogenetically related to one or more "classical" viruses, then it may be useful in this context to examine viruses having similar morphology. It has been argued previously that the baculoviruses bear at least a superficial resemblance to the bracoviruses (Stoltz and Vinson, 1979a). It is therefore conceivable that genes determining the structure/packaging of bracovirus nucleocapsids may share homology with cognate baculovirus genes. It will be of particular interest in this regard to examine an unusual baculovirus that replicates in the ovary of the ichneumonid parasitoid *Mesoleius tenthredinis* (Stoltz, 1981).

2. What Are the Functional Roles of Polydnavirus Gene Products?

This section should of course be prefaced with a cautionary note to the effect that what are referred to here as polydnavirus functions can just as legitimately be considered to represent parasitoid functions. As discussed in the previous section, there is no easy distinction to be made here (perhaps, conceptually, none need be made). In any case, in keeping with the hypothesis that polydnaviruses are phylogenetically related to classical viruses, I will assume here—as I have earlier—that there are indeed polydnavirus-specific genes. Like other viruses, polydnaviruses presumably code for both structural and regulatory proteins; the former represent components of assembled virions, whereas the latter are typically required for replicative functions. However, unlike other viruses, polydnaviruses are likely to require *two* types of regulatory proteins: those operating in the parasitoid (e.g., DNA replication, etc.) and those expressed in the host (functions required for genetic colonization). The identification of at least some major structural proteins can be readily accomplished using techniques that are now standard in many laboratories; mapping these to particular genome segments should prove relatively straightforward. The identification of regulatory gene products required for replication may prove to be a more demanding task, especially since the usual requirements for studying replication—for example, an ability to grow the virus *in vitro*—are not met. Thus, progress here may occur more slowly. On the other hand, rapid progress in identifying proteins expressed in host animals should be expected. Indeed, preliminary work has already been reported (Blissard *et al.*, 1989); in addition, *in vitro* systems (e.g., Stoltz *et al.*, 1988) could be exploited for this purpose.

What remains to be answered is the very important question of determining what functions viral gene products are performing in the parasitized host. In some cases, obvious approaches will be suggested by the appearance of one or more major new polypeptides (e.g., Cook *et al.*, 1984; Beckage *et al.*, 1987); these presumably, but not necessarily, will be encoded by polydnavirus genes. It may be possible to purify the relevant protein and then use it directly in biological experiments: Will injection of such a protein, for example, mimic some known polydnavirus-induced change in host physiology? In most cases, however, novel methodologies may be required in order to correlate biological activity with a particular polydnavirus genome segment. Possible approaches here might include the use of nonlytic baculovirus expression vectors carrying individual polydnavirus genes or genome segments. It might also be possible to transfect newly parasitized host larvae using anti-sense constructs designed to abrogate the expression of genes required for successful parasitism; alternatively, similar constructs could be delivered into host eggs by microprojectile bombardment. Antisera could be raised against virus-specific gene products identified using cDNA expression libraries, and then assessed in terms of their ability to interfere with polydnavirus-induced activity in the host animal (Webb and Summers, 1990). None of this will be particularly easy, but the rewards may be considerable: polydnaviruses may encode a number of products having potential for practical application in the biopesticide industry.

There are of course a number of additional questions that relate to virus-specific expression in the parasitized host. We as yet know nothing about the tissue specificity, if any, of viral gene products affecting host physiology. Nor do we know which gene products promote the development of the parasitoid egg and which the larva. Host specificity, in terms of a requirement for polydnaviruses in successful parasitism, is also a largely unexplored area; of potential interest here is the observation that HfV DNA persists in *Lymantria dispar* larvae, but parasitism of that host is nevertheless unsuccessful (Stoltz *et al.*, 1986).

3. Why Is the Polydnavirus Genome Segmented?

This is one of the more interesting questions concerning the polydnaviruses and has yet to be addressed. Segmentation will perhaps ultimately be found to have something to do with the origin of these viruses; purely mechanistic explanations (e.g., segmentation could facilitate excision, or reduce mutational loads) seem overly simplistic. It could be argued that if polydnaviruses arose from the parasitoid genome, then their genes would most likely already have been separated from each other, in which case excision/encapsidation simply serves as a means to bring them together as a functioning whole. Alternatively, if one assumes that polydnaviruses are related to classical viruses, then it becomes reasonable to suppose that segmentation must have conferred some as yet unrecognized benefit to the establishment and continued evolution of polydnavirus/parasitoid/host complexes. Perhaps the polydnaviruses could have arisen from ancestral defective interfering particles; DI particles encapsidate subgenomic molecules and in addition may reduce the pathogenicity of the parental virion, which could well have been a prerequisite for the acquisition of polydnavirus progenitors by parasitoids. At present, it is difficult to envision how this issue might be addressed experimentally.

In the foregoing discussions, I have considered only a few of the many interesting questions facing students of polydnavirus biology. My exploration has been largely and necessarily speculative in nature since, in the absence of definitive information bearing on these questions, perhaps the best that can be hoped for is a set of working hypotheses. It is the author's hope that some of these will prove sufficiently attractive to stimulate further dialogue, and perhaps even some directed research.

Acknowledgments

I wish to acknowledge the contributions of a number of individuals who have made significant contributions over the years in the capacity of either graduate student (Peter Krell, David Guzo, Deming Xu) or research assistant (Doug Cook, Elizabeth Belland). Original research from my laboratory was funded in part by the Medical Research Council and in part by the Canadian Forestry Service.

References

Beckage, N. E. (1985). Endocrine interactions between endoparasitic insects and their hosts. *Annu. Rev. Entomol.* **30:**371–413.

Beckage, N. E., Templeton, T. J., Nielsen, B. D., Cook, D. I., and Stoltz, D. B. (1987). Parasitism-induced hemolymph polypeptides in *Manduca sexta* (L.) larvae parasitized by the braconid wasp *Cotesia congregata* (Say). *Insect Biochem.* **17:**439–455.

Beckage, N. E., Metcalf, J. S., Nesbit, D. J., Schleifer, K. W., Zetlan, S. R., and De Buron, I. (1990). Host hemolymph monophenoloxidase activity in parasitized *Manduca sexta* larvae and evidence for inhibition by wasp polydnavirus. *Insect Biochem.* **20:**285–294.

Berg, R., Schuchmann-Feddersen, I., and Schmidt, O. (1988). Bacterial infection induces a moth protein which has antigenic similarity to virus-like particle proteins of a parasitoid wasp. *J. Insect Physiol.* **34:**473–480.

Blissard, G. W., Vinson, S. B., and Summers, M. D. (1986). Identification, mapping, and *in vitro* translation of *Campoletis sonorensis* virus mRNAs from parasitized *Heliothis virescens* larvae. *J. Virol.* **57:**318–327.

Blissard, G. W., Smith, O. P., and Summers, M. D. (1987). Two related viral genes are located on a single superhelical DNA segment of the multipartite *Campoletis sonorensis* virus genome. *Virology* **160:**120–134.

Blissard, G. W., Theilmann, D. A., and Summers, M. D. (1989). Segment W of *Campoletis sonorensis* virus: Expression, gene products, and organization. *Virology* **169:**78–89.

Cook, D. I., Stoltz, D. B., and Vinson, S. B. (1984). Induction of a new haemolymph glycoprotein in larvae of permissive hosts parasitized by *Campoletis sonorensis*. *Insect Biochem.* **14:**45–50.

Dahlman, D. L., and Vinson, S. B. (1977). Effect of calyx fluid from an insect parasitoid on host hemolymph dry weight and trehalose content. *J. Invertebr. Pathol.* **29:**227–229.

Davies, D. H., Strand, M. R., and Vinson, S. B. (1987). Changes in differential haemocyte count and *in vitro* behaviour of plasmatocytes from host *Heliothis virescens* caused by *Campoletis sonorensis* polydnavirus. *J. Insect Physiol.* **33:**143–153.

Dover, B. A., Davies, D. H., Strand, M. R., Gray, R. S., Keeley, L. L., and Vinson, S. B. (1987). Ecdysteroid-titre reduction and developmental arrest of last-instar *Heliothis virescens* larvae by calyx fluid from the parasitoid *Campoletis sonorensis*. *J. Insect Physiol.* **33:**333–338.

Dover, B. A., Davies, D. H., and Vinson, S. B. (1988a). Degeneration of last instar *Heliothis virescens* prothoracic glands by *Campoletis sonorensis* polydnavirus. *J. Invertebr. Pathol.* **51:**80–91.

Dover, B. A., Davies, D. H., and Vinson, S. B. (1988b). Dose-dependent influence of *Campoletis sonorensis* polydnavirus on the development and ecdysteroid titers of last-instar *Heliothis virescens* larvae. *Arch. Insect Biochem. Physiol.* **8:**113–126.

Dover, B. A., Strand, M. R., Davies, D. H., and Vinson, S. B. (1989). Ultrastructure of host tissues exposed to the calyx fluid of the parasitoid, *Campoletis sonorensis* (Cameron) (Hymenoptera: Ichneumonidae). *Int. J. Insect Morphol. Embryol.* **18:**47–57.

Edson, K. M., and Vinson, S. B. (1977). Nutrient absorption by the anal vesicle of the braconid wasp, *Microplitis croceipes*. *J. Insect Physiol.* **23:**5–8.

Edson, K. M., Vinson, S. B., Stoltz, D. B., and Summers, M. D. (1981). Virus in a parasitoid wasp: Suppression of the cellular immune response in the parasitoid's host. *Science* **211:**582–583.

Fleming, J.G.W., and Summers, M. D. (1986). *Campoletis sonorensis* endoparasitic wasps contain forms of *C. sonorensis* virus DNA suggestive of integrated and extrachromosomal polydnavirus DNAs. *J. Virol.* **57:**552–562.

Fleming, J.G.W., and Summers, M. D. (1990). The integration of the genome of a segmented DNA virus in the host insect's genome. *In* "Molecular Insect Science" (H. H. Hagedorn, J. G. Hildebrand, M. G. Kidwell, and J. H. Law, eds.). pp. 99–105. Plenum, New York.

Fleming. J.G.W., Blissard, G. W., Summers, M. D., and Vinson, S. B. (1983). Expression of *Campoletis sonorensis* virus in the parasitized host, *Heliothis virescens*. *J. Virol.* **48:**74–78.

Guzo, D. (1988). Biological functions, genome organization, and transmission of a polydnavirus associated with the braconid endoparasitoid, *Cotesia melanoscela*. Ph.D. Thesis, Dalhousie University, Halifax.

Guzo, D., and Stoltz, D. B. (1985). Obligatory multiparasitism in the tussock moth, *Orgyia leucostigma*. *Parasitology* **90:**1–10.

Guzo, D., and Stoltz, D. B. (1987). Observations on cellular immunity and parasitism in the tussock moth. *J. Insect Physiol.* **33:**19–31.

Hayakawa, Y. (1990). Juvenile hormone esterase activity repressive factor in the plasma of parasitized insect larvae. *J. Biol. Chem.* **265:**10813–10816.

Junnikkala, E. (1985). Testis development in *Pieris brassicae* parasitized by *Apanteles glomeratus*. *Entomol. Exp. Appl.* **37:**283–288.

Kitano, H. (1982). Effect of the venom of the gregarious parasitoid *Apanteles glomeratus* on its hemocytic encapsulation by the host, *Pieris*. *J. Invertebr. Pathol.* **40:**61–67.

Kitano, H. (1986). The role of *Apanteles glomeratus* venom in the defensive response of its host, *Pieris rapae crucivora*. *J. Insect Physiol.* **32:**369–375.

Krell, P. J., and Stoltz, D. B. (1980). Virus-like particles in the ovary of an ichneumonid wasp: Purification and preliminary characterization. *Virology* **101:**408–418.

Krell, P. J., Summers, M. D., and Vinson, S. B. (1982). Virus with a multipartite superhelical genome from the ichneumonid parasitoid, *Campoletis sonorensis*. *J. Virol.* **43:**859–870.

Kurata, S., Komano, H., and Natori, S. (1989). Dissociation of *Sarcophaga peregrina* (flesh fly) fat body by pupal haemocytes *in vitro*. *J. Insect Physiol.* **35:**559–565.

Lawrence, P. O. (1986). Host–parasite hormonal interactions: An overview. *J. Insect Physiol.* **32:**295–298.

Norton, W. N., and Vinson, S. B. (1977). Encapsulation of a parasitoid egg within its natural host: An ultrastructural investigation. *J. Invertebr. Pathol.* **30:**55–67.

Norton, W. N., and Vinson, S. B. (1983). Correlating the initiation of virus replication with a specific pupal developmental phase of an ichneumonid parasitoid. *Cell Tissue Res.* **231:**387–398.

Norton, W. N., Vinson, S. B., and Stoltz, D. B. (1975). Nuclear secretory particles associated with the calyx cells of the ichneumonid parasitoid *Campoletis sonorensis* (Cameron). *Cell Tissue Res.* **162:**195–208.

O'Reilly, D. R., and Miller, L. K. (1989). A baculovirus blocks insect molting by producing ecdysteroid UDP-glucosyl transferase. *Science* **245:**1110–1112.

Osheim, Y. N., and Miller, O. L., Jr. (1983). Novel amplification and transcriptional activity of chorion genes in *Drosophila melanogaster* follicle cells. *Cell (Cambridge, Mass.)* **33:**543–553.

Reed-Larsen, D. A., and Brown, J. J. (1990). Embryonic castration of the codling moth, *Cydia pomonella,* by an endoparasitoid, *Ascogaster quadridentata. J. Insect Physiol.* **36:**111–118.

Rizki, R. M., and Rizki, T. M. (1990). Encapsulation of parasitoid eggs in phenoloxidase-deficient mutants of *Drosophila melanogaster. J. Insect Physiol.* **36:**523–529.

Ross, D. R., and Dunn, P. E. (1989). Effect of parasitism by *Cotesia congregata* on the susceptibility of *Manduca sexta* larvae to bacterial infection. *Dev. Comp. Immunol.* **13:**205–216.

Schmidt, O., and Theopold, U. (1991). Immune defense and suppression in insects. *BioEssays* **13:**343–346.

Stark, G. R., Debatisse, M., Giulotto, E., and Wahl, G. M. (1989). Recent process in understanding mechanisms of mammalian DNA amplification. *Cell (Cambridge, Mass.)* **57:**901–908.

Stoltz, D. B. (1981). A putative baculovirus in the ichneumonid parasitoid, *Mesoleius tenthredinis. Can. J. Microbiol.* **27:**116–122.

Stoltz, D. B. (1990). Evidence for chromosomal transmission of polydnavirus DNA. *J. Gen. Virol.* **71:**1051–1056.

Stoltz, D. B., and Cook, D. I. (1983). Inhibition of host phenoloxidase activity by parasitoid hymenoptera. *Experientia* **39:**1022–1024.

Stoltz, D. B., and Guzo, D. (1986). Apparent haemocytic transformations associated with parasitoid-induced inhibition of immunity in *Malacosoma disstria* larvae. *J. Insect Physiol.* **32:**377–388.

Stoltz, D. B., and Vinson, S. B. (1977). Baculovirus-like particles in the reproductive tracts of female parasitoid wasps. II. The genus *Apanteles. Can. J. Microbiol.* **23:**28–37.

Stoltz, D. B., and Vinson, S. B. (1979a). Viruses and parasitism in insects. *Adv. Virus Res.* **24:**125–171.

Stoltz, D. B., and Vinson, S. B. (1979b). Penetration into caterpillar cells of virus-like particles injected during oviposition by parasitoid ichneumonid wasps. *Can. J. Microbiol.* **25:**207–216.

Stoltz, D. B., Krell, P., Summers, M. D., and Vinson, S. B. (1984). Polydnaviridae—A proposed family of insect viruses with segmented, double-stranded, circular DNA genomes. *Intervirology* **21:**1–4.

Stoltz, D. B., Guzo, D., and Cook, D. (1986). Studies on polydnavirus transmission. *Virology* **155:**120–131.

Stoltz, D. B., Guzo, D., Belland, E. R., Lucarotti, C. J., and MacKinnon, E. A. (1988). Venom promotes uncoating *in vitro* and persistence *in vivo* of DNA from a braconid polydnavirus. *J. Gen. Virol.* **69:**903–907.

Strand, M. R., and Noda, T. (1991). Alterations in the haemocytes of *Pseudoplusia includens* after parasitism by *Microplitis demolitor. J. Insect Physiol.* **37:**839–850.

Tanaka, T. (1986). Effects of the calyx and venom fluids of *Apanteles kariyai* Watanabe (Hymenoptera:Braconidae) on the fat body and hemolymph protein contents of its host *Pseudaletia separata* (Walker) (Lepidoptera:Noctuidae). *Appl. Ent. Zool.* **21:**220–227.

Tanaka, T. (1987a). Effect of the venom of the endoparasitoid, *Apanteles kariyai* Watanabe, on the cellular defence reaction of the host, *Pseudaletia separata* Walker. *J. Insect Physiol.* **33:**No. 6, 413–420.

Tanaka, T. (1987b). Morphological changes in haemocytes of the host, *Pseudaletia separata,* parasitized by *Microplitis mediator* or *Apanteles kariyai. Dev. Comp. Immunol.* **11:**57–67.

Tanaka, T., and Vinson, S. B. (1991a). Depression of prothoracic gland activity of *Heliothis virescens* by venom and calyx fluids from the parasitoid, *Cardiochiles nicriceps. J. Insect Physiol.* **37:**139–144.

Tanaka, T., and Vinson, S. B. (1991b). Interactions of venoms with the calyx fluids of three parasitoids, *Cardiochiles nigriceps, Microplitis croceipes* (Hymenoptera:Braconidae), and *Cam-*

poletis sonorensis (Hymenoptera:Ichneumonidae) in effecting a delay in the pupation of *Heliothis virescens* (Lepidoptera:Noctuidae). *Ann. Entomol. Soc. Am.* **84:**87–92.

Tanaka, T., Agui, N., and Hiruma, K. (1987). The parasitoid *Apanteles kariyai* inhibits pupation of its host, *Pseudaletia separata,* via disruption of prothoracicotropic hormone release. *Gen. Comp. Endocrinol.* **67:**364–374.

Theilmann, D. A., and Summers, M. D. (1986). Molecular analysis of *Campoletis sonorensis* virus DNA in the lepidopteran host *Heliothis virescens. J. Gen. Virol.* **67:**1961–1969.

Theilmann, D. A., and Summers, M. D. (1988). Identification and comparison of *Campoletis sonorensis* virus transcripts expressed from four genomic segments in the insect hosts *Campoletis sonorensis* and *Heliothis virescens. Virology* **167:**329–341.

Vinson, S. B. (1972). Factors involved in successful attack on *Heliothis virescens* by the parasitoid *Cardiochiles nigriceps. J. Invertebr. Pathol.* **20:**118–123.

Vinson, S. B. (1974). The role of the foreign surface and female parasitoid secretions on the immune response of an insect. *Parasitology* **68:**27–33.

Vinson, S. B. (1977). *Microplitis croceipes:* Inhibition of the *Heliothis zea* defense reaction to *Cardiochiles nigriceps. Exp. Parasitol.* **41:**112–117.

Vinson, S. B., and Iwantsch, G. F. (1980). Host regulation by insect parasitoids. *Q. Rev. Biol.* **55:**143–165.

Vinson S. B., and Stoltz, D. B. (1986). Cross-protection experiments with two parasitoid (Hymenoptera:Ichneumonidae) viruses. *Ann. Entomol. Soc. Am.* **79:**216–218.

Vinson, S. B., Edson, K. M., and Stoltz, D. B. (1979). Effect of a virus associated with the reproductive system of the parasitoid wasp, *Campoletis sonorensis,* on host weight gain. *J. Invertebr. Pathol.* **34:**133–137.

Wago, H., and Tanaka, T. (1989). Synergistic effects of calyx fluid and venom of *Apanteles kariyai* Watanabe (Hymenoptera: Braconidae) on the granular cells of *Pseudaletia separata* Walker (Lepidoptera: Noctuidae). *Zool. Sci.* **6:**691–696.

Webb, B. A., and Summers, M. D. (1990). Venom and viral expression products of the endoparasitic wasp *Campoletis sonorensis* share epitopes and related sequences. *Proc. Natl. Acad. Sci. U.S.A.* **87:**4961–4965.

Webb, B. A., and Summers, M. D. (1992). Stimulation of polydnavirus replication by 20-hydroxyecdysone. *Experientia* **48:**1018–1022.

Xu, D., and Stoltz, D. B. (1991). Evidence for a chromosomal location of polydnavirus DNA in the ichneumonid parasitoid, *Hyposoter fugitivus. J. Virol.* **65:**6693–6704.

Zhang, D., and Dahlman, D. L. (1989). *Microplitis croceipes* teratocytes cause developmental arrest of *Heliothis virescens* larvae. *Arch. Insect Biochem. Physiol.* **12:**51–61.

Chapter 9

Polydnavirus Genome Organization

Jo-Ann G. W. Fleming
Department of Animal Science
Texas A&M University
College Station, Texas

Peter J. Krell
Department of Microbiology
University of Guelph
Guelph, Ontario, Canada

I. Introduction

Polydnaviruses play an unusual role in parasitoid life cycles because the viruses appear to be symbiotically associated with the wasp but induce significant physiological changes in lepidopteran host larvae parasitized by the wasps. These changes have been considered evidence of "host regulation" by the parasitoid or factors derived from it (Vinson and Iwantsch, 1980). Polydnaviruses are implicated as one of the factors that may be needed to secure the successful development of the parasitoid in its host and therefore are believed to be involved in determining the species specificity of host–parasitoid interactions. This aspect of the biology of polydnaviruses is covered in more detail in the preceding chapter by D. Stoltz and in other reviews (Stoltz and Vinson, 1979; Krell, 1991a,b; Fleming, 1992).

The most distinctive characteristic of viruses in the family Polydnaviridae is the unusual organization of the genomes as collections of several supercoiled DNAs (Stoltz *et al.,* 1984; Francki *et al.,* 1991). The structural and functional organization of these complex viral genomes is the primary focus of this review. The genetic relationship of the viruses with the wasp host also will be considered because of the implications that relationship has for virus replication and the evolution of the virus and parasitoid (Fleming, 1991, 1992; Krell, 1991a). The discussion also will include some speculations in the hope that they will stimulate debate and help to identify problems for future investigations of these important viruses.

II. Polydnaviruses as Part of Biological Systems

All polydnaviruses have segmented genomes of multiple circular DNAs (see Section III). Two major subgroups of polydnaviruses, the bracoviruses and the ichnoviruses, are recognized (Francki *et al.,* 1991). The bracoviruses have been isolated from wasps in the family Braconidae, whereas ichnoviruses have been isolated from wasps in the family Ichneumonidae. In addition to differences in their host ranges, these two subgroups can be differentiated on the basis of their significantly different morphologies (Stoltz and Vinson, 1979; Krell and Beveridge, 1987; Krell, 1991a,b). To date the ichnovirus *Campoletis sonorensis* virus (CsV) is the polydnavirus that has been most intensively studied at the molecular level. The remarks in this review are based heavily, but *not* exclusively, on studies of that virus system. Future investigations of other polydnaviruses will be necessary before it will be clear whether the physical and genetic organization of CsV is typical of ichnoviruses and how similar bracoviruses and ichnoviruses are at the molecular level.

Polydnaviruses appear to replicate primarily, and possibly exclusively, in the wasp host (Stoltz and Vinson, 1979). Available Southern and dot blot data suggest that the amount of *Campoletis sonorensis* virus-specific DNA remains at relatively constant levels in parasitized or virus-injected *Heliothis virescens* larvae, one habitual host of *C. sonorensis* wasps (Theilmann and Summers, 1986). In addition, electron microscopic studies have not reported virus particles in parasitized host tissues at later times after oviposition. A significant level of viral replication thus does not appear to occur in the parasitized host, although *in vivo* labeling studies are needed to confirm that low to moderate levels of replication do not occur in one or more tissues of the lepidopteran host. The observations that the virus replicates primarily in the wasp host but causes pathological changes only in the parasitized host were the first indication that polydnaviral genes involved in the different functions might be regulated in a species-specific manner. Available data on CsV gene transcription are consistent with the idea that at least some polydnaviral genes are expressed in a species-specific manner (see Section VI). However, because the functions of the viral gene products are not yet known, it is not clear to what extent particular processes like replication are restricted to only one of the host species of polydnaviruses, that is, wasp or (parasitized) lepidopteran larva.

The gross features of viral replication and morphogenesis have been shown in numerous electron microscopic studies, even though the molecular mechanisms involved are not known (summarized in Stoltz and Vinson, 1979; Krell, 1991a). Viral DNA synthesis and virus maturation begin in the late pharate female wasp and continue in the adult for at least several days after eclosion (Norton *et al.*, 1975; Norton and Vinson, 1983; Fleming and Summers, 1986; Theilmann and Summers, 1986). Viral replication appears to be restricted to the nuclei of epithelial cells in the calyx region of the oviducts (Stoltz and Vinson, 1979). The newly synthesized viral DNAs are packaged in structurally complex nucleocapsids that acquire an envelope from membranes that appear to be free within the nucleus itself (Stoltz and Vinson, 1979). This method of envelopment, which also is seen during baculovirus maturation, has been interpreted as evidence for *de novo* intranuclear membrane synthesis (Stoltz *et al.*, 1973; Federici, 1986). No biochemical evidence to suggest new pathways for membrane biosynthesis in the nucleus exists, but some additional research to determine the origin of the intranuclear membranes may be warranted since this method of envelopment is not generally observed in other animal virus families. The lenticular ichnovirus nucleocapsids are individually enveloped in this membrane and then acquire a second membrane as they bud through the nuclear envelope and the plasmalemma (Norton *et al.*, 1975; Stoltz and Vinson, 1979; Krell and Stoltz, 1980; Krell, 1987; Krell and Beveridge, 1987). In contrast, one or more cylindrical bracovirus nucleocapsids may be enveloped in each membrane found in the nucleus. The bracovirus virions are released without acquiring additional membranes when the cells lyse (Norton *et al.*, 1975; Krell and Stoltz, 1979; Stoltz and Vinson, 1979; Krell and Beveridge, 1987). In both cases, the virions accumulate to very high densities in the lumen of the lateral oviducts. The viruses thus are ideally positioned both temporally and physically for transfer into lepidopteran host larvae when the female wasp oviposits eggs.

III. Multipartite Genome Structure

The unique feature of all polydnaviruses is the multipartite or segmented nature of the supercoiled genome. Although other DNA viruses contain covalently closed circular, double-stranded (ds) DNA (e.g., one DNA molecule per baculovirus) or are segmented (e.g., the two single-stranded, circular

DNAs of the geminivirus subgroup with bipartite genomes), polydnaviruses are unusual in having multiple covalently closed circular, ds DNAs in the genome of a given virus. Viruslike particles reported in some braconid and ichneumonid wasps cannot legitimately be considered polydnaviruses until a multipartite, supercoiled DNA genome has been demonstrated (reviewed in Fleming, 1992; Krell, 1991a). Polydnaviral DNAs, whose covalently closed circular nature has been demonstrated by molecular and biophysical methods, are resolved into supercoiled and relaxed circular fractions by isopycnic centrifugation in cesium chloride–ethidium bromide gradients (Krell *et al.*, 1982; Theilmann and Summers, 1986; Blissard *et al.*, 1987). It is probable that virions contain the superhelical form of the DNAs because the supercoiled form would be expected to be favored thermodynamically and the supercoiled form predominates when nicking due to handling is minimized.

Polydnaviruses from different wasp species differ in terms of segment number, DNA size range, segment molar ratios, and total genome size. Although these characteristics are quite reproducible for a given polydnavirus "species" isolated from a particular wasp host species, the heterogeneity in these characteristics is greater among viruses in the family Polydnaviridae than among viruses in other virus families with segmented genomes (Murphy and Kingsbury, 1990).

The number of DNA segments per genome can be estimated from the number of bands detected in agarose gels, but such estimates should be regarded only as approximations for several reasons. Both superhelical and relaxed circular forms of the DNAs are present in agarose gels and are resolved from one another to varying extents, depending on the range of DNA sizes, agarose concentration, field strength, etc. The two forms often can be differentiated in regions where the resolution is good if the supercoiled and relaxed circular fractions are run in parallel. However, in areas of the gel where the two forms of different DNAs comigrate significantly, this distinction is not always clear-cut. As expected for large circular molecules, the extraction method, storage conditions, and electrophoresis methods can profoundly affect the number of bands resolved and/or the relative proportion of supercoiled molecules and the corresponding relaxed circular forms detectable per sample (Stoltz *et al.*, 1981; Jones *et al.*, 1986). The total number of bands (open circular and supercoiled) detectable in *Hyposoter exiguae* virus, for example, increased from approximately 9 to 30 when the extraction protocol was modified (Stoltz *et al.*, 1981). Similarly, the number of segments in the CsV genome increased from approximately 4 to 25 when larger amounts of material were examined after isolation by more rigorous methods (Stoltz *et al.*, 1981; Krell *et al.*, 1982).

Estimates are further complicated by the fact that the various DNA segments appear to be present in nonequimolar ratios. The DNAs in the CsV genome appear to be present in at least three "classes" based on relative abundance (Fig. 1, and Blissard *et al.*, 1986a; Krell *et al.*, 1982). The molecular basis and biological significance of the nonequimolarity of the genomic segments are not known. DNAs present in more limited amounts may be undetectable under certain experimental conditions. Three additional DNA segments, A^2, L^2, and O^2, were visualized in ethidium bromide-stained gels of the CsV genome after electrophoresis of higher amounts of DNA (Blissard *et al.*, 1986a). More sensitive techniques like autoradiography can result in the detection of other DNAs present in even lower amounts, for example, C^2 in CsV (Theilmann and Summers, 1987). The apparent differences in relative abundance may be partly due to the comigration of two or more unrelated DNAs of the same size. Comigration of unrelated superhelical DNAs can result in the underestimation of the total number of segments and may be undetected except by cloning and/or physical mapping. The CsV band designated DNA G (8.4 kbp) contains two forms that could be distinguished by SalI and HindIII digestion (Krell *et al.*, 1982). These forms may represent unrelated DNAs, but because cloning, detailed physical mapping, and reciprocal hybridization of these two forms of superhelix G were not undertaken, the possibility that the two forms represent two restriction fragment length polymorphisms (RFLPs) on essentially the same DNA cannot be excluded.

Alternatively, the viral DNAs present in apparently "hypermolar" amounts relative to the other segments may be single DNA species that are synthesized more efficiently. Whether the greater abundance of these DNAs is functionally significant is not known, but it is tempting to speculate that the nonequimolarity of the DNAs reflects a gene dosage effect by which the copy number of segments containing essential viral genes for proteins required in high amounts is increased. The apparently "hypermolar" CsV DNA W (15.8 kbp), for example, contains two genes that are expressed in *Heliothis* larvae within 2 hr after parasitization by *C. sonorensis* wasps and that continue to be among the most highly expressed viral genes for about 9 days while the wasp develops as an endoparasite (Blissard *et al.*, 1986b, 1987). If the nonequimolarity is not an artifact, the different relative abundances of the DNAs raise significant questions about how the synthesis of the DNAs is differentially regulated or what packaging signals are involved in their differential encapsidation.

Whether the reported DNA profiles give a representative picture of the genome of a given polydnavirus "species" is also open to question. The amount of viral DNA per female is usually significantly less than 1 μg per pair of lateral oviducts, for example, 150 ng CsV DNA per *C. sonorensis* female (Fleming and Summers, 1986). Many of the reported DNA electrophoretic profiles therefore were determined with viral DNA isolated from several females in a single laboratory population. Because of the difficulty of rearing

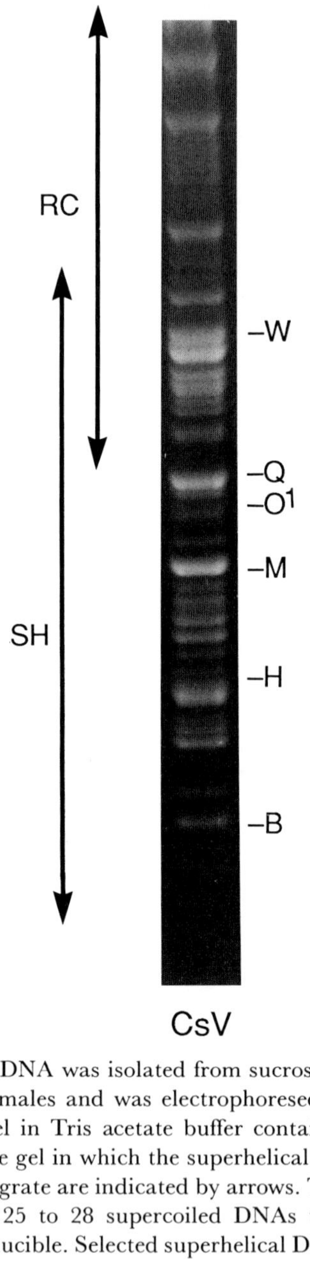

Figure 1 CsV DNA. CsV DNA was isolated from sucrose gradient-purified virions from pooled *C. sonorensis* females and was electrophoresed in its undigested, native form in a 0.7% agarose gel in Tris acetate buffer containing 0.5μg ethidium bromide/ml. The regions of the gel in which the superhelical (SH) and relaxed circular (RC) forms of the DNAs migrate are indicated by arrows. The differences in intensity among the approximately 25 to 28 supercoiled DNAs that migrate in the lower portion of the gel are reproducible. Selected superhelical DNA bands are indicated by letter designations according to the scheme used in the original description of CsV (Krell *et al.*, 1982). Sizes of the selected DNAs (in kilobase pairs) are: B (6.7), H (8.5), M (10.4), O^1 (11.3), Q (12.2), and W (15.8).

the parasitoids, other reported profiles represent the DNA obtained from a single feral female (Stoltz *et al.*, 1981). Apart from the problem of detectability where limited amounts of material were used, the number of insects sampled to produce a DNA profile may be important because genetic variation in the number of segments can occur within a given virus. The number of DNAs in the genomes of the viruses from some geographically isolated populations of *Cotesia melanoscela* (CmV) or from individual females from laboratory colonies of *Hyposoter lymantriae* or *C. sonorensis* varied (Stoltz *et al.*, 1986; Stoltz and Xu, 1990; J. Fleming, unpublished). Although the similarities between viral profiles of isolates from individual females within the same wasp species exceeded the differences, several segments within a virus "species" were variable in some cases (Stoltz and Xu, 1990).

The observed intraspecific variations suggest that the use of viral DNA from pooled wasps may result in an overestimation of the genomic complexity of these viruses whereas use of material from single females may cause one to underestimate the potential complexity. The variations observed in viral DNAs from individual females within a wasp species underscore the apparent plasticity of polydnavirus genomes. The apparent absence of whole viral genomic segments in some individuals within a species also suggests that the viral genes on the missing segment may be nonessential for viability in either the wasp or the parasitized host. Alternatively, if an essential gene were located on a superhelix that is deleted in some strains, a duplicate or closely related gene located on another viral genomic segment might supply the necessary genetic information. The presence of viral gene families in polydnaviruses like CsV and the multipartite structure of the genomes thus could potentially increase the apparent plasticity of polydnavirus genomes with respect to genomic segment size and number.

Interspecific differences in viral DNA profiles generally exceed intraspecific differences. Jones and his associates (1986) present evidence that the polydnaviruses of the congeneric species *Chelonus* near *curvimaculatus* and *C. insularis* can be distinguished by these criteria. The DNA profiles of polydnaviruses from more distantly related wasps in different genera or families are less similar in terms of the numbers and sizes of the DNAs than the viruses of the two *Chelonus* species are. Unlike the limited range of 10–12 ds RNAs per genome of a typical reovirus, for example, different polydnaviruses can have from fewer than 10 (*Hyposoter annulipes* virus) to more than 25 (CsV) circular DNA molecules (Stoltz *et al.*, 1981; Krell *et al.*, 1982; Blissard *et al.*, 1986a). Polydnaviral genomes often have 15 or more segments (Stoltz *et al.*, 1981; Stoltz and Xu, 1990; Xu and Stoltz, 1991). The DNAs tend to be fairly large but can vary within a fairly broad size range. The CsV genome, for example, is composed of DNAs of approximately 5–21 kbp, and larger DNAs appear to be present in the genome of *Cotesia melanoscela* virus (Krell *et al.*, 1982; Blissard *et al.*, 1986a; Stoltz and Xu, 1990).

The variability in viral DNA number, size, and molarity makes it difficult to estimate an aggregate genome size accurately for any polydnavirus. Available estimates are based on the sums of the sizes of the circular DNA molecules or all restriction fragments in digested viral DNA and should be regarded only as preliminary estimates. Sizes can range from 75 kbp for *Heliothis exiguae* virus (HeV) to 250 kbp or more for CsV, indicating that polydnaviral genomes thus tend to be large and vary within a wide range (Krell *et al.*, 1982; Krell, 1991a).

Although differences in the number and sizes of DNAs in polydnavirus genomes historically have been used as a preliminary method of distinguishing the viruses from different wasp species, these criteria clearly are inadequate for phylogenetic comparisons. Both experimental methods and the level of genetic polymorphisms within a polydnavirus "species" (or even a population within a "species") can significantly influence the DNA profile observed. More significantly, these traits do not address the genetic relatedness of the viral genomes in a meaningful way. DNA hybridization and nucleotide sequence data will provide a sounder basis for future comparisons.

IV. Genome Packaging

The DNA profiles of given polydnavirus "species" are quite reproducible, except for the fairly limited variations in DNA segment number noted earlier. This appears to be true whether the viral DNA is obtained from different individual females, wasp generations, or wasp populations (Krell *et al.*, 1982; Stoltz *et al.*, 1986; J. Fleming, unpublished data). The reproducibility suggests that the number and ratio of segments in a virion are regulated at the level of either DNA replication and/or packaging. However, because an electrophoretic profile of a given polydnaviral genome reflects all the DNAs found in a population of virus particles that are present in one (or more) oviducts, it is not clear whether each nucleocapsid (and/or virion) contains all viral DNAs, specific subsets of the genome, random collections of the viral DNAs, or individual DNAs. If each nucleocapsid (or virion) were to contain less than the full complement of viral DNAs, the observed reproducibility would suggest that the ratio of different viral subclasses containing varying numbers of DNAs per particle also is regulated in each female. Precedents for the various possible patterns of packaging exist in other virus families characterized by segmented genomes (Belloncik, 1989; Murphy and Kingsbury, 1990; Matthews, 1991).

Apart from its potential interest from a molecular biological standpoint, the packaging strategy adopted by polydnaviruses could theoretically affect the probability that a particular viral gene is expressed in a given cell of a parasitized host. Available data suggest that polydnaviral genes are segregated among the genomic segments (see Section VI). Assuming that each virion is equally capable of infecting a given host cell, the probability that a particular viral gene is expressed in the cell is theoretically greater if each infecting virion contains the entire viral genome than if each virion contains only a subset of the genome. In the latter case, the probability that a particular gene is expressed in the cell would be proportional to the percentage of the injected inoculum virus particles that contained the genomic segment on which the gene is located. The packaging strategy would be less important from this standpoint if the gene of interest were duplicated on more than one genomic DNA segment, more than a single virion were to infect each cell, or specific viral subclasses with different affinities for receptors on different host cell types were shown to exist.

Little direct experimental work has addressed the problem of polydnaviral genome packaging. The ratio of the nucleocapsid size (volume) to the viral genome size can provide a rough estimate of the proportion of the genome that a typical virus particle could contain. Nucleocapsids of ichnoviruses are of uniform size and tend to be large, for example, 85 nm × 330 nm for CsV (Norton *et al.*, 1975; Stoltz and Vinson, 1979). Comparison of the particle volume : genome size ratios of CsV or HeV with the ratios for baculoviruses suggests that each ichnovirus virion is sufficiently large to encapsidate the entire genome and could possibly contain multiple copies of at least some of the viral DNAs (Stoltz and Vinson, 1979; Krell and Stoltz, 1980; Krell *et al.*, 1982). Attempts to fractionate CsV into putative subclasses containing different subsets of DNAs by sucrose or CsCl gradient ultracentrifugation were unsuccessful (Krell *et al.*, 1982). Thus, either subclasses of virions containing different complements of viral DNA do not exist for CsV or these classes have densities that are too similar to each other to be separated by the techniques used.

Bracoviruses have rod-shaped nucleocapsids with fixed diameters but variable lengths, but whether they are multicomponent viruses in which different portions of the genome are distributed among different nucleocapsids that are physically distinct from one another is not clear. Whether the larger nucleocapsids could contain the entire genome has not been determined. If one compares a histogram of the frequency of different-sized nucleocapsids in CmV with the frequency of the different sizes of circular CmV DNAs (as determined by electron microscopy of Kleinschmidt spreads), the two histograms are very similar (Krell and Stoltz, 1979). In other words, the proportions of long nucleocapsids and large DNAs, for example, were similar. Suggestive as these data are, more detailed biochemical and electron microscopic studies are required to determine whether the viral DNAs are singly encapsidated.

If each bracovirus nucleocapsid eventually were shown to contain only one or a few DNAs, the portion of the viral genome present in any infected lepidopteran host cell would then depend on the number of virions that infect the host cell. However, in several bracovirus species, for example, CmV, more than one nucleocapsid is enveloped in each virion (Stoltz and Vinson, 1979). A single infecting virion thus could introduce more than one genomic segment even if each nucleocapsid contains a single DNA. It will be of interest to determine whether bracovirus nucleocapsids are co-enveloped at random or whether specific subsets of nucleocapsids are enveloped in particular virus particles, because a nonrandom envelopment strategy hypothetically would be an efficient mechanism to ensure that specific subsets of viral DNAs would be present in a given infected cell.

V. Intraspecific Genomic Complexity

Preliminary analysis of the genome of HeV by Southern blot hybridization demonstrated that the HeV DNAs present in higher molar amounts did not cross-hybridize with other HeV DNAs under the high stringency conditions used (Krell and Stoltz, 1980). Similar results were reported for the "hypermolar" CsV DNAs (Krell *et al.*, 1982). These studies suggested that the more abundant HeV or CsV DNAs were unrelated to other DNAs in the respective viral genomes and therefore might be unique. However, subsequent Southern blot analyses of CsV with cloned viral DNA probes under similar high stringency conditions consistently detected significant cross-hybridization of the probe DNAs with one or more viral DNAs in addition to the expected hybridization to the superhelical and relaxed circular forms of the segment from which the probe DNAs were cloned (Fleming and Summers, 1986; Blissard *et al.*, 1987; Theilmann and Summers, 1987). Each of the cloned DNA probes reproducibly hybridized with only a subset of the genomic segments. The subsets of hybridizing viral DNAs differed for the various probes tested. None of the probes tested hybridized with all genomic segments or detected a "nested set" of segments, that is, a probe that hybridized with a given superhelix did not hybridize with all successively larger segments (Blissard *et al.*, 1986b, 1980; Fleming and Summers, 1986; Theilmann and Summers, 1987, 1988). The hybridization patterns argue *against* CsV being a defective virus. The data support the interpretation that the multiple DNAs in a polydnavirus reflect a segmented genome.

Part of the apparent cross-hybridization may reflect multiple relaxed circular forms of a given circular molecule, as has been suggested for CsV DNA B (Theilmann and Summers, 1987). More importantly, several gene families exist in the CsV genome (see Section VI), and at least in the case of the 540-bp repeat element gene family, the members have been shown unambiguously to exist on more than one superhelical DNA (Blissard *et al.,* 1986b, 1987; Theilmann and Summers, 1987). Segregation of the gene family members among the genomic segments suggests that fairly extensive intragenomic recombination has occurred during the evolution of the virus. A third possible explanation, which has not yet been tested experimentally, may be more applicable to cases where virus from two wasp populations contain slightly different numbers of cross-hybridizing superhelices, for example, DNAs D and G of the (Truro) Nova Scotia and Connecticut strains of CmV (Stoltz *et al.,* 1986). Duplication of a gene or insertion of random viral or wasp sequences into a viral superhelix could theoretically give rise to two or more differently sized genomic segments that have the same sequence apart from the inserted or duplicated fragments. Deletions could similarly result in two closely related molecules of different size.

The genomic complexity of the CsV genome was corroborated by Southern cross-blot hybridization analyses in which each EcoRI fragment hybridized with one or more EcoRI fragments in addition to itself (Theilmann and Summers, 1987). The data are consistent with the existence of gene families and/or repetitive DNA within the viral genome. However, because the origin of each fragment in the (digested) genome is not known, at least some of the cross-hybridizing sequences may represent restriction fragment length polymorphisms, which have been detected in the CsV genome (Fleming *et al.,* 1990; Fleming, 1991; J. Fleming, unpublished data). [RFLPs also have been detected in the genomes of other ichnoviruses and bracoviruses (Stoltz and Xu, 1990).] Because RFLPs reflect the gain or loss of a restriction endonuclease site on different copies of a DNA that exist in a population, a given probe will hybridize with several differently sized fragments, part of which are identical. The existence of RFLPs in a virus population thus could result in an exaggerated estimate of intragenomic complexity in some experimental designs.

VI. Transcription and Gene Families

A. Functional Organization of the Genome

One of the earliest suggestions that polydnavirus genomes are transcriptionally active was the observation that ultraviolet irradiation of *C. sonorensis* "calyx fluid" (the virus and other substances in the lateral oviducts) reduced its biological activity in injected lepidopteran larvae in a dosage-dependent manner (Vinson *et al.*, 1979). To date, transcription has been studied in detail only in the CsV system. Hybridization of cDNAs of the polyadenylated mRNAs from parasitized *H. virescens* larvae or *C. sonorensis* pupae to the CsV genome provided the first proof that polydnavirus genes are transcribed (Fleming *et al.*, 1983). The data indicated that CsV transcription in the parasitized host occurs within 2 hr after the female wasp oviposits eggs and polydnavirus and continues at readily detectable levels for approximately 9 days, when the final *C. sonorensis* instar emerges from the moribund host. Some of the viral gene products are believed to alter the physiology of the parasitized host. These studies also indicated that several, but not all, of the CsV genomic segments contain sequences that are transcribed in the parasitized host and suggested that different viral sequences were expressed in the wasp and the lepidopteran host.

Later investigations significantly extended and refined the early expression studies (Blissard *et al.*, 1986a, 1987, 1989; Theilmann and Summers, 1987; Theilmann and Summers, 1988). At least 12 mRNAs are detectable in *H. virescens* larvae parasitized by *C. sonorensis* wasps or injected with purified CsV, indicating that neither wasp tissue nor other nonviral factors are required for CsV transcription (Blissard *et al.*, 1986b). These experiments also provided the supporting molecular data for the pivotal experiments of Edson and her coworkers (1981), who demonstrated that gradient-purified CsV alone was sufficient to prevent the encapsulation of *C. sonorensis* eggs in host larvae and endoparasitic stages of the wasps. Although additional factors, for example, venoms, also are necessary in some bracovirus systems, the data from the CsV system strongly suggest that the alteration of the physiology of the parasitized host depends primarily on viral gene expression.

The probable location of the genes in the viral genome and the size of the transcripts encoded by particular viral DNA fragments were determined by blot analyses (Blissard *et al.*, 1986a,b). Cloned viral DNAs containing the expressed sequences or the flanking, nontranscribed sequences were used separately as probes with Southern blots of undigested CsV genomic DNA and Northern blots of total polyadenylated mRNAs from parasitized host larvae, respectively (Blissard *et al.*, 1986b). The results of these experiments provided the first indication of several of the unusual features of the functional organization of polydnavirus genomes. First, typical of what one might expect for a virus with a segmented genome, the different probes hybridized with different genomic segments. However, each probe expectedly hybridized with more than one viral DNA. Unrelated probe DNAs hybridized with different subsets of the viral genomic segments, but a given genomic segment sometimes hybridized with more than one of the unrelated probes. In other words, different probes identified different, but sometimes partially overlapping, subsets of the viral genomic DNA segments. The apparent cross-hybridization was much too extensive and too complex to be wholly explained by the presence of multiple topological forms of the circular DNAs. Second, probes that were unrelated (as determined by physical mapping and the Southern blot data) hybridized with different mRNAs. The majority of the probes each hybridized with one mRNA apiece. Unexpectedly, two of the gene-specific probes each hybridized with two abundant viral messages apiece, one to transcripts of 1.0 and 1.6 kb and the other to mRNAs of 1.1 and 1.4 kb. The related mRNAs could be interpreted as transcripts of members of a multigene family, alternatively spliced mRNAs, etc. Third, hybrid-selected mRNAs could be translated *in vitro*, indicating that the coding sequences were not pseudogenes.

The model of the CsV genome that emerged from these studies is one in which the polydnaviral genome is both functionally and physically multipartite. Individual genes, including members of gene families, are segregated among the genomic segments. The viral superhelical DNAs are not identical but may share one or more sequences. Available data from studies of three CsV gene families (discussed in the following) support this model. However, none of the CsV genomic segments has been sequenced in its entirety. Additional molecular comparisons of different genomic segments are needed to determine whether each superhelix is a mosaic of unique and shared sequences. Such comparisons will demonstrate (1) how extensive the shared sequences are, that is, are they restricted to the genes in multigene families or do they also include other, nonexpressed sequences, and (2) how similar the related sequences on different DNAs are.

B. Known CsV Gene Families

The preliminary identification of the genes encoding the 1.0- and 1.6-kb mRNAs as members of a gene family was confirmed by nucleotide (nt) sequence analysis of the related sequences (Blissard *et al.*, 1987). cDNAs corresponding to the 1.6- and 1.0-kb transcripts were mapped to genes on different regions of (cloned) superhelix W (15.8 kbp). It is of interest that the cDNA for the 1.6-kb mRNA hybridized CsV genomic DNAs W, R (13.3 kbp), and M (10.4 kbp), whereas the cDNA for the 1.0-kb transcript hybridized only with DNA W, consistent with the concept of the segregation and sequence divergence of related genes in the genome. Comparison of the nucleotide sequences of the cDNA for the 1.6-kb transcript and the homologous genomic sequence on DNA W indicated that the two sequences are identical except for two introns that are present in the genomic sequence. (The genomic copy of the gene for the 1.0-kb mRNA has not yet been sequenced.) The two cDNAs of the 1.0- and 1.6-kb mRNAs showed that the genes share five regions (approximately 120–170 nt each) of significant sequence similarity (68–88% identity) that were interspersed with regions of significant sequence divergence. Based on the predicted amino acid sequences, the amino and carboxy termini of the two proteins are similar. An identical 26-amino acid sequence is centrally located in each gene and may represent a functionally important domain. Using the baculovirus expression vector system, it was shown that the 1.0-kb transcript coded for a 24-kDa polypeptide that was glycosylated to a 26-kDa form whereas the 1.6-kb transcript coded for a nonglycosylated 23-kDa polypeptide (Blissard *et al.*, 1989). Both gene products were secreted from *Spodoptera frugiperda* cells.

Much less is known about a second gene family that contains the genes for the 1.1- and 1.4-kb transcripts that also are abundant in the parasitized host (Blissard *et al.*, 1986b). The function of these genes is not known, but the fragment containing one of the two genes was recently shown to hybridize with *C. sonorensis* venom gland cDNAs (Webb and Summers, 1990b). Whether these genes code for proteins related to the venom proteins per se or other nonvenom wasp proteins found in the venom gland tissues has not been determined.

A third CsV gene family, termed the 540-bp repeat element gene family, was identified in the course of studies to determine whether repetitive sequences were common in the genome and responsible for much of the observed cross-hybridization (Theilmann and Summers, 1987, 1988). DNA O^1 (11.3 kbp) hybridized with most of the viral superhelices under the lowest stringency condition tested whereas the other CsV DNA probes hybridized with much smaller subsets of the genome (Fig. 2, and Theilmann and Summers, 1987). Further analyses revealed that the sequences on DNA O^1 responsible for the high level of intragenomic hybridization mapped to two discrete regions of the superhelix. Related sequences also were mapped by hybridization to discrete fragments on most, but not all, of the selected cloned genomic superhelices that were examined in detail (Theilmann and Summers, 1987).

Nucleotide sequence analysis of the cross-hybridizing regions revealed that related fragments are composed of an imperfectly conserved, repeated sequence element of approximately 540 bp (Theilmann and Summers, 1987). A single copy of the repeat element exists on DNA B (6.7 kbp) whereas 4.5 and 2.5 copies of it occur as direct tandem arrays on DNAs H (8.5 kbp) and O^1, respectively. The nine copies of the element that have been sequenced to date are on average approximately 60–70% similar at the nucleotide level, although short patches of higher similarity occur.

Subsequent comparison of the viral genomic copy and a homologous cDNA confirmed that the repeat element on DNA B is transcribed in the parasitized host, but Northern blot analyses suggested that this element is not expressed in the wasp's oviducts (Theilmann and Summers, 1988). In contrast, a 540-bp related sequence identified on DNA W appears to be transcribed only in the wasp (Blissard *et al.*, 1989). Northern blot analyses indicated that the repeat related sequences on DNAs H and O[1] are transcribed in both parasitized host larvae and *C. sonorensis* oviducts (Theilmann and Summers, 1988). Although the repeat element related sequences on CsV DNAs B and H each hybridized with one transcript in the mRNAs of the host in which the sequence is expressed, the repeat element DNAs on O[1] or W each hybridized with two or more transcripts (Theilmann and Summers, 1988; Blissard *et al.*, 1989). The relative abundance of the four transcripts detected with the DNA O[1] repeat element related probes varied depending on the host and the particular DNA O[1] fragment used. Whether some of the multiple transcripts detected by the DNA O[1] probes are alternatively spliced mRNAs or reflect cross-hybridization with mRNAs encoded by related sequences on other superhelices is not known. Whether the 540-bp element related mRNAs can be translated has not yet been determined by *in vitro* or *in vivo* studies, but open reading frames have been identified (Theilmann and Summers, 1987).

A limited number of other CsV genes have been identified and partially characterized. A second gene that codes for a spliced mRNA expressed only in the wasp was identified on DNA B (Theilmann and Summers, 1988). This gene does not appear to be closely related to the 540-bp repeat element gene. Further study of CsV DNA W also identified three to five wasp-specific genes in addition to the two previously characterized genes for the 1.0- and 1.6-kb transcripts present in the parasitized host (Blissard *et al.*, 1989). One of these wasp-specific genes maps to the region containing 540-bp repeat related sequences. Whether the other wasp-specific genes on DNA W belong to multigene families is not clear from the limited extant data.

Very few polydnaviral genes have been sequenced, but the available data suggest that the structure of polydnaviral genes is typical of eukaryotic genes in general. At least two of the CsV genes are spliced, and nucleotides consistent with the 5′ and 3′ splice junction consensus sequences occur at the exon–intron junctions (Blissard *et al.*, 1987; Theilmann and Summers, 1988). Sequences resembling typical transcriptional and translational signals appear to be used, and the sequence context around the translational start site is consistent with the translation initiation consensus sequence of Kozak (Blissard *et al.*, 1987; Theilmann and Summers, 1988). The temporal pattern of expression of genes involved in virus replication in the female wasp suggests that future investigations may identify unique regulatory sequences, for example, hormone-responsive elements, in the 5′ regions of some polydnaviral genes (see Section VII).

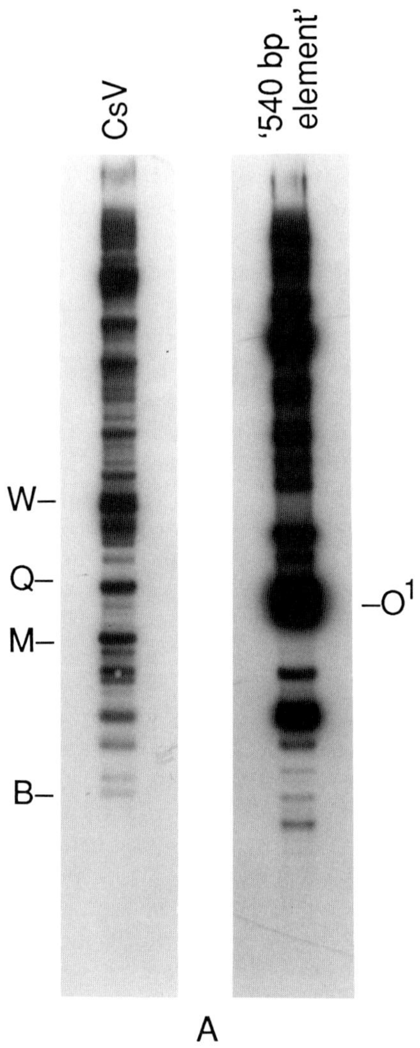

Figure 2 Distribution of 540-bp repeat element related sequences in the CsV genome. (A) Duplicate lanes of undigested CsV DNA were hybridized with ^{32}P-labeled CsV DNA under high (50% formamide) stringency conditions or with vector-free 540-bp repeat element sequences from DNA O^1 (pO^1-HC1185) under lower (30% formamide) stringency conditions. Equivalent autoradiographic exposures are shown. Although 540-bp repeat element related sequences are widespread in the genome, not all CsV DNAs hybridize detectably with the 540-bp repeat element probe under the stringency conditions used. The genomic distribution of the 540-bp repeat element related sequences detected in these Southern blots is equivalent to that reported in the original description of the gene family by Theilmann and Summers (1987).

(continues)

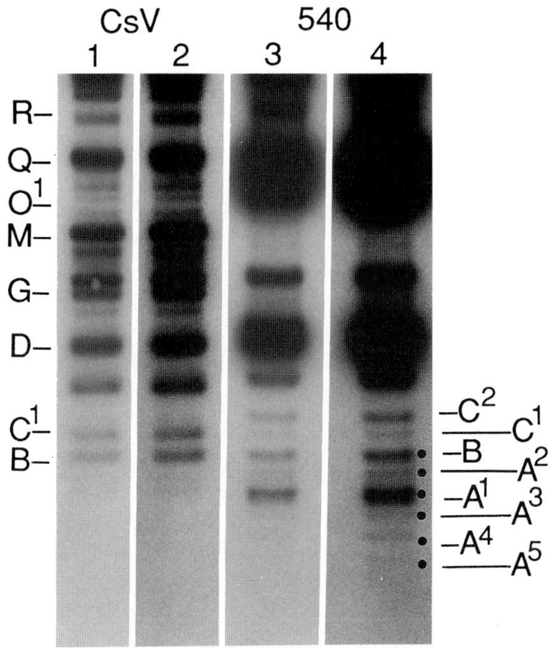

B

Figure 2 *(continued)* (B) The region containing predominately superhelical forms of the viral DNAs is enlarged. To illustrate both faint and intense bands more clearly, lands 1 and 2 show short and long photographic exposures, respectively, of the autoradiograph of viral DNA hybridized with the CsV DNA probe. Lanes 3 and 4 are comparable short and long photographic exposures, respectively, of lanes hybridized with the 540-bp repeat element probe. Both the identity and complexity of probes can significantly affect the hybridization results in studies of polydnavirus genomes. Note that some CsV DNA bands, for example, A^1–A^5 and C^2, are more readily detectable with the 540-bp repeat element probe than with the more complex (total) viral genomic DNA probe. The DNA bands designated A^2–A^5or C^2 were identified after the initial characterization, illustrating the difficulty in accurately estimating the total number of DNAs in a polydnavirus genome. During the last decade the original scheme for designating bands has been modified to accommodate the additional bands as they were discovered. Additonal DNA band with the most similar size and distinguished from the DNA band observed in the initial study by numerical superscripts or subscripts greater than one, for example, C^2. DNA bands designated by the same letter are of similar size but are *not* necessarily related at the molecular level. As the complexity of polydnavirus genomes has become more apparent with time, other designation schemes are being considered by investigators in the field.

C. Temporal Patterns of CSV Gene Expression

Limited data about the temporal regulation of CsV genes are available. The most striking feature of the presently available, semiquantitative Northern blot data is the similarity of the steady-state levels of many of the transcripts from approximately 12 hr to at least 6 days postoviposition (Blissard *et al.*, 1986b). Some minor differences in the temporal expression of the genes for the 1.0-, 1.6-, and a related 1.3-kb mRNA have been noted. However, these differences were not consistent in the several studies of this gene family and cannot be clearly distinguished from the differences in the amounts of the three related transcripts relative to one another at given sampling times (Blissard *et al.*, 1986b, 1987, 1989). The 540-bp repeat element genes, especially the genes located on CsV DNAs B and O^1, appear to differ from the pattern described here. These genes are expressed maximally and at similar levels between 2 and 24 hr postoviposition and then remain at slightly decreased but relatively constant levels until levels increased at 8 days postoviposition (Theilmann and Summers, 1988). Whether the variations in the mRNA levels noted in the published studies are evidence of temporal regulation is unclear because the observed differences were slight and internal controls (e.g., a constitutively expressed host gene) that would allow unambiguous comparisons of samples taken at different time points were not included.

The relatively constant steady-state levels suggest that several CsV gene products may be required at approximately similar levels immediately after oviposition and at later times to prevent the encapsulation of the wasp's egg and larval instars. This interpretation should be considered tentative until more detailed quantitative studies are done because the reported work used complex probes (e.g., total CsV genomic DNA) or examined the transcription of members of gene families in a single tissue and/or developmental stage of the insects. Genes in multigene families often are coordinately and developmentally regulated, as are the hemaglobin or chorion genes (Li, 1983; references in Hibner *et al.*, 1991; Xiong *et al.*, 1988). Greater differences in the temporal pattern of expression of single-copy viral genes might be expected. What effect polydnaviruses have on host transcription is not known but also should be considered in future studies.

D. Significance of the Gene Families

The presence of more than one gene family in the CsV genome is important for several reasons. First, the CsV gene families, especially the 540-bp repeat element gene family, are evidence for past sequence duplication and intersegmental recombination (Blissard *et al.*, 1987; Theilmann and Summers, 1987).

The multiple gene copies can alternatively be interpreted as possible sites for future intermolecular or intramolecular crossovers in the continued evolution of the virus. The concomitant segregation of related genes within a polydnaviral genome is consistent with the model that each polydnaviral genomic segment may be a chimera of unique and shared sequences. As noted previously, such a model could account for much of the apparent intragenomic complexity of CsV. The segregation of related genes on different viral superhelical DNAs could be important if less than the entire viral genome were encapsidated in each virion. This would be especially true if the viral genes were expressed in a tissue-specific manner.

Second, it may be significant that the four CsV genes initially identified as being members of gene families appear to be among the viral genes that are the most highly expressed in host larvae. Gene duplication is a mechanism for increasing gene dosage, thereby influencing the mRNA levels of genes with critical functions. This could be especially important for genes that are expressed immediately after oviposition. Whether the CsV genes being discussed play critical roles is conjectural at this time because it is not known what the function of *any* CsV or other polydnaviral gene is.

Third, gene families provide a mechanism for regulating gene expression in a tissue-specific manner, such as the human collagen and chicken troponin gene families (Breitbart and Nadal-Ginard, 1989; Vuorio and de Crombrugghe, 1990; Marchant *et al.,* 1991). The tissue specificity of CsV genes is not known. Related but variant forms of a gene may encode products that differ functionally to varying degrees, for example, the collagens specific to the cornea or muscle, respectively. Altered tissue specificities among members of a gene family may be one mechanism of refining a system physiologically. This may be particularly important where more than one tissue appears to be infected, as in parasitized hosts.

Fourth, the studies on the 540-bp repeat element gene family confirmed that polydnaviral genes in some cases are regulated in a species-specific manner, that is, the genes are expressed exclusively in either the wasp or the parasitized host. The finding that some viral genes are expressed in both insects, however, was somewhat surprising in view of the significant biological differences in the two insects (wasp and lepidopteran larvae) and the interaction of the virus with the two species. Polydnaviral gene regulation has not been investigated systematically, but future investigations of the differential regulation of the three subclasses of viral genes (wasp specific, lepidopteran specific, and genes expressed in both species) may provide information useful to the study of insect gene regulation in general. In the case of genes expressed in both host species, what biological activity the protein product(s) have (if any) in each species also will be of interest.

Fifth, the studies demonstrated that a given genomic segment can contain *both* genes expressed exclusively in the parasitized host and genes that are wasp specific. That genes expressed solely in either the wasp or the lepidopteran host are interspersed on a DNA segment and are not clustered together on different viral genomic segments is of interest with regard to the origin and evolution of polydnaviruses (see Section VIII). The interspersion of the two species-specific classes of viral genes on a given superhelix is of less interest from the standpoint of gene transcription because data from other eukaryotic systems suggest that tissue- or species-specific expression depends on specific regulatory elements of the genes themselves, and the genes therefore would be expected to be regulated independently.

VII. Polydnavirus Transmission, Integration, and Replication

Evidence from studies of the host–parasitoid interaction supports the theory that polydnaviruses play an essential role in determining whether the eggs and immature stages of certain ichneumonid or braconid wasps can complete endoparasitic development in the lepidopteran host. Given the critical function of polydnaviruses in the life cycles of these parasitoids and the generally accepted notion that these viruses occur in the oviducts of *all* females within species in which a polydnavirus has been detected, it is important to know how the viruses are transmitted between wasp generations with sufficient efficiency to ensure that all progeny acquire the virus. The issue of how a given polydnavirus is maintained within a wasp species becomes doubly important in light of the apparent species specificity of the viruses (see Section VIII). Maintaining the "correct" polydnavirus, that is, the virus that appears to be specific for the given wasp species, could be critical to the reproductive success of the parasitoid because not all polydnaviruses are equally effective in altering the physiology of a given lepidopteran host (Vinson, 1977; Vinson and Stoltz, 1986). In nature, larval hosts can be parasitized by wasps of more than one species, and thus endoparasitoids potentially could come into contact with the virus from more than one wasp species. Postulated mechanisms therefore must account for the efficiency and specificity of transmission.

Insect viruses typically are transmitted by *per os* acquisition or by transovarial transmission. Baculoviruses are well-known examples of insect viruses transmitted by ingestion of virus-infested material. Transovarial transmission, a mechanism in which ovarial tissues are infected, has been demonstrated for bunyaviruses (e.g., La Crosse virus), flaviviruses (e.g., dengue virus), and rhabdoviruses (e.g., the sigma virus of *Drosophila melanogaster*) (Gonzalez-Scarano and Nathanson, 1990; Monath, 1990; Rosen, 1984). Either of these mechanisms appeared possible when the studies on transmission were initiated. Larval endoparasitoids consume host tissue or hemolyph, either of which could be infected or contaminated with virus injected by the female wasp parent during oviposition. A parasitized host contains a fairly significant amount of viral DNA (Theilmann and Summers, 1986). It is unclear, however, whether the viral DNA present in a parasitized host is in a form that would be infectious for the developing parasitoid. The species specificity of virus acquired *per os* could be explained by arguing that the wasp tissues were permissive for replication of only the virus typically associated with that wasp species. (Although this argument is vague and ultimately unsatisfactory, it often is invoked for lack of a more definitive explanation due to our comparatively poor understanding of permissiveness and host range in most viral systems.) In the second model (transovarial transmission), infection of the egg would provide a ready source of inoculum that is compatible with the wasp host.

A third mechanism, vertical transmission through the germline as proviruses, is important in the biology of many animal viruses such as retroviruses. The covalent linkage of the viral genome to host cellular DNA sequences distinguishes proviral transmission from transovarial transmission in which the viral genome is extrachromosomal and can self-replicate under appropriate conditions. No precedent for proviral germline transmission of insect viruses existed. Although retroviruslike particles containing RNA related to *copia* have been detected in *Drosophila melanogaster,* this generally is not considered an example of germline transmission of an insect virus (Heine *et al.,* 1980; Shiba and Saigo, 1983). The particles are noninfectious, and the retrotransposon *copia,* which lacks an *env* (envelope) gene, is genetically distinct from true retroviruses (Echalier, 1989).

The physical and genetic data supporting the probable transmission of polydnaviruses through the germline come from the ichnoviruses CsV and *Hyposoter fugitivus* virus (HfV) and the bracovirus CmV. Virions have not been detected in male wasps or in nonoviduct tissues of female wasps. It therefore was surprising that each body segment, that is, head, thorax, and abdomen, of each male or female *C. sonorensis* wasp sampled contained low but reproducible amounts of CsV-specific DNA (Fleming and Summers, 1986). Subsequent physical analysis of the virus-specific DNA in male *C. sonorensis* wasps revealed two differently organized forms. One fraction was identical to the DNA isolated from virions when analyzed either in its native, undigested state or after digestion with restriction endonucleases. This episomal fraction appeared to contain all the segments of the viral genome but was present in male somatic tissues in substantially lower amounts than the second form.

Southern blot analyses of digested male wasp cellular DNA and comparably treated CsV DNA demonstrated that the second, predominate form of virus-specific DNA in wasp somatic tissues is consistently detectable as offsize (anomalously sized) restriction fragments when compared to the fragments in comparably digested viral DNA controls (Fig. 3, and Fleming and Summers, 1986). The offsize restriction fragments detected in digested cellular DNA are reproducible when DNA from different individuals or genera-

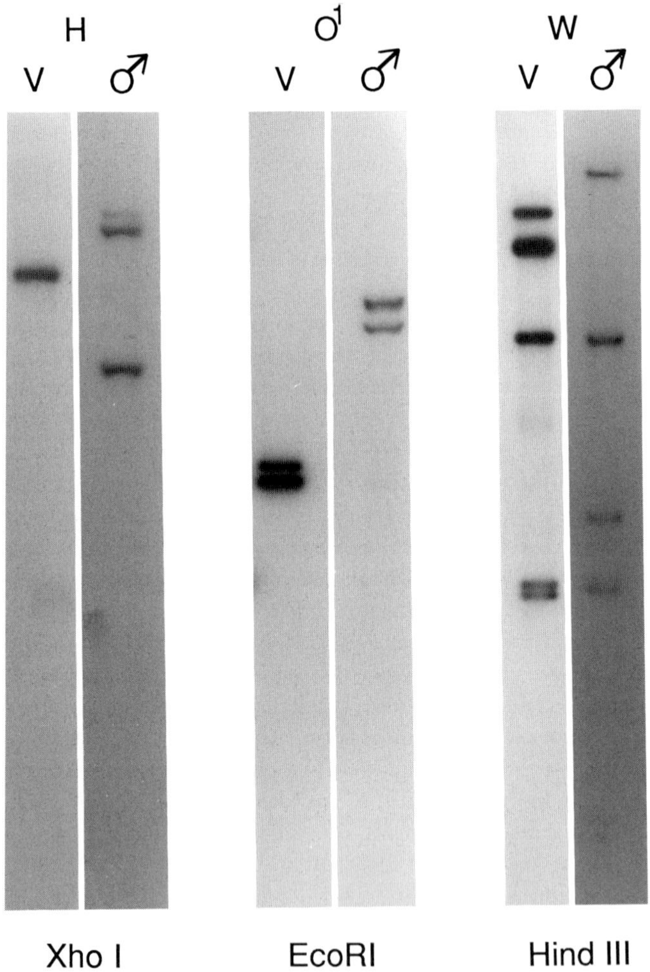

Figure 3 Offsize restriction fragments. Southern blots of cellular DNA from pooled male *C. sonorensis* wasps or CsV DNA (V) digested with the indicated restriction endonucleases were hybridized with cloned sequences from viral DNAs H, O^1, or W under stringent conditions. Each lane of male DNA contains one or more offsize restriction fragments not present in the viral DNA control.

tions is used (J. Fleming, unpublished data). Identical offsize restriction fragments are detectable in the nonoviduct somatic tissue (head and thorax) DNA of *C. sonorensis* female wasps. When wasp DNA is digested with enzymes that do not cut the viral superhelical DNA used as a probe, single high-molecular fragments that exceed the size of the viral genomic segment used as a probe are detectable in the cellular DNA (Fleming and Summers, 1986). This is consistent with results indicating that the virus-specific DNA is associated with the high-molecular-weight fraction of wasp DNA purified on cesium chloride–ethidium bromide gradients (Fleming and Summers, 1986). A single, discrete fragment on each viral genomic segment tested is consistently homologous to the offsize restriction fragments detected by that probe, and the remainder of each genomic segment appears to be unaltered in wasp cellular DNA. These findings suggested that the viral DNAs are integrated.

Comparison of cloned wasp DNA related to CsV DNA B and cloned extrachromosomal DNA B (6.7 kbp) by detailed physical mapping, reciprocal Southern blot hybridization, and nucleotide sequence analysis indicates that the CsV DNA B-specific sequences in *C. sonorensis* cellular DNA are colinear with the extrachromosomal sequences. Relative to the extrachromosomal circular molecule, the homologous cellular copy is linearized at a site within the fragment associated with the offsize restriction fragments. Rearrangement of the viral molecule in somatic tissues can be ruled out as an explanation for the offsize restriction fragments. The viral DNA in wasp cellular DNA is covalently linked to nonviral sequences, that is, DNA B is integrated in the wasp genome (Fleming and Summers, 1990, 1991). All the CsV segments that have been examined (to date, six superhelices or approximately 25% of the CsV genome) also are detectable as offsize restriction fragments (Fig. 3, Fleming and Summers, 1986; J. Fleming, unpublished data). The entire CsV genome thus appears to be integrated, suggesting that CsV may be the first documented example of an integrated DNA virus in insects (Fleming and Summers, 1990, 1991). Physical mapping and Southern blot data also have demonstrated that HfV-specific DNA is integrated in the cellular DNA of male *H. fugitivus* ichneumonid wasps (Xu and Stoltz, 1991).

The sequences abutting the junctions of the viral and wasp sequences are structurally complex and suggest that the integrated CsV DNA is not closely related to the characterized classes of transposable elements. The integrated copy of DNA B terminates in short (59 nt) imperfect direct repeats, but a single copy of the terminal repeat sequence is found in the circular, extrachromosomal DNA B molecule isolated from virions (Fig. 4). Sequences near each terminus of integrated DNA B form imperfect inverted repeats with sequences in the 540-bp repeat element that is located centrally on DNA B (Fleming and Summers, 1991). Base pairing between the terminal and internal sequences that form these inverted repeats are hypothesized to result in a

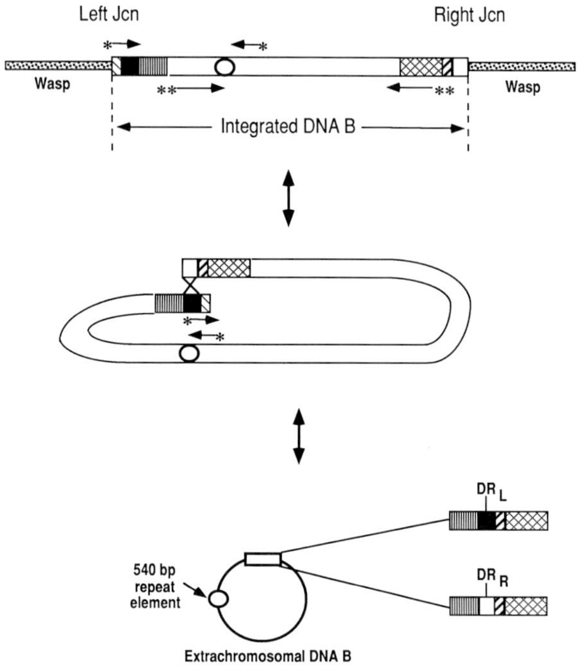

Figure 4 Integrated and extrachromosomal CsV DNA B. The integrated copy of CsV DNA B (wide bars) flanked by wasp sequences (narrow speckled bars) is shown schematically at the top. The terminal imperfect direct repeats (DR_R or DR_L, open or solid squares at the right or left junction, respectively) are immediately flanked by short sequences that are related to each other but contain more mismatches than the DR_R or DR_L sequences (rectangles with heavy or narrow diagonal lines that are located immediately to the left of DR_R or DR_L in the figure). Two sets of sequences close to the terminal repeats form imperfect inverted repeats (sequence *pairs* marked by short or long arrows with one or two asterisks, respectively) with sequences near the 5′ and 3′ termini of the single 540-bp repeat element (small open circle) that is located centrally in integrated CsV DNA B. The inverted repeats have been hypothesized to contribute to the stabilization of a conformation that facilitates recombination between the terminal direct repeats during replication, shown schematically in the middle figure (Fleming and Summers, 1991). The episomal form of DNA B isolated from purified CsV virions is shown schematically at the bottom and contains a single copy of the direct repeat sequence. Different (cloned) molecules of extrachromosomal DNA B contain *either* DR_R or DR_L. Other viral sequences immediately adjacent to the direct repeat sequences are identical in both episomal DNA B subclasses.

conformation in which the ends of the virus-specific sequences are brought into proximity. Recombination between the direct repeats then could occur during replication (Fleming and Summers, 1990, 1991).

The genetic analyses of the bracovirus CmV support the hypothesis of transmission of polydnaviruses through the wasps' germline, but the mechanism utilized by bracoviruses presently is unclear. The initial analyses took advantage of the polymorphic genomic segments present in geographically isolated populations of *C. melanoscela*. A genomic segment designated D is present in the virus particles in the Connecticut strain but is absent from the virions in the (Truro) Nova Scotia strain. When Connecticut males were mated with Nova Scotia females, DNA D was found in the DNA isolated from virions in F1 females (Stoltz *et al.*, 1986). These data indicate that polydnaviral DNA can be transmitted in wasp sperm. What form of CmV DNA is transmitted in sperm, however, is unclear from the accompanying physical analyses. CmV probes failed to hybridize with the *C. melanoscela* male wasp chromosomal DNA, and only episomal forms of CmV-specific DNA were detectable in male wasp DNA (Stoltz *et al.*, 1986). These data are inconsistent with the integration of the bracovirus DNA and suggested that the viral genome could be transmitted as episomal DNAs. However, in later genetic studies of the same two *C. melanoscela* populations, three HindIII fragments that distinguish the viruses in the wasp populations segregated in ratios consistent with Mendelian inheritance, suggesting that the viral DNAs are inherited as part of the wasp chromosomal genome (Stoltz, 1990). Further physical analyses are necessary to resolve the discrepancy between the physical and genetic data from the bracovirus system.

The Mendelian segregation ratios observed in the bracovirus system are strong evidence against either *per os* or transovarial transmission. Both episomal and integrated polydnaviral DNAs are present in (*C. sonorensis*) wasp somatic tissues (Fleming and Summers, 1986), but how important the episomes are for virus transmission is questionable in view of the Mendelian inheritance patterns that have been observed for CmV. (The alternative interpretation that the circular DNAs in males result from low levels of virus replication from the integrated viral DNA templates is of considerable interest from the standpoint of gene regulation because significant levels of virus replication are restricted to the female reproductive tract.) Inheritance as (integrated) endogenous proviruses appears to be the most probable mechanism for the transmission of polydnaviruses of ichneumonids, for example, CsV and HfV. Additional work, however must be done to determine more clearly what molecular mechanism is used by bracoviruses like CmV.

The genetic relationship of the parasitoid and an endogenous polydnavirus is a close one. It therefore is reasonable to use polydnavirus DNAs as genetic markers for intraspecific genetic analyses (reviewed in Fleming,

1991). Either polymorphic genomic segments or RFLPs in the extrachromosomal DNA in virus from females can be used. The predicted higher frequency of RFLPs and the decreased number of artifacts due to DNA topology in digested DNA suggest that RFLPs will be more useful. This simple approach is limited by the fact that virus is present in significant amounts only in the female wasp. The integrated viral DNAs, however, appear to be stably integrated in the chromosomal DNA of both male and female wasps and do not appear to be repetitive, that is, each viral DNA appears to exist at a single locus or at only a few loci. Thus, the integrated viral DNAs probably will ultimately prove to be more important in the genetic analysis of parasitoids, but more sensitive and sophisticated methods, for example, polymerase chain reaction technology and pulsed field electrophoresis, (PCR), will be required.

The replication strategy of polydnaviruses is not known, and a number of models can be hypothesized. Two models that explain multipartite genomes in some other systems probably can be dismissed for polydnaviruses because no evidence for a large "master molecule" or a tandem array of several different genomic segments has been found in the studies of either the extrachromosomal or integrated viral DNAs (Blissard *et al.*, 1986b; Fleming and Summers, 1991). The integrated CsV DNA B, for example, is flanked by wasp sequences, and the mapping data for the other viral DNAs suggest that the viral genomic segments exist at different, noncontiguous loci. Unlike plant mitochondrial genomes, the individual segments therefore are unlikely to result from the recombination between repeated sequences on a large circular molecule composed of several segments (Palmer and Shields, 1984). The data from the CsV system also suggest that the superhelices are unlikely to be synthesized as part of a long, linear concatamer that subsequently is cleaved to yield different individual genomic segments.

If, as seems likely, the integrated viral DNAs are used as templates, postulated mechanisms must account for both the high level of amplification of the DNAs and the change from a linear molecule to a circular one. It remains to be determined whether the viral DNAs are excised from the chromosome and whether the viral DNAs are amplified as extrachromosomal elements. The single copy of the terminal direct repeat sequence in CsV extrachromosomal DNA B suggests that the two direct repeats at the ends of the linear (integrated) copy recombine during replication, resulting in a circular molecule (Fig. 4 and Fleming and Summers, 1991). Recombination between directly repeated sequences can result in excision of the intervening DNA that would be present in the resultant circular molecule, as has been demonstrated unambiguously in *in vitro* studies of a mouse retrotransposon (Edelmann *et al.*, 1989). Alternatively, it is also possible that the viral sequences are synthesized by a copy choice or other mechanism, and termini of the extrachromosomal, linear product recombine to form a circular DNA without the excision of the genomic sequences. Although retrovirus DNAs generated by reverse transcription prior to integration in retrovirus-infected cells are not strictly analogous to polydnaviral DNAs, the circles with a single long terminal repeat (LTR) suggest that homologous recombination between the termini of linear DNA can result in a circular molecule with a single copy of the direct repeat (Varmus and Brown, 1989).

A biphasic replication strategy also is possible and might better explain the logarithmic increase of viral DNA in newly eclosed adults. Viral replication begins in late pharate females but continues in the adult female wasp for several days after eclosion (Norton and Vinson, 1983; Fleming and Summers, 1986; Theilmann and Summers, 1986). The initial event may generate circular molecules from the integrated viral DNAs by one of the mechanisms outlined earlier. The circular DNAs generated in the first phase might then act as templates for DNA synthesis in the second phase. Continued amplification of the episomal DNAs by a rolling circle or other mechanism could account for the rapid increase in the viral superhelical DNAs in newly eclosed adult females. Different episomes presumably would be templates for the synthesis of the different DNA genomic segments.

Whatever replication mechanism is used, the restriction of high levels of replication to late pharate and adult female wasps suggests that replication is developmentally regulated (Norton and Vinson, 1983). Altered hormonal status in the pupating female is an obvious possible mechanism of regulating the wasp and/or viral genes involved in virus replication. Ecdysteroids have been reported to influence the level of CsV replication (Webb and Summers, 1990a, 1993). What other (wasp) host factors are involved and whether replication in the pupal and adult female is similarly regulated are not known.

VIII. Polydnavirus Evolution

A. Gene Divergence

The most obvious features of polydnavirus genomes, the segmentation and large size, are of interest from an evolutionary point of view. Viruses have evolved several strategies (alternative splicing, frameshifting, etc.) to increase the coding capacity of their genomes without significant concomitant increases in size. The aggregate genome sizes of polydnaviruses suggest that they are among the viruses (e.g., baculoviruses, poxviruses, herpesviruses, etc.) with the largest genomes (Murphy and Kingsbury, 1990). The large genomes partly may reflect the structural complexity of the virions [e.g., 25 polypeptides in CsV (Krell *et al.*, 1982)] and the multiple functions that polydnaviruses may play in the wasp and in the one or more lepidopteran hosts of the wasp. This explanation, however, may be insufficient because broad host ranges do not necessarily require genomes as large as those of polydnaviruses. Flaviviruses, for example, have both invertebrate and mammalian hosts but have RNA genomes of approximately 10 kb (Murphy and Kingsbury, 1990).

The multipartite nature of the genome may have reduced the selection pressure for small genome size. Although the aggregate genomes are large, each segment typically is less than 15–20 kbp and could presumably accommodate multiple insertions without adversely affecting the efficiency of replication. The integration of the polydnaviral DNAs also may have reduced the selection for small genome size because the viral DNA would be propagated as part of the chromosomes during part of the viral life cycle.

The number of multigene families and the divergence within the viral gene families are noteworthy. Viral gene families are not unique to polydnaviruses and have been reported for African swine fever and myxoma virus (Almendral *et al.*, 1990; Gonzalez *et al.*, 1990; Upton *et al.*, 1990). Multigene families, however, are not a typical feature of viral genomes, and the presence of more than two families in a given virus, that is, CsV, is unusual. The integration of polydnaviruses may have influenced both the number of gene families and the apparent number of members within particular (CsV) multigene families. Ohta (1988b) suggests that evolution through gene duplication occurs more rapidly during interchromosomal unequal crossing-over during sexual reproduction.

Viral gene families can reflect the acquisition of host sequences (e.g., the serpins of myxoma virus) or the duplication of a viral gene with no detectable homology with a host gene (Upton *et al.*, 1990). Western blot data suggest that some CsV envelope proteins and *C. sonorensis* venom gland proteins share common epitopes (Webb and Summers, 1990b). To what extent the cross-reactive proteins are evolutionarily related is unclear because neither the wasp genes for the venoms nor the viral genes for the cross-reactive envelope proteins have been identified. Amino acid or nucleotide sequence data should help to clarify the evolutionary relationship of the genes, but the serological cross-reactivity of the proteins is at least consistent with the possibility that CsV acquired an ancestral venom gene(s) during the coevolution of the wasp and virus. The possible selective advantage of acquiring wasp genes with biological activity in the lepidopteran host(s) is obvious.

Duplication of (viral) genes not derived from the wasp genome also could provide a selective advantage to the virus by increasing the copy number of a gene or a functional domain within a gene. The tandemly repeated 540-bp elements probably resulted from unequal crossing-over and significant sequence divergence among the elements. Published dot plot analyses suggest that certain sequences within the tandemly repeated 540-bp elements on CsV DNA H are more highly conserved than the other sequences in those elements (Theilmann and Summers, 1987). Whether the more highly related sequences within the DNA H 540-bp repeats represent functional domains presently is unclear.

Gene duplication can be indirectly important to an organism because the existence of multiple copies of a gene permits mutations to accumulate in one or more copies of the gene without detriment to the organism (Li, 1983; Ohta, 1988a). Alternatively, the divergence of related genes can help to refine the physiology of a system by encoding variant proteins with slightly modified functions or by coding for related peptides that form multimeric proteins that are functionally superior to the monomers (Li, 1983). The three similar regions in the predicted amino acid sequences for the 1.0- and 1.6-kb mRNAs suggest that the proteins have similar, but not necessarily identical, biological activities. Whether any related CsV genes encode subunits of multimeric proteins cannot be addressed at the present time because little is known about the physical and functional characteristics of the protein products of the related genes or the interaction of the proteins when they are coexpressed in cells of the wasp or parasitized host.

The level of divergence among members of some CsV gene families studied to date appears to be fairly high compared to the relative homogeneity within gene families such as the *Bombyx* chorion genes (Xiong *et al.*, 1988; Hibner *et al.*, 1991). Because the 540-bp repeat element sequences were initially identified by Southern blot hybridization, a minimum level of 60–70% similarity could be predicted from the hybridization stringency conditions that were used. It is unclear why the level of variation among the members is so uniform. The elements on different genomic segments are approximately 60–70% similar to one another (Theilmann and Summers, 1987). Adjacent elements in a tandem array might be expected to be more highly related to one another than the more distal elements, but the elements that are tandemly arrayed on a given genomic segment also are only approximately 60–70% similar (or less in a few cases). The observed divergence can be interpreted as an indication that strong selection pressure for variant genes existed prior to unequal crossing-over. As noted earlier, short sequences within most of the elements on CsV DNA H appear to be similar. The patterns can be alternatively interpreted as suggesting that the changes outside of these short regions of similarity were selectively neutral while these conserved regions were under positive selection and diverged less. Krell (1991a) has proposed that the 540-bp repeat elements represent remnants of duplicated but nonessential pseudogenes that are selectively neutral and hence were not eliminated during viral replication. A less extreme interpretation is that one or more pseudogenes may exist in this or other polydnavirus multigene families in addition to functional genes, as typically occurs in other eukaryotic gene families (Li, 1983). Which of the four explanations applies to the 540-bp repeat element gene family cannot be evaluated now because it is not known whether these sequences can be translated or what biological activity the products have. However, investigators of polydnavirus genomic complexity and gene families in the future should consider the possibility that functional and nonfunctional related sequences can coexist in a genome.

B. Species Specificity

Polydnaviruses generally are considered to be species specific with respect to the wasp host in which they replicate. A given polydnavirus "species," as judged by its DNA electrophoretic profile, can be reproducibly isolated from a given wasp species. The similarities in the electrophoretic profiles of undigested DNAs from individual *Hyposoter lymantriae* females in a single population outweighed the observed differences (Stoltz and Xu, 1990). The overall similarity of the electrophoretic profiles of the undigested genomic DNAs of individual females from geographically isolated populations of *C. melanoscela* also is consistent with this conclusion, even though some polymorphic segments were identified (Stoltz *et al.*, 1986). Conversely, the viruses from different wasp species can be distinguished from one another by their DNA profiles and by their biological activities in different lepidopteran species. In spite of the problems associated with these criteria, the interpretation that the viruses of different wasp species are themselves distinct species appears to be valid for the polydnaviruses that have been studied so far.

The apparent species specificity of the viruses and the mutualistic interdependence of wasp and polydnavirus for successful reproduction and survival suggest that the wasp and virus may have coevolved along parallel paths. Integration and germline transmission of the viruses are not necessary for the postulated parallel coevolution to have occurred. However, the apparently dual life cycle of the viruses as alternately extrachromosomal (and extracellular) virions and endogeneous proviruses provides an efficient mechanism for selecting and maintaining viral variants that result in improved survival of wasp progeny. Coevolution of wasp and virus suggests that polydnaviruses from parasitoid species that are phylogenetically related may be more related to each other than viruses from less closely related wasps. *Hyposoter fugitivus* virus genomic DNA hybridized with viral DNAs of six *Hyposoter* or *Diadegma* species but not with the viral DNAs of *Campoletis*, *Campoplex*, or *Glypta* species (Krell, 1991a). Stoltz and Xu (1990) also present Southern blot data suggesting that viral DNAs from several *Hyposoter* species are related. These data and the limited available serological comparisons of polydnaviruses suggest that viruses of congeneric wasp species or species in more closely related genera are more highly related than viruses from more distantly related wasps (Cook and Stoltz, 1983).

Which polydnaviruses represent distinct viral species as opposed to pathovars or biotypes of a single virus and how species specific polydnaviruses are with respect to the wasp will become clearer as more detailed molecular data become available. Whether the various polydnaviruses have diverged at the same rates as the wasp species within a genus (or other taxon) is not clear. First, either the virus or the wasp may be under stronger selection than the other symbiont. Selection for viral variants would not necessarily result in the selection of wasp variants, and vice versa. Second, the influence of the lepidopteran host as a selective force also must be considered when making phylogenetic comparisons of the wasps and the viruses. Wasps that parasitize the same host or closely related hosts presumably must overcome the same or similar host defense mechanisms. More distantly related hosts might be expected to exert different selection pressures on the wasp. Because viral genes appear to be involved in the abrogation of host immunity, viruses of related wasps that parasitize the same or related lepidopteran hosts may be more similar than viruses from related wasps that have become adapted to more distantly related hosts.

How related two viruses (or wasps) appear to be can depend on what genes are studied. Not all genes within the genome of either the virus or the wasp would be expected to evolve at the same rate. Viral genes involved in an essential function such as viral DNA synthesis might change more slowly than genes for structural proteins. The envelope genes of human and simian immunodeficiency viruses, for example, are notoriously variable relative to some other genes in the respective viral genomes. (The questions of whether the polydnaviral structural genes are under similar selection pressure and, if so, which of the many structural genes are variable and which are less subject to change are currently unanswered.) Nucleotide sequence comparisons of particular viral genes of similar function thus appear more likely to yield useful information than hybridization analyses using entire genomic segments or uncharacterized viral DNA fragments. Not all portions of a gene are likely to evolve at similar rates, and this must be taken into account when analyzing the data. Functional domains may be more highly conserved if they are critical to the function of the gene product, as in the case of enzymes, or they may be somewhat more variable if they are selected against, as in the case of epitopes recognized by the immune system (Good *et al.*, 1988; Hughes, 1991; Thomas and Wilson, 1991).

Comparisons of polydnaviruses from congeneric species or from closely related genera are likely to be more easily interpretable from the standpoint of systematics, but comparisons of viruses from more distantly related species, for example, those in different wasp subfamilies or families, also may be useful in tracing the evolution of polydnaviruses. Although it is reasonable to expect that ichnoviruses and bracoviruses will be substantially less related to one another than viruses of congeneric wasp species, it cannot be predicted from extant data to what extent the two viral subgroups are genetically related. Molecular analyses will help to determine whether the similar multipartite genomic organization of polydnaviruses in the two subgroups reflects convergent evolution or evolution from a common ancestral virus or wasp sequence.

C. Virus Origin

Molecular analyses may help to trace the paths of polydnavirus evolution, but the origin of polydnaviruses is likely to remain obscure. Theories about the origin of the viruses must account for the fact that, unlike most viruses, polydnaviruses have distinctly different ecological relationships with the two insect hosts. The wasp–virus relationship appears to be mutualistic, whereas the interaction of the virus with the parasitized host can be interpreted as pathogenic. Polydnaviruses might be viewed as the progeny of an ancestral virus or as wasp (host), DNA sequences that have gained a limited degree of autonomy (Strauss *et al.*, 1990; Whitfield, 1990).

Pathogenicity can be considered as selectively disadvantageous to a virus if it results in the premature death of the host prior to viral replication and dissemination (Temin, 1989). From this standpoint, partial or selective attenuation of an ancient pathogenic virus might have been selectively favorable to it if the loss of virulence to one of its hosts (the wasp) increased the probability of virus replication and transmission to progeny (wasp) hosts. The virus's virulence for the lepidopteran host also may have decreased if such changes decreased premature host mortality relative to wasp emergence. This is consistent with observations that injection of polydnaviruses can result in supernumerary instars or psuedoparasitized larvae and rarely causes a fatal infection in the *habitual* host in the absence of the parasitoid or other microbial infections. Alternatively, selection for increased virulence for the lepidopteran host may have occurred if debilitation of the host increased the survival of the alternate (wasp) host.

In a second hypothesis, wasp sequences may have been transferred to the parasitized host. If the wasp DNAs were expressed and the products were biologically active in the host and indirectly increased the fitness of the wasp, the sequences could have become fixed in the wasp population. The origin of the multiple viral genomic segments can be explained simply if they are derived from ancient dispersed repetitive elements. Acquisition of the ability to replicate autonomously in selected tissues or cells of the wasp, for example, the calyx epithelial cells, would be associated with increased copy number, which would increase the probability of transmission. A viral origin explains the consistent transmission of biologically active DNA between the two insect species more simply than a wasp origin, but several possible mechanisms can be speculated. Analogous to transformation, the simplest hypothetical mechanism may be lysis of wasp cells followed by DNA uptake by cells of the parasitized host. In a second scenario, extrachromosomal wasp sequences may have become entrapped in vacuoles or other unit membranes that subsequently were sloughed off from the wasp cell and fused with the host's cell membranes.

IX. Conclusions

Today polydnaviruses are increasingly recognized as being remarkably genetically complex biological entities that occupy an unusual ecological niche in the insect world. Many questions about them are presently unanswered, but it is becoming more apparent that the function of these viruses in the host–parasitoid interaction cannot be fully understood until their molecular genetics is better investigated. In addition to their importance to insect science, their genome organization and genetic complexity make polydnaviruses unique among viruses, also warranting further molecular study of them for what it can tell us about the diversity of genetic strategies that have evolved in the biological world.

Acknowledgments

We thank Brent Graham and Robert Harrison for valuable discussions and comments on the manuscript and Bruce Webb for providing a preprint of a manuscript on ecdysteroid effects. We thank David Theilmann for providing clone pO[1]-HC1185 and Loyd Sneed for advice on photography. This work was supported by grants from the USDA Competitive Grants Program to J.G.W.F. and from the Natural Sciences and Engineering Research Council of Canada to P.J.K.

References

Almendral, J. M., Almazan, F., Blasco, R., and Vinuela, E. (1990). Multigene families in African swine fever virus: Family 110. *J. Virol.* **64:**2064–2072.

Belloncik, S. (1989). Cytoplasmic polyhedrosis virus-reoviridae. *Adv. Virus Res.* **37:**173–209.

Blissard, G. W., Fleming, J.G.W., Vinson, S. B., and Summers, M. D. (1986a). *Campoletis sonorensis* virus: Expression in *Heliothis virescens* and identification of expressed sequences. *J. Insect Physiol.* **32:**351–359.

Blissard, G. W., Vinson, S. B., and Summers, M. D. (1986b). Identification, mapping, and *in vitro* translation of *Campoletis sonorensis* virus mRNAs from parasitized *Heliothis virescens* larvae. *J. Virol.* **57:**318–327.

Blissard, G. W., Smith, O. P., and Summers, M. D. (1987). Two related viral genes are located on a single superhelical DNA segment of the multipartite *Campoletis sonorensis* virus genome. *Virology* **160:**120–134.

Blissard, G. W., Theilmann, D. A., and Summers, M. D. (1989). Segment W of *Campoletis sonorensis* virus: Expression, gene products, and organization. *Virology* **169:**78–89.

Breitbart, R. E., and Nadal-Ginard, B. (1989). Tissue specific alternative splicing in the tropinin T multigene family. *In* "Tissue Specific Gene Expression" (R. Renkawitz, ed.), pp. 199–215. VCH Publishers, New York.

Cook, D. I., and Stoltz, D. B. (1983). Comparative serology of viruses isolated from ichneumonid parasitoids. *Virology* **130:**215–220.

Echalier, G. (1989). *Drosophila* retrotransposons: Interactions with the genome. *Adv. Virus Res.* **36:**33–15.

Edelmann, W., Kroger, B., Goller, M., and Horak, I. (1989). A recombination hotspot in the LTR of a mouse retrotransposon. *Cell (Cambridge, Mass.)* **57:**937–946.

Edson, K. M., Vinson, S. B., Stoltz, D. B., and Summers, M. D. (1981). Virus in a parasitoid wasp: Suppression of the cellular immune response in the parasitoid's host. *Science* **211:**582–583.

Federici, B. A. (1986). Ultrastructure of baculoviruses. *In* "The Biology of Baculoviruses" (R. R. Granados and B. A. Federici, eds.), Vol. 1, pp. 61–88. CRC Press, Boca Raton, FL.

Fleming, J.G.W. (1991). The integration of polydnavirus genomes in parasitoid genomes: Implications for biocontrol and genetic analyses of parasitoid wasps. *Biol. Control* **1:** 127–135.

Fleming, J.G.W. (1992). Polydnaviruses: Mutualists and pathogens. *Annu. Rev. Entomol.* **37:**401–425.

Fleming, J.G.W., and Summers, M. D. (1986). *Campoletis sonorensis* endoparasitic wasps contain forms of *C. sonorensis* virus DNA suggestive of integrated and extrachromosomal polydnaviral DNAs. *J. Virol.* **57:**552–562.

Fleming, J.G.W., and Summers, M. D. (1990). The integration of the genome of a segmented DNA virus in the host insect's genome. *In* "Molecular Insect Science" (H. H. Hagedorn, J. G. Hildebrand, M. G. Kidwell, and J. H. Law, eds.), pp. 99–105. Plenum Press, New York.

Fleming, J.G.W., and Summers, M. D. (1991). Polydnavirus DNA is integrated in the DNA of its parasitoid wasp host. *Proc. Natl. Acad. Sci. U.S.A.* **88:**9770–9774.

Fleming, J.G.W., Blissard, G. W., Summers, M. D., and Vinson, S. B. (1983). Expression of *Campoletis sonorensis* virus in the parasitized host, *Heliothis virescens*. *J. Virol.* **48:**74–78.

Fleming, J.G.W., Gibbs, M. C., and Summers, M. D. (1990). Heterogeneous forms of selected *Campoletis sonorensis* virus superhelical DNAs: Implications for polydnavirus replication and genome complexity. *Annu. Meet. Am. Soc. Virol.* (Abstr.), Salt Lake City, Utah.

Francki, R.I.B., Fauquet, C. M., Knudson, D. L., and Brown, F. (1991). Classification and nomenclature of viruses: Fifth report of the International Committee on Taxonomy of Viruses. *Arch. Virol., Suppl.* **2:**129–131.

Gonzalez, A., Calvo, V., Almazan, F., Almendral, J. M., Ramirez, J. C., de la Vega, I., Blasco, R., and Vinuela, E. (1990). Multigene families in African swine fever virus: Family 360. *J. Virol.* **64:**2073–2081.

Gonzalez-Scarano, F., and Nathanson, N. (1990). Bunyaviruses. *In* "Fields Virology" (B. N. Fields, D. M. Knipe, R. M. Chanock, M. S. Hirsch, J. L. Melnick, T. P. Monath, and B. Roizman, eds.), Vol. 1, pp. 1195–1228. Raven Press, New York.

Good, M. F., Pombo, D., Quakyi, I. A., Riley, E. M., Houghten, R. A., Menon, A., Alling, D. W., Berzofsky, J. A., and Miller, L. H. (1988). Human T-cell recognition of the circumsporozoite protein of *Plasmodium falciparum:* Immunodominant T-cell domains map to the polymorphic regions of the molecule. *Proc. Natl. Acad. Sci. U.S.A.* **85:**1199–1203.

Heine, C. W., Kelly, D. C., and Avery, R. J. (1980). The detection of intracellular retrovirus-like entities in *Drosophila melanogaster* cell cultures. *J. Gen. Virol.* **49:**385–395.

Hibner, B. L., Burke, W. D., and Eickbush, T. H. (1991). Sequence identity in an early chorion multigene family is the result of localized gene conversion. *Genetics* **128:**595–606.

Hughes, A. L. (1991). Circumsporozoite protein genes of malaria parasites (*Plasmodium* spp.): Evidence for positive selection on the immunogenic regions. *Genetics* **127:**345–353.

Jones, D., Sreekrishna, S., Iwaya, M., Yang, J.-N., and Eberely, M. (1986). Comparison of viral ultrastructure and DNA banding patterns from the reproductive tracts of eastern and western hemisphere *Chelonus* spp. (Braconidae: Hymenoptera). *J. Invertebr. Pathol.* **47:**105–115.

Krell, P. J. (1987). Polydnavirus replication and tissue organization of infected cells in the parasitic wasp *Diadegma tenebrans*. *Can. J. Microbiol.* **33:**176–183.

Krell, P. J. (1991a). The polydnaviruses: Multipartite DNA viruses from parasitic hymenoptera. *In* "Viruses of Invertebrates" (E. Kurstak, ed.), pp. 141–177. Dekker, New York.

Krell, P. J. (1991b). Polydnaviridae. *In* "Atlas of Invertebrate Viruses" (J. R. Adams and J. R. Bonami, eds.). CRC Press, Boca Raton, FL.

Krell, P. J., and Beveridge, T. J. (1987). The structure of bacteria and molecular biology of viruses. *Int. Rev. Cytol. Suppl.* **17:**15–88.

Krell, P. J., and Stoltz, D. B. (1979). Unusual baculovirus of the parasitoid wasp *Apanteles melanoscelus:* Isolation and preliminary characterization. *J. Virol.* **29:**1118–1130.

Krell, P. J., and Stoltz, D. B. (1980). Virus-like particles in the ovary of an ichneumonid wasp: Purification and preliminary characterization. *Virology* **101:**408–418.

Krell, P. J., Summers, M. D., and Vinson, S. B. (1982). Virus with a multipartite superhelical DNA genome from the ichneumonid parasitoid *Campoletis sonorensis*. *J. Virol.* **43:**859–870.

Li, W.-H. (1983). Evolution of duplicate genes and pseudogenes. *In* "Evolution of Genes and Proteins" (M. Nei and R. K. Koehn, eds.), pp. 14–37. Sinauer Associates, Sunderland, MA.

Marchant, J. K. Linsenmayer, T. F., and Gordon, M. K. (1991). cDNA analysis predicts a cornea-specific collagen. *Proc. Natl. Acad. Sci. U.S.A.* **88:**1560–1564.

Matthews, R.E.F. (1991). "Plant Virology." Academic Press, San Diego, CA.

Monath, T. P. (1990). Flaviviruses. *In* "Fields Virology" (B. N. Fields, D. M. Knipe, R. M. Chanock, M. S. Hirsch, J. L. Melnick, T. P. Monath, and B. Roizman, eds.), Vol. 1, pp. 763–814. Raven Press, New York.

Murphy, F. A., and Kingsbury, D. W. (1990). Virus taxonomy. *In* "Fields Virology" (B. N. Fields, D. M. Knipe, R. M. Chanock, M. S. Hirsch, J. L. Melnick, T. P. Monath, and B. Roizman, eds.), Vol. 1, pp. 9–35. Raven Press, New York.

Norton, W. N., and Vinson, S. B. (1983). Correlating the initiation of virus replication with a specific pupal developmental phase of an ichneumonid parasitoid. *Cell Tissue Res.* **231:**387–398.

Norton, W. N., Vinson, S. B., and Stoltz, D. B. (1975). Nuclear secretory particles associated with the calyx cells of the ichneumonid parasitoid *Campoletis sonorensis* (Cameron). *Cell Tissue Res.* **162:**195–208.

Ohta, T. (1988a). Evolution by gene duplication and compensatory advantageous mutations. *Genetics* **120:**841–847.

Ohta, T. (1988b). Time for acquiring a new gene by duplication. *Proc. Natl. Acad. Sci. U.S.A.* **85:**3509–3512.

Palmer, J. D., and Shields, C. R. (1984). Tripartite structure of the *Brassica campestris* genome. *Nature (London)* **307:**437–440.

Rosen, L. (1984). Ovarian infection and transovarial transmission of viruses in insects. *In* "Concepts in Viral Pathogenesis" (A. L. Notkins and M.B.A. Oldstone, eds.), Vol. 1, pp. 194–198. Springer-Verlag, New York.

Shiba, T., and Saigo, K. (1983). Retrovirus-like particles containing RNA homologous to the transposable element *copia* in *Drosophila melanogaster. Nature* (*London*) **302:**119–124.

Stoltz, D. B. (1990). Evidence for chromosomal transmission of polydnavirus DNA. *J. Gen. Virol.* **71:**1051–1056.

Stoltz, D. B., and Vinson, S. B. (1979). Viruses and parasitism in insects. *Adv. Virus Res.* **24:**125–171.

Stoltz, D. B., and Xu, D. (1990). Polymorphism in polydnavirus genomes. *Can. J. Microbiol.* **36:**538–543.

Stoltz, D. B., Pavan, C., and DaCunha, A. B. (1973). Nuclear polyhedrosis virus, a possible example of de novo intranuclear membrane morphogenesis. *J. Gen. Virol.* **19:**145–150.

Stoltz, D. B., Krell, P. J., and Vinson, S. B. (1981). Polydisperse viral DNA's in ichneumonid ovaries: A survey. *Can. J. Microbiol.* **27:**123–130.

Stoltz, D. B., Krell, P. J., Summers, M. D., and Vinson, S. B. (1984). Polydnaviridae—A proposed family of insect viruses with segmented, double-stranded, circular DNA genomes. *Intervirology* **21:**1–4.

Stoltz, D. B., Guzo, D., and Cook, D. (1986). Studies on polydnavirus transmission. *Virology* **155:**120–131.

Strauss, E. G., Strauss, J. H., and Levine, A. J. (1990). Virus evolution. *In* "Fields Virology" (B. N. Fields, D. M. Knipe, R. M. Chanock, M. S. Hirsch, J. L. Melnick, T. P. Monath, and B. Roizman, eds.), Vol. 1, pp. 167–190. Raven Press, New York.

Temin, H. M. (1989). Evolution of retroviruses. *In* "Concepts in Viral Pathogenesis" (A. L. Notkins and M.B.A. Oldstone, eds.), Vol. 3, pp. 189–195. Springer-Verlag, New York.

Theilmann, D. A., and Summers, M. D. (1986). Molecular analysis of *Campoletis sonorensis* virus DNA in the lepidopteran host *Heliothis virescens. J. Gen. Virol.* **67:**1961–1969.

Theilmann, D. A., and Summers, M. D. (1987). Physical analysis of the *Campoletis sonorensis* virus multipartite genome and identification of a family of tandemly repeated elements. *J. Virol.* **61:**2589–2598.

Theilmann, D. A., and Summers, M. D. (1988). Identification and comparison of *Campoletis sonorensis* virus transcripts expressed from four genomic segments in the insect hosts *Campoletis sonorensis* and *Heliothis virescens. Virology* **167:**329–341.

Thomas, W. K., and Wilson, A. C. (1991). Mode and tempo of molecular evolution in the nematode *Caenorhabditis:* Cytochrome oxidase II and calmodulin sequences. *Genetics* **128:**269–279.

Upton, C., Macen, J. L., Wishart, D. S., and McFadden, G. (1990). Myxoma virus and malignant rabbit fibroma virus encode a serpin-like protein important for virus virulence. *Virology* **179:**618–631.

Varmus, H., and Brown, P. (1989). Retroviruses. *In* "Mobile DNA" (D. E. Berg and M. M. Howe, eds.), pp. 53–108. Am. Soc. Microbiol., Washington, DC.

Vinson, S. B. (1977). *Microplitis croceipes:* Inhibitions of the *Heliothis zea* defense reaction to *Cardiochiles nigriceps. Exp. Parasitol.* **41:**112–117.

Vinson, S. B., and Iwantsch, G. F. (1980). Host regulation by insect parasitoids. *Q. Rev. Biol.* **55:**143–165.

Vinson, S. B., and Stoltz, S. B. (1986). Cross-protection experiments with two parasitoid (Hymenoptera: Ichneumonidae) viruses. *Ann. Entomol. Soc. Am.* **79:**216–218.

Vinson, S. B., Edson, K. M., and Stoltz, D. B. (1979). Effect of a virus associated with the reproductive system of the parasitoid wasp, *Campoletis sonorensis*, on host weight gain. *J. Invertebr. Pathol.* **34:**133–137.

Vuorio, E., and de Crombrugghe, B. (1990). The family of collagen genes. *Annu. Rev. Biochem.* **59:**837–872.

Webb, B. A., and Summers, M. D. (1990a). Developmental regulation of the replication of *Campoletis sonorensis*. *Annu. Meet. Entomol. Soc. Am.* (Abstr.), New Orleans, Louisiana.

Webb, B. A., and Summers, M. D. (1990b). Venom and viral expression products of the endoparasitic wasp *Campoletis sonorensis* share epitopes and related sequences. *Proc. Natl. Acad. Sci. U.S.A.* **87**:4961–4965.

Webb, B. A., and Summers, M. D. (1993). *Experientia* (in press).

Whitfield, J. B. (1990). Parasitoids, polydnaviruses and endosymbiosis. *Parasitol. Today* **6**:381–384.

Xiong, Y., Sakaguchi, B., and Eickbush, T. H. (1988). Gene conversion can generate sequence variants in the late chorion multigene families of *Bombyx mori*. *Genetics* **120**:221–231.

Xu, D., and Stoltz, D. B. (1991). Evidence for a chromosomal location of polydnavirus DNA in the ichneumonid parasitoid *Hyposoter fugitivus*. *J. Virol.* **65**:6693–6704.

Chapter 10

Venoms of Parasitic Hymenoptera as Investigatory Tools

Davy Jones
Graduate Center for Toxicology
University of Kentucky
Lexington, Kentucky

Thomas Coudron
Biological Control of Insects Lab
USDA-ARS
Columbia, Missouri

Hymenopteran venoms hold great promise as useful and effective tools for addressing fundamental questions in many areas of biological science. They also have use in practical biological control and classroom instruction. In this chapter we review ways in which hymenopteran venoms have or may facilitate investigations on aspects of insect bionomics, physiology, evolution, taxonomy, behavior, pharmacology, molecular biology, biological control, and curricular instruction. A goal of this review is to foster attention to the numerous opportunities awaiting in the use of these venoms. Examples are given of new, testable hypotheses following the application of venoms as research tools.

I. Introduction

The toxinology of venoms is a rapidly growing field, especially with the application of molecular techniques that facilitate an understanding of the thousands of different naturally occurring venomous compounds. The forces of survival and food procurement have placed strong selective pressure on the composition of the venoms. As a result, the components of hymenopteran venoms represent a wide spectrum of chemical activities. An increased understanding of the diversity, mechanisms of action, and structures of venom components will permit their use as investigative tools in a wide variety of disciplines. In some situations, use of venoms may offer the means to address fundamental questions that have been refractory to other methods.

The study of venoms has much to offer in practical biological control. Recent efforts utilizing effects of venoms on a wide range of insects have delineated less empirical methods for predicting the success of given biological control efforts (Jones, 1986; Coudron and Puttler, 1988). In the arena of the classroom, the wide diversity of hymenopteran venoms offers numerous instructional possibilities for demonstrating various biological, evolutionary, or molecular principles.

The potential for major advances and uses of parasitic wasp venoms has not been widely realized because much less effort has been devoted to the biochemistry and molecular biology of these agents, as compared with other aspects of host–parasite interactions. The paucity of case examples forces us to examine hymenopteran venoms from a different perspective than is usually done in reviewing various parasite interactions or regulatory mediators. Reviews of wasp venoms based on group of wasp or kind of effect caused in the host are available (Piek and Spanjer, 1986; Coudron, 1991). In this review, we assess the status of research on wasp venoms from the perspective of how such venoms can be used to address questions in different scientific fields. By following such a format, we feel that we can provide the reader with information and new perspectives on how researchers can apply the growing body of information on hymenopteran venoms to their own research needs.

II. Uses of Venoms in Addressing Questions in Various Scientific Fields

A. Inferences on the Bionomics of the Host and Parasite

Faunal surveys frequently yield new species of insects about which little is known, other than the date and place of collection. Many species of parasitic wasps are known only from museum specimens, with little information on the natural biology of the wasp or the identity of its hosts. In such cases, the investigator who desires to identify the unknown host may not have a basis from which to focus additional field studies (see Whitfield and Wagner, 1988, for discussion). Fortunately, it is becoming apparent that the venoms of different wasp species are tailored to accommodate the physiology of their host species. Components of venoms are becoming identified that are diagnostic of interaction with specific physiological pathways found in certain host taxa. Thus, analysis of the venom of a new, or little known, wasp species can assist the investigator in focusing initial research efforts toward identifying the most likely candidate hosts.

1. Biology of Host

Investigators of the bionomics of host–parasite interactions have long recognized certain correlates between the accessibility or mobility of hosts and the occurrence of permanent paralysis of the host caused by the wasp. Specific cases of permanent paralysis of the concealed host include *Coeloides* spp. (Braconidae) that permanently paralyze their cryptic curculionid hosts (DeLeon, 1935), and the permanent paralysis induced in the codling moth by *Mastrus carpocapsae* (McClure, 1933). Hosts that are permanently paralyzed are sometimes "preserved" *in situ* by the agents injected into the host. The host of *M. carpocapsae* is preserved up to 73 days, and the host of *Mellitoba acasta* can be preserved up to 9 months (Steiner, 1986). Preservation of exposed hosts *in situ* occurs, but is much less common since there is a greater likelihood of predation on these hosts compared with a concealed host. These considerations can be informative in initial surveys on hymenopteran fauna, as well as in ecological studies defining niche partitioning and food webs. If analysis shows that a wasp specimen's venom contains components for both paralysis and preservation, it is more probable that the wasp either (1) interacts with a concealed host or (2) makes the host concealed after injecting it with these factors. Thus, consideration of the wasp's venom components will generate testable hypotheses about the taxonomic group and biology of the wasp's host. Diagnostic components of the venom may provide initial clues as to the niches that should be examined first to identify an unknown host. As wasp venom components become better understood, it should be possible to identify specific molecular components that are directed toward particular host templates (e.g., moth versus butterfly, or Lepidoptera versus Coleoptera).

Similar considerations can be used with respect to sessile hosts, which are usually not paralyzed. For example, venom from *Chelonus* wasps (egg-larval parasite) does not paralyze the host (Taylor and Jones, 1989). *Nasonia vitripennis,* a pupal parasite, does not paralyze its host (Ratcliffe and King, 1967). It would be predicted that in species that sting sessile hosts the paralyzing components are absent, or have lost or changed function. Future research on venoms will show whether this hypothesis is correct. If correct, then in the absence of other information the presence of a paralyzing venom would indicate that the (poorly known) wasp species does not attack a sessile host (or sessile host stage). Polydnaviruses occur primarily in wasps that attack lepidopterans or symphytans (Stoltz and Vinson, 1979). The presence of both polydnaviruses and a paralyzing venom then narrows the host range to the larval stage of a lepidopteran or symphytan.

We believe that uses of venom components to draw inferences on parasite–host biology will be important in limiting critical situations such as disappearing tropical rain forests. Indeed, it is likely that there will be extinct parasite species from such areas that will be known only from museum specimens taken in current faunal surveys. Our primary and most direct source of information about their hosts in the future may come from analysis of the venom that interfaced the wasp with its host.

2. Biology of Parasite

Venom components may provide leads to the biology of the parasite. For example, endoparasites seldom paralyze their host, or, if they do, it is usually temporary. *Cotesia congregata, Cotesia glomerata,* and *Microplitis croceipes* are all examples of endoparasites that inject into the mobile host a venom that does not have paralyzing activity (Kitano, 1986; Beckage *et al.*, 1987; Tanaka and Vinson, 1991). Conversely, those parasites that do paralyze their host, permanently or temporarily, are usually ectoparasites, especially those that cause permanent paralysis (e.g., *Bracon hebetor,* Beard, 1952; *Clinocentrus gracilipes*, Shaw, 1981). Thus, analysis of the venom from field-collected adult wasps can provide clues on their biology that cannot be inferred from morphological information. As the biochemistry of parasite venoms becomes better understood, it will be possible to use molecular markers that distinguish between venoms that cause temporary versus permanent paralysis. In situations where now we can only say "biology unknown," information in the future, based on venom analysis, may permit inferences such as: "The wasp most likely permanently paralyzes lepidopteran larval hosts that occur in, or which the wasp puts in, concealed places, in a preserved state (i.e., the host larval stage can be found weeks or months out of season), and upon which the parasite larvae feed as ectoparasites." Conceivably other aspects of the parasite biology, such as solitary versus gregarious parasitism, will be diagnosable on the basis of key venom components.

B. Use of Venoms in Physiological Studies

1. Use as Probes of Normal Host Physiology

a. Timing of developmental events. The ability of venom components to exert an effect only on certain developmental stages of the host suggests changes in aspects of the host template that might otherwise remain uninvestigated. For example, venom proteins from *Chelonus* sp. are relatively stable during the first 48 hr of host embryonic development, but are rapidly degraded by a serine protease activity appearing late in embryonic development (Leluk and Jones, 1989). Larvae of *Paramyelois transitella* that are paralyzed by the venom from the bethilid wasp *Goniozus emigratus* immediately prior to pupation will proceed to complete the pupation process. However, parasitism earlier in the larval stadium causes a host paralysis that results in death (Gordh and Hawkins, 1981).

An example of use of venom to uncover hormonally driven events was provided by Coudron (1991), who showed that injection of venom from *Euplectrus plathypenae* sufficiently prior to the next molt blocks apolysis in the host. Material injected by female *C. gracilipes* (Braconidae) will prevent the host from molting if injected sufficiently in advance of the next molt (Shaw, 1981). Temporary paralysis of spiders by wasps (e.g., *Polysphincta eximia*, Clausen, 1940) can also be associated with inhibition of molting.

These examples illustrate how the action of venoms can reveal developmental windows that occur during arthropod development that may not be accessible or detectable with present tools. In fact, some venoms produce a set of symptoms in the host that have not been produced by conventional physiological probes (e.g., Coudron, 1991, described in the foregoing). We predict that these venoms will be found to interact with parts of the cognate physiological machinery not presently known to exist.

b. Host immune response. Melanization of the hemolymph is considered to be a part of the host immune response that is suppressed by the polydnavirus of braconid and ichneumonid wasps (Beckage *et al.*, 1990). In an interesting study by Stoltz *et al.* (1988) it was shown that without coinjection of the venom, the wasp's (*Cotesia melanoscela*) polydnavirus did not uncoat at the nuclear pores of host tissue. Leluk *et al.* (1988) showed that injection of the venom from *Chelonus*, which also has a polydnavirus (Chelliah and Jones, 1990), at the proper time was necessary for the melanization response of host hemolymph to be suppressed. No known agents can arrest the infection process of insect viruses. Therefore, experimental use of venoms in systems such as those cited earlier could reveal pathways involved in insect immune response and virus infection.

c. Physiology of nervous system. Venom components can be very powerful tools in analysis of the nervous system of insects, and as more insect venom proteins become well studied, their use in such investigations will probably expand. Paralyzing components from parasitic wasps have already been used to assess pre- and postsynaptic aspects of the nervous system of their hosts (Walther et al., 1976). Scorpion venom proteins have been used to isolate insect sodium channel components to which they bind (de Lima *et al.*, 1988) and we can predict that similar uses will be found for some paralyzing venoms of parasitic wasps.

2. Use in Studies on Biochemical Pathways in the Wasp

The venom gland is a specialized tissue that produces a limited number of proteins in high relative abundance. These characteristics make the venom gland a simple and amenable system with which to study the physiological regulation of tissue activity. For example, Jones and Wozniak (1991) have assessed the ontogeny of biosynthesis of venom components in development of wasp pupae and adults, and have identified specific physiological events correlated with induced expression of the venom proteins. Webb and Summers (1990) analyzed polydnavirus structural proteins and venom proteins with common immunodeterminants, and further determined that the expression of these and other polydnavirus proteins was regulated by pupal ecdysteroids. Data from Piek and Spanjer (1986) suggest developmentally regulated production of paralyzing activity in the venom gland of *Venturia canescens*. Thus, wasp venom proteins, in themselves, make useful models in studies on physiological pathways and regulation of the developmental expression of proteins.

C. Use of Venoms in Behavioral Studies

Wasp venoms provide useful perspectives with which to study the behavior of insects. For example, researchers have inquired as to the behavior of parasitic wasps as adapted for effective use of venom. Askew (1971) reported that some braconids place their eggs in nerve ganglia, and predatory wasps inject venom into specific ganglia where it can act. Wilbert (1964) determined that *Aphelinus semiflavus* first stings the host leg to paralyze the host, and only then inserts its sting further into the host to oviposit. *Euplectrus* spp. use their ovipositor to first sting, injecting their venom into the hemocoel of the host, followed by a series of shallow punctures of the host cuticle and deposition of eggs on each puncture site (Coudron and Kelly, 1985).

Wasp behaviors can also assist in biochemical studies on venom proteins. The ovipositional behavior of a wasp was utilized in studies on the venom of a *Chelonus* species. Jones (1987) found that *Chelonus* near *curvimaculatus* wasps prefer to oviposit into young host embryos much more than into old embryos, and that oviposition into older host embryos results in the death of the endoparasite. It was hypothesized that something about the template of older embryos prevented the venom from acting. Leluk and Jones (1989) tested this hypothesis and in the process discovered and characterized a large increase in serine protease activity in older embryos that rapidly degraded the venom. The degradation of the venom by the serine protease activity leads to the death of the endoparasite.

One can also modify behavioral events of the parasite to probe the action of its venom. For example, Leluk and Jones (1989) sought to determine the order of entry of several parasite-derived factors during oviposition. Using female wasps calibrated to spend an average of 20 seconds for oviposition, they interrupted oviposition at successive intervals. Antibodies against the venom proteins were then used to determine that the female injects the venom first during oviposition, followed by the egg and polydnavirus. Carefully timed disruption of wasp oviposition behavior allowed the authors to obtain hosts injected only with natural amounts of venom, and no other regulatory materials. Thus, it was possible for these authors to probe host biochemical processes as influenced solely by natural amounts of injected venom. Strand *et al.* (1983) similarly used female oviposition behavior as a tool in examining the effects of specific injected materials on host development. Shaw (1981) utilized the behavior of an ectoparasite to assess the action of its regulatory materials on the host. The parasite placed its eggs on the surface of the host after injection of venom, enabling their subsequent removal to obtain a host only influenced by the injected material. Thus, combined use of both biochemistry and behavior can permit experimental manipulations and tests that could not be accomplished with either approach alone. At present, we have limited example studies combining behavioral aspects and venom studies. However, many new and different ways exist in which the disciplines of ethology and biochemistry can become coupled through studies on and uses of parasite venoms.

D. Use of Venoms in Evolutionary Studies

Parasitic wasp venoms can be used to address questions on the evolutionary forces bearing upon parasites that are in a coevolutionary state with their hosts, and in speciation of parasites to exploit new hosts. In many wasp–host systems, the venom is the first chemical contact from the wasp that is received by the host. The success of that venom in altering the host is the pivotal point that determines whether the utilization of the host by the parasite will proceed. Therefore, one would predict that venoms would be under very strong pressures of molecular evolution for successful intercession in the host's physiological pathways. There is essentially no biochemical framework of parasitic wasp venoms from which hypotheses can be generated concerning this aspect of wasp–host coevolution. Presently investigations cannot pre- or postdict convergent evolution in parasites for analogous effects versus homologous effects on similar versus dissimilar host templates. For example, there are any number of pathways by which, say, blockage of a molt can be caused by a venom. The venom may block apolysis independent of the host ecdysteroid titer (Kelly and Coudron, 1990; Coudron, 1991), or the parasite's regulatory mediator may actually suppress the ecdysteroid titer (Jones *et al.*, 1986). It is of fundamental interest to understand how selective pressures on unrelated parasites would result in convergent evolution of two different mechanisms of blocking host molting, and when the homologous venom molecules would be independently selected toward the same function. As speciation events partition various resources (host templates), information may be extrapolated concerning advantageous evolutionary strategies. For example, does a very late lineage that "encounters" a template niche similar to that of an early lineage readapt the same previous molecular mechanism? Alternatively, does that late lineage alter the function of a different molecule used for a different purpose by an immediately ancestral lineage? Analysis of the actual venom components causing apparently homologous effects will reveal which venom components of unrelated or distantly related parasites have converged to have the same function. Evolutionary rates may be ascertained by comparing sequence data for specific venom components and the frequency of conserved sequences.

An example from studies on snake venoms illustrates how such considerations as described here can provide insights on molecular evolution. Jeng *et al.* (1979) found evidence that one type of venom protein functioning in high-affinity membrane binding had evolved from phospholipase A2 enzymes (also present in snake venom), without retention and expression of phospholipase activity. Similar mechanisms may operate in hymenopteran venoms. Leluk *et al.* (1989) showed immunological conservation of epitopes in venoms from parasitic Hymenoptera with epitopes in venom proteins from higher social or predatory Hymenoptera in which the venom serves a different function. Webb and Summers (1990) showed that epitopes from wasp venom proteins produced in the venom gland were also found on proteins synthesized in the ovary that serve as polydnavirus structural proteins.

With respect to comparisons across wider phylogenetic distances, Leluk *et al.* (1989) found several trends in hymenopteran venoms (Fig. 1). Braconid

Figure 1 Sodium dodecyl sulfate polyacrylamide gel electrophoresis of venom proteins from different species of Hymenoptera. Lane assignments: (1) *Chelonus insularis,* (2) *Chelonus* near *curvimaculatus,* (3) *Dinoponera grandis,* (4) *Diacamma* sp., (5) *Paraponera clavata,* (6) Ponerine sp., (7) *Pogonomyrmex rugosus,* (8) *Pogonomyrmex maricopa,* (9) *Dasymutilla foxi,* (10) *Dasymutilla nogalensis,* (11) *Pepsis chrysothemis,* (12) *Sphecius grandis,* (13) *Apis mellifera,* (14) *Bombus sonorus,* (15) *Xylocopa varipuncta,* (16) molecular size markers, in kilodaltons, (17) *Microplitis* sp., (18) *Cotesia congregata,* (19) Ichneumonidae sp., (20) Ichneumonidae sp., (21) *Dasymutilla nogalensis* (same as 10), (22) *Pepsis chrysothemis* (same as 11), (23) *Sphecious grandis* (same as 12), (24) *Polistes arizonensis,* (25) *Paravespula pensylvanica,* (26) *Vespa luctuosa,* (27) *Brachygastra mellifica,* (28) *Synoeca septentrionalis,* (29) *Parachartegus fraternus,* (30) *Apis mellifera* (same as 13), (31) *Bombus sonorus* (same as 14). Figure is adapted from Leluk *et al.* (1989), with permission of the publisher.

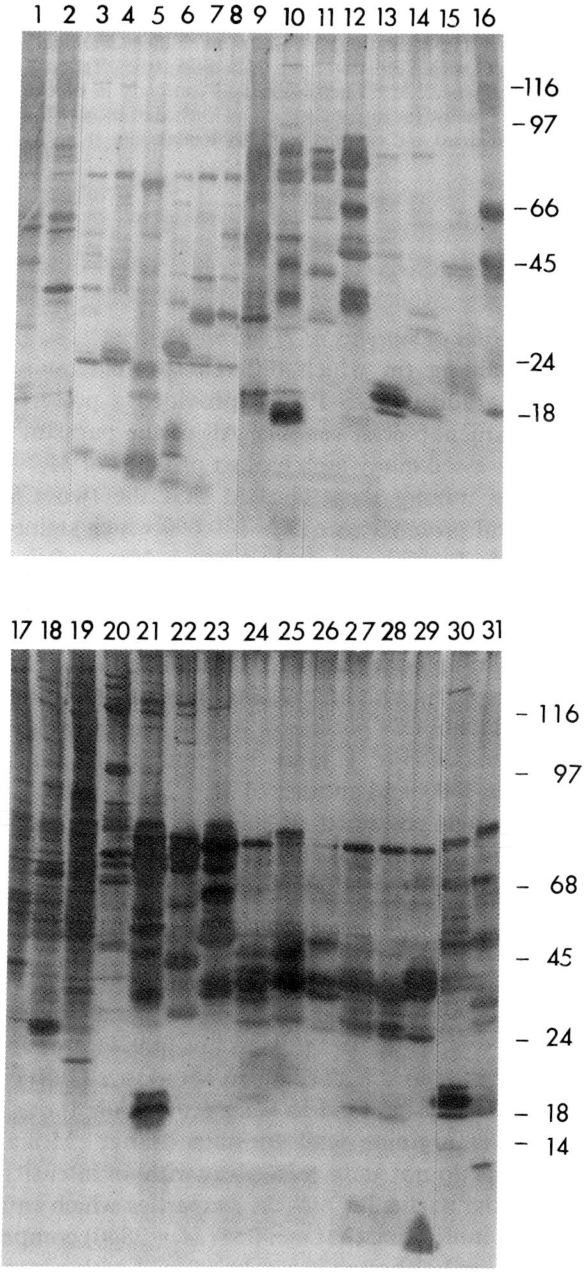
1 2 3 4 5 6 7 8 9 10 11 12 13 14 15 16
–116
–97
–66
–45
–24
–18
17 18 19 20 21 22 23 24 25 26 27 28 29 30 31
– 116
– 97
– 68
– 45
– 24
– 18
– 14

and ichneumonid venoms have an abundance of high-molecular-weight proteins over 100 kDa, whereas most of such proteins are missing from the venoms of higher Hymenoptera such as bees and predatory wasps. Further, the venoms of braconids and ichneumonids are missing the very small, but abundant proteins found in venoms of bees and predatory wasps such as melittin (Fig. 1). We have now examined a parasitic wasp from the family Eulophidae and find that it is also missing most of the abundant, high-molecular-weight proteins that typify braconid and ichneumonid wasps (Fig. 2). Another parameter involved in this comparison is that the venom proteins that were examined previously from identified wasps were all from endoparasitic species, whereas *Euplectrus plathypenae* is an ectoparasite.

Venom proteins are clearly responsive to, and can serve as subjects of study of, forces of molecular evolution. The field of molecular evolution of venoms is in its infancy. Given the tens of thousands of parasitic Hymenoptera that are thought to exist, and the even greater number of hosts against which their venoms act, there clearly exists a huge area of molecular research that is essentially an open field for investigators interested in molecular evolution.

E. Use of Venoms in Taxonomic Studies

Many hymenopteran phylogenies are unclear, with the taxonomic placement of many groups uncertain. Although no single feature of an insect can serve as the definitive taxonomic criterion, the intense selection pressure that the venom is under (discussed in the foregoing) should make it an extremely sensitive indicator of the phylogenic distance between species. Since change in a single protein in the complex mixture of venom proteins (Fig. 1) can conceivably confer access to an altered host range, the venom could be a sensitive detector of sibling species. Again, rigorous use of molecular analysis of venom proteins for such taxonomic purposes at this early stage has very few examples. However, on those occasions in which properties of venoms have been tested in drawing phylogenetic inferences, it has yielded useful results. For example, Whitfield (1992) found a useful application in the use of the presence or absence of a permanently paralyzing venom as a taxonomic character within the Rogadini. Van Achterberg (1988) and Quicke and van Achterberg (1990) found that the morphology of the venom apparatus itself was a useful character toward solutions to such taxonomic questions.

Leluk and Jones (1989) found that the venom of species in the genus *Chelonus* has a characteristic group of high-molecular-weight, yellow-staining proteins (Fig. 1). Other genera in the subfamily Cheloninae (e.g., *Ascogaster*) share many immunologically conserved venom epitopes with *Chelonus* (Fig. 2; Jones *et al.*, 1990). One group whose relationship to the Cheloninae on the

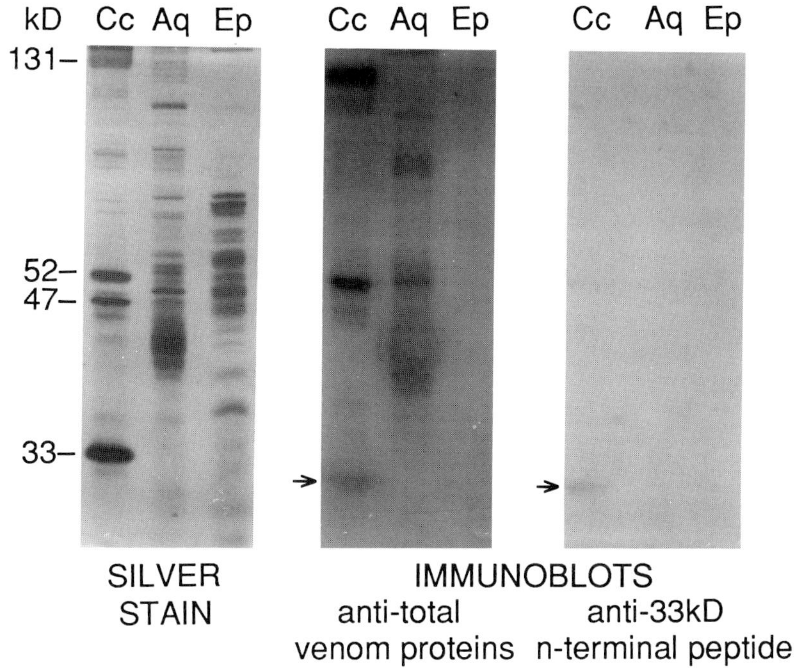

Figure 2 Sodium dodecyl sulfate polyacrylamide gel electrophoresis of venom proteins from *Chelonus* near *curvimaculatus* (Cc), *Ascogaster quadridentata* (Aq), and *Euplectrus plathypenae* (Ep). Left, the proteins visualized by silver staining; right, two panels show the result of immunoblotting similar gels (see Leluk *et al.*, 1989, for methods) using antiserum against total venom proteins from *C.* near *curvimaculatus* and antiserum against an N-terminal octapeptide for the 33-kDa protein from *C.* near *curvimaculatus* conjugated to keyhole limpet hemocyanin. These results show strong cross-reactivity in general between the venom proteins of *C.* near *curvimaculatus* and *A. quadridentata.* The venom from *E. plathypenae,* which blocks host apolysis independently of ecdysteroid suppression (Coudron, 1991), did not show detectable cross-reactivity with either antiserum.

basis of classical morphological characters was for many years uncertain is *Sigalphus* (Muesebeck *et al.,* 1951; Krombein *et al.,* 1979). Species in this group are very rare, and very little bionomic information about them is available that might otherwise assist in their taxonomic placement. However, the results obtained on yellow-staining proteins in both Old and New World *Chelonus,* and the very strong immunological conservation of epitopes between *Chelonus* and *Ascogaster,* suggest that such tests on *Sigalphus* can resolve questions on its phylogenetic association with *Chelonus*. Should the venom of *Sigalphus* possess the high-molecular-weight, immunoreactive yellow-staining proteins, the results would favor the interpretation that *Sigalphus* is closely related to the Cheloninae in general, and *Chelonus* in particular. The potential for extension of this kind of approach to taxonomic questions on other groups of parasitic wasps is immediately evident.

F. Use of Venoms to Study Molecular Biology of Proteins

A classic example of the contributions that studies on hymenopteran venoms can make to molecular biology is the work on the expression of honeybee melittin. Studies on melittin processing from prepromelittin to promelittin to final secreted melittin were pioneering efforts that led to generally applicable concepts on protein secretion (Kreil *et al.*, 1980). With respect to molecular architecture itself, the early studies on melittin led to its description in the Merck index (Windholz, 1983) as the "first polypeptide whose biological effects can be understood on the basis of its primary structure."

There have been few published reports of the sequence of any venom protein from parasitic wasps. Thus, insect venomology has not yet had an opportunity to contribute to general principles in molecular biology and gene expression. Recent studies on cloning and sequencing of venom proteins in *Chelonus* parasitic wasps (Jones *et al.*, 1992) and other nonparasitic Hymenoptera (Vlasak *et al.*, 1983; Vlasak and Kreil, 1984; Fang *et al.*, 1988) have led to the realization that signal peptides in hymenopteran venom proteins often have the standard basic amino acids at each end replaced with or followed by glutamic acid residues (Jones *et al.*, 1992). These results suggest the testable hypothesis that specialized tissues of the hymenopteran venom gland use a form of signal peptide machinery that is different than the classical mechanism. In addition, those authors found that a 33-kDa venom protein was primarily composed of a unique series of 12 tandem repeats. These and other data prompted their hypothesis that some critical venom proteins evolved by duplication/deletion of such repeats, during speciation or molecular coevolution. That the first sequencing of cloned cDNAs for parasite venom proteins would yield such intriguing outcomes, and generate such hypotheses, provides great optimism that future research into this area will be particularly fruitful. Thus, venom proteins offer model systems for elucidation of both generally applicable molecular principles and unique exceptions.

G. Use of Venoms as Pharmacological Probes of Heterologous Systems

Hymenopteran venoms from bees and predatory wasps have been very useful as experimental probes of heterologous systems. Use of heterologous neurotoxins from bees and predatory wasps were used to assess biochemical pathways in nontarget insects (Quistad *et al.*, 1988). Honeybee melittin has been used as a probe of membrane structure (Banks and Shipolini, 1986). A recent report describes the use of melittin to destabilize cell membranes in the isolation of cell nuclei (Smith *et al.*, 1988). Schmidt (1986a,b) found it useful to use venom components to study mechanisms by which allergenic reactions are elicited. With respect to parasitic wasps, Piek and Spanjer (1986) used a venom neurotoxin from one wasp to selectively block excitatory neuromuscular transmission, prior to probing the nerve preparation with the venom of a different wasp. We can anticipate that parasitic wasp venoms will also become useful for medicinal studies. The venom of *Microbracon* has effects that parallel the botulism toxin (Thesleff, 1960), the latter of which has been a popular pharmacological probe in studies of vertebrate systems.

The numerous advances that have been made using toxins and venoms attest to the power of such heterologous probes in dissecting molecular phenomena (e.g., actinomysin, cordycepin, tetradotoxin, etc.). We can anticipate that as the potential of parasitic wasp venoms becomes better understood and investigated, these agents will also become popular and effective pharmacological probes. Many of these efforts will be assisted with studies of the effect of venom components on cells in culture.

H. Use of Venoms in Practical Biological Control

Venoms have uses in biological control from a number of different perspectives. For example, genes for scorpion toxins that are selectively active against arthropods are being tested as a means to increase the effectiveness of baculoviruses (Carbonell *et al.*, 1988; Dee *et al.*, 1990; Coudron, 1991). The tremendous diversity of both parasitic wasp venoms and the host templates on which they act offers an unparalleled resource of single-gene products for engineering of insect pathogens (see also Beard, 1971). These are genes whose products are already evolutionarily designed to intercede in the biochemical pathways of our most important pest insects (Jones, 1986). Baculoviruses can act as the vector to deliver toxins to targeted insects that are not natural hosts of a venom-producing predator or parasite. For example, the venom from *Euplectrus plathypenae* is effective on over 50 factitious hosts (Coudron and Puttler, 1988). Thus, the limits on pathogenicity of a slow-acting baculovirus due to natural limits on host range can be overcome by insertion of the gene for this venom protein into the virus. Simultaneously, the virus provides a means of entry of the venom into hosts that the parasite would normally not sting.

In developing a less empirical basis for predicting the success of some biological control efforts, it is necessary to be able to predict whether the parasite will be able to regulate, and thus survive in, the physiology of a new target host (Jones, 1986). The venom plays a major role in successful utilization of the host by some groups of parasites. Therefore, as additional studies define the specific venom components that regulate specific physiological pathways, it will become possible to predict whether a proposed biological control effort will fail due to inability of the parasite's venom to regulate the potential host.

From a different approach, it has been proposed that a number of unusual host records occurring in over 70 years of biological control literature arise from "obligate mutual parasitism," in which the stinging of a host by one wasp species enables the survival of a second wasp species in that otherwise nonpermissive host (Guzo and Stoltz, 1985). The basis of this phenomenon had not been previously understood, and thus biological control researchers have not had a mechanism to anticipate which host records are of this type. It is now apparent that the venom from one wasp species can enable the survival of a second wasp species in an otherwise inhospitable physiological environment (Guzo and Stoltz, 1985; Vinson and Stoltz, 1986). Now that it is known that the venom participates in immunosuppression of the host and parasite survival in a number of systems (Kitano, 1986; Taylor and Jones, 1989; Tanaka and Vinson, 1991; other chapters in this volume), and that under certain conditions the action of the venom in a normally permissive host can be abrogated, the theoretical framework is present to develop predictive rules on records of obligate mutual parasitism.

In the future, germline transformation procedures for Hymenoptera will inevitably be developed. A theoretical framework established on the action of parasitic wasp venoms will permit selection of the appropriate venom gene(s) that will enable the transformed wasp to redirect the physiology of nonpermissive hosts that are not affected by the wasp's natural venom.

I. Use of Venoms as Classroom Tools

Venoms from Hymenoptera have a number of applications in the classroom or laboratory exercises for illustration or exercise in studying biological, evolutionary, or molecular principles. For example, melittin is a reagent used to lyse red blood cell membranes, and can be easily used to illustrate properties of cell membranes. Commercially available spider venom can be used to illustrate neurophysiological principles in physiology laboratory exercises. In medical school instruction, venoms of social hymenopteran species are used to illustrate principles in allergenic reactions. It is common for laboratory exercises on insect physiology to use ligation of various body regions to illustrate physiological principles. Piek (1966) and Coudron (1991) showed how this format can be easily extended to use of parasitic wasp venoms with different actions via different body parts. We have also utilized wasp venoms in our own classroom instruction in biological control topics. For example, the ability to observe and manipulate *Chelonus* stinging behavior for venom entry, and to observe whether the hosts become truly parasitized (surviving parasite) or pseudoparasitized (dead parasite), has enabled students in lab exercises to couple the disciplines of insect behavior and insect physiology. These real examples provide confidence that the many additional possibilities that can be envisioned could easily be adapted for various instructional purposes.

III. Conclusions

In this chapter we have articulated a number of ways in which venoms of parasitic Hymenoptera can serve as effective tools in investigations in diverse scientific disciplines. As illustrations of such use, we examined a variety of research areas from the different perspectives offered by utilization of wasp venoms. A number of new, testable hypotheses were generated as a consequence of such perspectives. For example, we have proposed that wasp venoms will be found to contain components that are diagnostic of taxa of hosts with distinctive bionomics and/or inferable physiology. We also propose that certain venom proteins have evolved during hymenopteran evolution by duplication or deletion of a core tandem repeat. We also hypothesize that the cellular machinery involved in protein secretion by the venom gland is different from, or is a distinct modification of, the orthodox machinery used in protein secretion from other tissues. We also predict that, as a general rule, the venoms of higher Hymenoptera contain few proteins with a molecular weight above 100 kDa, whereas the venoms of lower, parasitic Hymenoptera contain many relatively abundant proteins above that size.

As the biochemistry and molecular biology of venoms become better understood, we anticipate that the list of examples of such uses of parasite venoms will increase dramatically. The relatively small size of the list at present attests both to how much this area has been heretofore overlooked and to the lucrative research potential that awaits investigators attracted to the study of parasitic wasp venoms. There is every reason that major advances are in the offing for this field in itself, and for other disciplines toward which parasite venoms can be experimentally applied.

Acknowledgments

The authors gratefully recognize Drs. M. Chippendale, W. Steiner, and D. Stoltz for the review of an early draft. This work was supported, in part, by NIH GM 33995 and a Biomedical

Research Support Grant (D. Jones), the USDA ARS (T. Coudon), and the Kentucky Agricultural Experiment Station (D. Jones).

References

Askew, R. R. (1971). "Parasitic Insects." American Elsevier, New York.

Banks, B. E. C., and Shipolini, R. A. (1986). Chemistry and pharmacology of honey-bee venom. *In* "Venoms of the Hymenoptera" (T. Piek, ed.), pp. 330–416. Academic Press, Orlando, FL.

Beard, R. L. (1952). The toxicology of Habrobracon venom: A study of a natural insecticide. *Bull.—Conn. Agric. Exp. Stn. New Haven* **562:**346–351.

Beard, R. L. (1971). Arthropod venoms as insecticides. *In* "Naturally Occurring Insecticides" (M. Jacobson and D. G. Crosby, eds.), pp. 213–270. Dekker, New York.

Beckage, N. E., Templeton, T. J., Nielsen, B. D., Cook, D. I., and Stoltz, D. B. (1987). Parasitism-induced hemolymph polypeptides in *Manduca sexta* (L.) larvae parasitized by the braconid wasp *Cotesia congregata* (Say). *Insect Biochem.* **17:**439–455.

Beckage, N. E., Metcalf, J. S., Nesbit, D. J., Schleifer, K. W., Zetlan, S. R., and de Buron, I. (1990). Host hemolymph monophenoloxidase activity in parasitized *Manduca sexta* larvae and evidence for inhibition by wasp polydnavirus. *Insect Biochem.* **20:**285–294.

Carbonell, L. F., Hodge, M. R., Tomalski, M. D., and Miller, L. K. (1988). Synthesis of a gene coding for an insect-specific scorpion neurotoxin and attempts to express it using baculovirus vectors. *Gene* **73:**409–418.

Chelliah, J., and Jones, D. (1990). Biochemical and immunological studies of proteins from polydnavirus *Chelonus* sp. near *curvimaculatus. J. Gen. Virol.* **771:**2353–2359.

Clausen, C. P. (1940). "Entomophagous Insects," McGraw-Hill, New York.

Coudron, T. A. (1991). Host-regulating factors associated with parasitic Hymenoptera. *ACS Symp. Ser.* **449:**41–65.

Coudron, T. A., and Kelly, T. J. (1985). *Euplectrus plathypenae* parasitization of *Trichoplusia ni.* Effect of weight gain, ecdysteroid titer and molting. *ACS Symp. Ser.* **276:**319.

Coudron, T. A., and Puttler, B. (1988). Response of natural and factitious hosts to the ectoparasite *Euplectrus plathypanae* (Hymenoptera: Eulophidae). *Ann. Entomol. Soc. Am.* **81:**931–937.

Coudron, T. A., Kelly, T. J., and Puttler, B. (1990). Developmental responses of *Trichoplusia ni* (Lepidoptera: Noctuidae) to parasitism by the ectoparasite *Euplectrus plathypenae* (Hymenoptera: Eulophidae). *Arch. Insect Biochem. Physiol.* **13:**83–94.

Dee, A., Belagaje, R. M., Ward, K., Chio, E., and Lai, M.-H. T. (1990). Expression and secretion of a functional scorpion insecticidal toxin in cultured mouse cells. *Bio/Technology* **8:** 339–342.

DeLeon, D. (1935). The biology of *Coeloides dendroctoni* Cushman (Hymenoptera–Braconidae) an important parasite of the mountain pine beetle (*Dendroctonus monticolae* Hopk.). *Ann. Entomol. Soc. Am.* **28:**411–424.

de Lima, M. E., Couraud, F., Lapied, B., Pelhate, M., Diniz, C. D., and Rochat, H. (1988). Photoaffinity labelling of scorpion toxin receptors associated with insect synaptosomal Na^+ channels. *Biochem. Biophys. Res. Commun.* **151:**187–192.

Fang, K. S.-I., Vitale, M., Fehlner, P., and King, T. P. (1988). cDNA cloning and primary structure of a white-face hornet venom allergen, antigen 5. *Proc. Natl. Acad. Sci. U.S.A.* **85:**895–899.

Gordh, G., and Hawkins, B. (1981). *Goniozus emigratus,* a primary external parasite of *Paramyleois transitella,* and comments on bethylids attacking Lepidoptera. *J. Kans. Entomol. Soc.* **54:** 787–803.

Guzo, D., and Stoltz, D. B. (1985). Obligatory multiparasitism in the tussock moth, *Orgyia leucostigma*. *Parasitology* **90:**1–10.

Jeng, T.-W., Hendon, R. A., and Fraenkel-Conrat, H. (1979). Search for relationships among hemolytic phospholipolytic and neurotoxic activities of snake venoms. *Proc. Natl. Acad. Sci. U.S.A.* **75:**600–604.

Jones, D. (1986). Use of parasite regulation of host endocrinology to enhance the potential of biological control. *Entomophaga* **31:**153–161.

Jones, D. (1987). Material from adult female *Chelonus* sp. directs expression of altered developmental programme of host Lepidoptera. *J. Insect Physiol.* **33:**129–134.

Jones, D., Jones, G., Rudnicka, M., Click, A., Reck-Mallaczewen, V., and Iwaya, M. (1986). Pseudoparasitism of host *Trichoplusia ni* by *Chelonus* spp.: A new model system for parasite regulation of host physiology. *J. Insect Physiol.* **32:**315–328.

Jones, D., and Wozniak, M. (1991). Regulatory mediators in the venom of *Chelonus* sp.: Their biosynthesis and subsequent processing in homologous and heterologous systems. *Biochem. Biophy. Res. Commun.* **178:**213–220.

Jones, D., Taylor, T., Farkas, R., Chelliah, J., Haene, B., Brown, J., and Reed-Larsen, D. (1990). Intercession of parasitic wasps (Cheloniae) in host developmental and biochemical pathways. *Adv. Invertebr. Reprod.* **5:**157–162.

Jones, D., Sawicki, G., and Wozniak, M. (1992). Sequence, structure, and expression of a wasp venom protein with a negatively charged signal peptide and a novel repeating internal structure. *J. Biol. Chem.* **267:**14871–14878.

Kelly, T. J., and Coudron, T. A. (1990). Total and specific ecdysteroids in the hemolymph of *Trichoplusia ni* (Lepidoptera: Noctuidae) and its parasite, *Euplectrus plathypenae* (Hymenoptera: Eulophidae). *J. Insect. Physiol.* **36:**463–470.

Kitano, M. (1986). The role of *Apanteles glomeratus* venom in the defensive response of its host, *Pieris rapae crucivora*. *J. Insect. Physiol.* **32:**369–378.

Kreil, G., Haiml, L., and Suchanek, G. (1980). Stepwise cleavage of the pro part of promelittin by dipeptidlylpeptidase. IV. *Eur. J. Biochem.* **111:**49–58.

Krombein, K. V., Hurd, D. D., Jr., Smith, D. R., and Burks, B. D. (1979). "Catalog of Hymenoptera in America North of Mexico." Smithsonian Institution Press, Washington, DC.

Leluk, J., and Jones, D. (1989). *Chelonus* sp. near *curvimaculatus* venom proteins: Analysis of their potential role and processing during development of host *Trichoplusia ni*. *Arch. Insect. Biochem. Physiol.* **10:**1–12.

Leluk, J., Schmidt, J., and Jones, D. (1988). Characterization and functions of the proteins of hymenopteran venoms. *In* "Endocrinological Frontiers in Physiological Insect Ecology" (F. Sehnal, A. Zabza, and D. L. Denlinger, eds.), Vol. 1, pp. 457–460. Wroclaw Tech. Univ. Press, Wroclaw, Poland.

Leluk, J. Schmidt, J., and Jones, D. (1989). Comparative studies on the protein composition of hymenopteran venom reservoirs. *Toxicon* **27:**105–114.

McClure, (1933). The effectiveness of the sting of *Aenoplex carpocapsae* Cushman (Hymen.: Ichneumonidae). *Entomol. News* **44:**48–49.

Muesebeck, C.F.W., Krombein, K. V., and Townes, H. K. (1951). "Hymenoptera of America North of Mexico," U.S.D.A. Monogr. 2, U.S. Govt. Printing Off., p. 142.

Piek, T. (1966). Site of action of the venom of the digger wasp *Philanthus triangulum* F. on the fast neuromuscular system of the locust. *Toxicon* **4:**191–198.

Piek, T., and Spanjer, W. (1986). Chemistry and pharmacology of solitary wasp venoms. *In* "Venoms of the Hymenoptera" (T. Piek, ed.), pp. 161–308. Academic Press, Orlando, FL.

Quicke, D.L.J., and van Achterberg, C. (1990). Phylogeny of the subfamilies of the family Braconidae (Hymenoptera: Ichneumonoidea). *Zool. Verh.* **258:**1–95.

Quistad, G. B., Skinner, W. S., and Schooley, D. A. (1988). Venoms of the social Hymenoptera—Toxicity to the lepidopteran, *Manduca sexta*. *Insect. Biochem.* **18:**511–514.

Ratcliffe, N. A., and King, P. E. (1967). The venom system of *Nasonia vitripennis* (Walker) (Hymenoptera: Pteromalidae). *proc. R. Entomol. Soc. London, Ser. A* **42:**49–61.

Schmidt, J. O. (1986a). Chemistry, pharmacology, and ecology of ant venoms. *In* "Venoms of the Hymenoptera" (T. Piek, ed.), pp. 425–508. Academic Press, Orlando, FL.

Schmidt, J. O. (1986b). Allergy to Hymenoptera venoms. *In* "Venoms of the Hymenoptera" (T. Piek, ed.), pp. 509–546. Academic Press, Orlando, FL.

Shaw, M. R. (1981). Delayed inhibition of host development by the nonparalyzing venoms of parasitic wasps. *J. Invertebr. Pathol.* **37:**215–221, 285.

Smith, R. J., Friede, M. H., Scott, B. J., and von Holt, C. (1988). Isolation of nuclei from melittin destabilized cells. *Anal. Biochem.* **169:**390–394.

Steiner, A. L. (1986). Stinging behavior of solitary wasps. *In* "Venoms of the Hymenoptera" (T. Piek, ed.), pp. 63–160. Academic Press, Orlando, FL.

Stoltz, D. B., and Vinson, S. B. (1979). Viruses and parasitism in insects. *Adv. Virus Res.* **24:**125–171.

Stoltz, D. B., Belland, E. R., Lucarotti, C. J., and Mackinnon, E. A. (1988). Venom promotes uncoating *in vitro* and persistence *in vivo* of DNA from a braconid polydnavirus. *J. Gen. Virol.* **69:**903–907.

Strand, M. R., Rather, S., and Vinson, S. B. (1983). Maternally induced host regulation by the egg parasitoid *Telenomus heliothidis*. *Phys. Entomol.* **8:**469–475.

Tanaka, T., and Vinson, S. B. (1991). Interaction of venoms with the calyx fluids of three parasitoids, *Cardiochiles nigriceps, Microplitis croceipes* (Hymenoptera: Braconidae), *Campoletis sonorensis* (Hymenoptera: Ichneumonidae) in effecting a delay in the pupation of *Heliothis virescens* (Lepidoptera: Noctuidae). *Ann. Entomol. Soc. Am.* **84:**87–92.

Taylor, T., and Jones, D. (1989). Isolation and characterization of the 32.5 kDa protein from venom of an endoparasitic wasp. *Biochim. Biophys. Acta* **1035:**37–43.

Thesleff, S. (1960). Supersensitivity of skeletal muscle produced by *Botulinum* toxin. *J. Physiol. (London)* **151:**598–607.

Van Achterberg, C. (1988). Parallelisms in the Braconidae (Hymenoptera) with special reference to the biology. *In* "Advances in Parasitic Hymenoptera Research" (V. K. Gupta, ed.), pp. 85–115. E. J. Brill, New York.

Vinson, S. B., and Stoltz, D. B. (1986). Cross-protection experiments with two parasitoid (Hymenoptera: Ichneumonidae) viruses. *Ann. Entomol. Soc. Am.* **79:**216–218.

Vlasak, R., and Kreil, G. (1984). Nucleotide sequence of cloned cDNAs coding for preprosecapin, a major product of queen bee venom glands. *Eur. J. Biochem.* **145:**279.

Vlasak, R., Unger-Ullmann, Kreil, G., and Frishauf, A.-M. (1983). Nucleotide sequence of cloned cDNA coding for honeybee prepromelittin. *Eur. J. Biochem.* **135:**123.

Walther, C., Zlotkin, E., and Rathmayer, W. (1976). Action of different toxins from the scorpion *Androctonus australis* on a locust nerve-muscle preparation. *J. Insect Physiol.* **22:**1187–1194.

Webb, B. A., and Summers, M. D. (1990). Venom and viral expression products of the endoparasitic wasp *Campoletis sonorensis* share epitopes and related sequences. *Proc. Natl. Acad. Sci. U.S.A.* **87:**4961–4965.

Whitfield, J. B. (1992). The polyphyletic origin of endoparasitism in the cyclosteme lineages of Braconidae (Hymenoptera). *Syst. Entomol.* **17:**273–286.

Whitfield, J. B., and Wagner, D. L. (1988). Patterns of host ranges within the nearctic species of the parasitoid genus *Pholetesor* Mason (Hymenoptera: Braconidae) *Environ. Entomol.* **17:**608–615.

Wilbert, H. (1964). Das Auslesverhalten von *Aphelinus semiflavus* Howard und die Abwerrreaktionen seiner Wirte (Hymenoptera: Aphelinidae). *Beitr. Entomol.* **14:**159–221.

Windholz, M., ed. (1983). "Merck Index." Merck & Co., Rathway, NJ.

Chapter 11

Biochemical and Molecular Basis of Mosquito Susceptibility to *Plasmodium* and Filarioid Nematodes

Bruce M. Christensen
David W. Severson
Department of Animal Health
and Biomedical Sciences
University of Wisconsin
Madison, Wisconsin

I. Introduction

Mosquito-borne diseases are a major cause of morbidity and mortality in human populations throughout the world, but are especially important in the tropics and subtropics of both hemispheres. Malaria, lymphatic filariasis, yellow fever, and other arboviral diseases infect a significant percentage of the world's population and place an even larger percentage at risk. *Plasmodium*, the causative agent of malaria, is transmitted by *Anopheles* species and infects over 200 million individuals each year, with young children suffering the greatest mortality. The filarioid nematodes, *Wuchereria bancrofti, Brugia malayi*, and *B. timori*, are transmitted to humans by several species of mosquitoes, primarily in the genera *Culex, Aedes, Mansonia*, and *Anopheles*, and cause debilitating lymphatic filariasis in nearly 100 million people. In addition, yellow fever, dengue fever, Japanese encephalitis, and other mosquito-borne arboviruses infect thousands of people throughout the world. Despite the economic and health-related impact these diseases have on the world's popu-

Parasites and Pathogens of Insects
Volume 1: Parasites

lation, relatively little attention has been given to research aimed at clarifying the complex relationships that exist between these pathogens and the mosquitoes required for their development, maintenance, and subsequent transmission to humans. Although primary control of these diseases has been accomplished through mosquito control programs, rapidly increasing resistance to chemical insecticides, together with an awareness of the environmental insults caused by these chemicals, has significantly reduced the effectiveness of these traditional control efforts.

It is well known that only select species, or even select strains of a given species, of mosquitoes function as natural vectors of the above-mentioned pathogens. The majority of species do not support parasite development to the infective stage. We define a susceptible mosquito as one that supports the complete development of the parasite. A resistant mosquito species is one that kills the invading parasite by active immune mechanisms, whereas a refractory mosquito is defined as one in which there seems to be a physiological incompatibility between the parasite and the potential vector. In refractory situations, the parasite may successfully invade the hemocoel environment of the host and reach the appropriate tissue site for development, but then simply fails to develop. A thorough understanding of the mechanisms controlling susceptibility, resistance, and refractoriness of mosquitoes for the parasites they transmit is critical for the development of new and innovative vector control strategies, as well as to gain a complete understanding of the epidemiology of these mosquito-borne diseases; however, little research effort has been directed toward providing this fundamental information.

A significant body of literature, however, is beginning to accumulate regarding the immune response of insects and other arthropods against both microbial and metazoan pathogens. The number of books, symposia publications, and review articles on invertebrate immunity over the last five years support this statement (Brehélin, 1986; Dunn, 1986, 1990; Gupta, 1986; Lackie, 1986; Boman and Hultmark, 1987; Nappi and Christensen, 1987; Christensen and Nappi, 1988; Lackie, 1988; Christensen and Tracy, 1989; Law and Wells, 1989; Karp, 1990; Sugumaran, 1990). These research findings, together with the updated material reviewed elsewhere in this volume, have added significant insight to our understanding of resistance phenomena in select arthropods. But much of these data may be limited because the majority of studies have concerned only a limited number of insect species, with a major emphasis on select species of lepidopterans. It can be seriously questioned how applicable data from lepidopteran larvae really are to clarifying our understanding of mechanisms operating in adult, hematophagous dipterans; therefore, the amount of data presently available regarding the biological interplay between mosquitoes and the parasites they transmit is quite limited in relation to other model systems being used in insect immunity research. This lack of data is even more apparent for the medically important psychodids, simuliids, and ceratopogonids.

Recent events, however, have stimulated an increased research effort aimed at understanding the complex relationships that determine compatibility and incompatibility in vector–parasite models. One reason for this effort undoubtedly involves the publicized interest of private, national, and international funding agencies in the vector biology arena. In addition, the failure of high-visibility programs designed to produce effective vaccines, together with the continuing evolution of drug resistance by vector-borne parasites, has renewed interest in the vector as a logical focus for new control strategies. But perhaps more importantly, the development of contemporary technologies has provided the means whereby sophisticated molecular, cellular, and biochemical studies can be performed successfully using experimental animals of extremely small size. One hopes these incentives will encourage some of the outstanding researchers presently using model systems of little economic and no public health significance to test the findings from these model systems in vectors that cause morbidity and mortality for millions of humans throughout the world.

There have been some advances in our understanding of the biochemical and molecular basis of mosquito susceptibility to parasites during the last several years, and this review will attempt to update some specific findings during the last 3–4 years. Additional information is available in several other review articles dealing specifically with the response of mosquitoes to filarial worm infection (Christensen, 1986; Townson and Chaithong, 1991), mosquito influences on susceptibility to arboviruses (Bishop and Beaty, 1986), and immune mechanisms in arthropod vectors (Christensen and Tracy, 1989). The first section of this review will concern some of the cellular and biochemical aspects of the immune response of mosquitoes against parasites and the following section will address our current state of knowledge regarding the molecular basis for susceptibility and/or refractoriness in mosquito vectors. Because of the paucity of data available concerning either the genetic control of susceptibility or the defensive responses against arboviruses in mosquitoes, this review will concern only the relationships that mosquitoes have with *Plasmodium* and filarioid parasites.

II. Immune Mechanisms in Mosquitoes

A. Non-Self Recognition

Mosquitoes possess an arsenal of cellular and humoral components ostensibly designed to retain the integrity of self, and these components can be very effective in identifying and destroying a multitude of foreign insults. Active immune responses of mosquitoes against malaria parasites (Collins *et al.*, 1986) and filarioid nematodes (Beerntsen *et al.*, 1989) generally are characterized by the encasement of the parasite in a melanotic capsule. Although "melanization" or "melanotic encapsulation" has been the subject of intensive study in several species of insects representing several orders (see reviews in Gupta, 1986; Christensen and Nappi, 1988; Nappi and Christensen, 1987; Lackie, 1988; Christensen and Tracy, 1989; Sugumaran, 1990; Karp, 1990), very little specific information is available concerning the mechanisms controlling these reactions, or the cell biology and biochemistry involved. Virtually nothing is known concerning signal recognition and transduction required for immune recognition of non-self components and the initiation of effector mechanisms responsible for parasite destruction (see Christensen and Tracy, 1989). The suggestion that the prophenol oxidase activating system plays a major role in immune recognition (Ratcliffe *et al.*, 1984; Söderhäll and Smith, 1986) is not supported by appropriate experimental data. Although this cascade of reactions seems critical for the production of melanotic compounds (Ashida and Yamazaki, 1990; Sugumaran, 1990), there is no direct evidence that an immune response is initiated by or dependent on components of the prophenol oxidase activating cascade (Lackie, 1988; Bayne, 1990). Recent studies using tyrosine-deficient mutants of *Drosophila melanogaster* clearly showed this dipteran to be capable of recognition and encapsulation of parasitoids even though they were incapable of forming melanotic capsules (Rizki and Rizki, 1990). Bayne (1990) provides an excellent review of phagocytosis and non-self recognition in invertebrates that addresses many of the problems inherent in studies designed to determine mechanisms of signal recognition. It is apparent, however, that cellular components play key roles in both recognition and destruction of parasites in immunoresponsive mosquito hosts.

B. Cellular Aspects

Although numerous reports exist regarding humoral melanization of foreign invaders within the hemocoel of certain dipterans, without the participation of hemocytes (Götz, 1986), ultrastructural and *in vitro* studies with mosquitoes provide convincing evidence that hemocytes are critical for effective recognition and subsequent melanotic encapsulation (Chen and Laurence, 1985, 1987; Forton *et al.*, 1985; Christensen and Forton, 1986). In addition, hemocyte populations have been shown to increase significantly during immune responses, with this increase likely due to mitotic activity in circulating cells (Christensen *et al.*, 1989). Concomitant with the increase in hemocyte numbers, surface changes on immune-activated cells have been identified using fluorescein-labeled wheat germ agglutinin (WGA) (Nappi and Christensen, 1986; Li and Christensen, 1990). Results from these studies, and studies of *D. melanogaster* (Rizki and Rizki, 1983; Nappi and Silvers, 1984), suggest that surface binding of WGA provides a valid indicator of immunocompetence and/or immune activation in hemocyte populations. Because the identity of hemocyte surface glycoproteins binding WGA, or the potential role these molecules might play in recognition mechanisms, is not known, our laboratory has initiated studies designed to identify and characterize hemocyte surface membrane molecules that might be associated with immune activation or effector mechanisms.

Hemocytes from microfilariae-inoculated (immune-activated), saline-inoculated (wounding control), and uninoculated *A. aegypti* were compared using sodium dodecyl sulfate–polyacrylamide gel electrophoresis (SDS–PAGE), ^{125}I-labeling, and WGA binding techniques (Spray and Christensen, 1991). Both immune activation and wounding induced a significant increase in hemocyte polypeptide synthesis and resulted in the expression of a 200-kDa polypeptide that was not seen in cells from uninoculated mosquitoes. Between 15 and 20 hemocyte surface proteins were labeled with ^{125}I, with a significant increase in intensities seen with hemocytes from inoculated mosquitoes. One of the labeled surface proteins corresponded to the unique 200-kDa polypeptide visualized with SDS–PAGE. Two-dimensional electrophoresis also revealed several polypeptides unique to hemocytes from immune-activated or wounded mosquitoes, but certain polypeptides were characteristic only of cells from mosquitoes undergoing a melanotic encapsulation reaction. Radiolabeled WGA identified three surface protein–WGA complexes of 59, 73, and 107 kDa on the surface of hemocytes from all three groups of *A. aegypti*. This study is the first to evaluate polypeptide synthesis and surface protein changes in activated hemocytes from any insect species, and results obtained provide a firm starting point for the identification of specific hemocyte surface molecules that might be involved in wound healing and immune responses. Characterization of these surface molecules seems vital to our understanding of the initial events of the immune response in mosquitoes, and production of a bank of monoclonal antibodies against immune-activated hemocytes would likely provide the tools necessary to achieve this end. Likewise, the development of culture techniques for hemocytes would greatly facilitate these types of studies (Christensen and Tracy, 1989).

C. Humoral Aspects

Although understanding cellular aspects of immune responses in mosquitoes is strongly justified if we are to clarify recognition and effector events, there are undoubtedly humoral components that play equally critical roles. As it becomes more apparent that the immune responses of mosquitoes and other insects represent highly sophisticated and complex cellular and biochemical events, the dynamic interplay of humoral and cellular components must be considered when evaluating either of these hemolymph compartments in studies of insect immunity. Despite the identification of several "immune proteins" that function to protect hosts from microbial insult (e.g., lysozyme, diptericins, cecropins, attacins, defensins, hemolin, sarcotoxins, phormicins, and eicosanoids, see Dunn, 1990; Sun *et al.*, 1990; Stanley-Samuelson *et al.*, 1991), few studies have identified protein changes associated with melanotic encapsulation or the effect of specific "immune proteins" on mosquito-transmitted parasites.

Using *in vivo* labeling techniques with [^{35}S]methionine, we recently reported that an 84-kDa hemolymph plasma polypeptide is preferentially expressed in *A. aegypti* during wound healing or melanotic encapsulation reactions, and that the expression is greater and persists for a longer time following immune activation (Beerntsen and Christensen, 1990). Data from these studies also suggest that this polypeptide is closely associated with hemocytes, is not induced in response to bacterial contamination, and is not a stress-related (heat shock) protein. In experiments yet to be published, this polypeptide was electroblotted onto a polyvinylidene difluoride membrane for direct sequencing. Two internal sequences of 15 and 8 amino acids were obtained, but no conclusions could be drawn regarding the nature of this protein following analysis with the NBRF protein data base sequence system. To obtain enough material for cloning, we constructed degenerate oligonucleotide primers to each end of the 15-amino acid sequence and conducted PCR amplification as per Lee *et al.* (1988), using *A. aegypti* genomic DNA as a template. A clone was obtained and the deduced sequence matches the data obtained from our direct microsequencing. A full-length clone should provide valuable information as to the identity and function of this 84-kDa polypeptide. These new technologies now make it possible to obtain data for identification and characterization of polypeptides available only in extremely limited quantity, and therefore enable workers to answer questions in hematophagous dipterans that could not be adequately addressed only a few years ago. Insect host size is no longer a serious detriment to studies that address mechanisms of vector–parasite compatibility.

In addition to our studies (Beerntsen and Christensen, 1990), a report by Gwadz *et al.* (1989) confirms that identification and/or evaluation of these insect immune molecules in mosquitoes could provide significant insight into mechanisms of resistance. Gwadz and coworkers injected cecropins into *Anopheles* mosquitoes previously infected with *Plasmodium* and determined that cecropins prevented sporogony by aborting the normal development of oocysts. Although these results were derived from an artificial situation, they provide sound justification for studies designed to assess protein alterations in mosquito strains and species both resistant and susceptible to the parasites they transmit.

A long-standing question in insect immunity regards the potential for opsonic factors in hemolymph plasma (Bayne, 1990), and the role lectins might play in this process (Renwrantz, 1986). Although data regarding hemagglutinin activity in insect blood are equivocal, the series of studies by Ogura and coworkers (1985; Ogura, 1986, 1987) with *Armigeres subalbatus* and *Brugia* microfilariae suggest that a hemagglutinin in the plasma of this mosquito might function as a mediator for melanotic encapsulation reactions. This is especially relevant because *A. subalbatus* functions as a natural vector for *B. pahangi,* but kills *B. malayi* microfilariae in both the hemocoel and thoracic musculature by melanotic encapsulation (Yamamoto *et al.,* 1985; Beerntsen *et al.,* 1989). The discrete specificity of this response could be disrupted by the addition of specific sugars against the hemagglutinin or by the exposure of microfilariae to hemolymph samples from pupae, where the hemagglutinating activity is higher than in adult mosquitoes (Ogura, 1986). The *A. subalbatus–Brugia* spp. model system is deserving of further study and could prove valuable for addressing fundamental questions regarding recognition phenomena in mosquitoes.

D. Biochemical Aspects

Because "melanization" is the primary mechanism responsible for the destruction of filarial worm larvae and malarial organisms in resistant mosquitoes, the biochemistry responsible for the production of melanotic materials has received considerable attention in recent years. Critical to the success of these studies has been the development of highly sensitive and accurate assays for measuring the activity of enzymes involved in melanotic encapsulation reactions and the assessment of quantitative changes in levels of substrates and intermediates required for the production of protein–polyphenol complexes. It has become apparent in the last several years that the production of melanotic capsules is a complex biochemical process involving catecholamines, catecholamine derivatives, catecholamine-metabolizing enzymes, and likely other inducible hemolymph components (Christensen and Tracy, 1989).

Phenol oxidase (tyrosinase) is a copper-containing oxygenase that catalyzes the hydroxylation of tyrosine to *o*-diphenols (monophenol oxidase activity), and the oxidation of *o*-diphenols to their corresponding *o*-quinones (diphenol oxidase activity). The *o*-quinones are chemically reactive and can both polymerize, if structurally appropriate, and bind covalently to nucleophilic protein residues to form protein–polyphenol complexes, commonly referred to as "melanin" or "sclerotin" (Christensen and Tracy, 1989; Li and Nappi, 1991). Consequently, phenol oxidases have been the most intensively studied component of cellular immune responses in arthropods. Because it is critical to assess the augmentation of enzyme activity *in vivo* to clearly delineate the function of these molecules in the production of melanotic compounds, new approaches were necessary in order to accurately measure phenol oxidase activity in the minute hemolymph samples available from mosquitoes. Previously, phenol oxidase activity was measured by monitoring formation of dopachrome from L-3,4-dihydroxyphenylalanine (L-Dopa) at 475 nm. Although useful for *in vitro* assays with large amounts of sample, this assay is fairly insensitive because of the low molar absorption coefficient of dopachrome. In addition, dopachrome is not an end product and is further oxidized to other intermediates and eventually to melanin, thereby resulting in erroneously low estimations of enzymatic activity (Vachtenheim *et al.*, 1985).

Nappi *et al.* (1987) used a radiometric assay, originally described by Townsend *et al.* (1984), to assess monophenol oxidase activity in hemolymph plasma from immune-reactive *A. aegypti.* This assay is based on the quantitation of tritiated water formed during the hydroxylation of L-[3,5-^{3}H]tyrosine to L-dopa. Subsequently, Li *et al.* (1989) used this assay to evaluate monophenol oxidase activity in hemocytes collected from immune-reactive and control *A. aegypti* and *Aedes trivittatus.* They were able to assay enzyme activity from hemocytes perfused from only three mosquitoes, and reported that enzyme activity in hemocytes likely plays a significant role in the production of melanotic capsules around microfilariae.

Although mammalian tyrosinase can clearly catalyze three reactions in the pathway of melanin biosynthesis—the hydroxylation of tyrosine, the dehydrogenation of L-dopa, and the dehydrogenation of 5,6-dihydroxyindole (Korner and Pawelek, 1982)—distinct monophenol and diphenol oxidases have been reported in *D. melanogaster* (Rizki and Rizki, 1983; Pentz *et al.*, 1986). To accurately assess diphenol oxidase activity, we developed an improved assay based on the report by Sugumaran (1986) that mushroom tyrosinase catalyzes the oxidative decarboxylation of 3,4-dihydroxymandelic acid to form 3,4-dihydroxybenzaldehyde (Li *et al.*, 1990). Because these are catechols and the product of the reaction is stable, a quantitative assay was developed using high-pressure liquid chromatography with electrochemical detection (HPLC-ECD), and the applicability of this assay for measuring diphenol oxidase activity in microgram quantities of crude, cell-free hemolymph from mosquitoes was demonstrated (Li *et al.*, 1990).

The availability of highly sensitive and specific assays for both monophenol and diphenol oxidase activity enabled Li *et al.* (1992) to critically assess these catalytic activities in hemolymph compartments of *A. aegypti* during melanotic encapsulation reactions. No significant differences in either monophenol or diphenol oxidase activity in hemocytes, cell-free plasma, or complete hemolymph (plasma + hemocytes) were noted between samples from uninoculated (naive) or saline-inoculated (wounded) mosquitoes, although baseline levels of both activities were detected. Both enzyme activities were significantly elevated in hemocyte and complete hemolymph samples following immune activation by the inoculation of microfilariae, but no significant increase in enzyme activity was detected in cell-free plasma. Increases in the activity of both enzymes were proportional in all samples, thereby suggesting that a single enzyme might be reacting with both monophenols and *o*-diphenols within the hemolymph of this mosquito. These studies also suggested that the increased phenol oxidase activity noted during melanotic encapsulation reactions is associated primarily with the hemocytes. This augmented enzyme activity was a result of *in vivo* activation by parasites, and this activity could not be increased by the application of the commonly used phenol oxidase activators zymosan and laminarin (Li *et al.*, 1992).

A monphenol oxidase assay recently was developed that is much more sensitive than the specific radioenzymatic assay (Li and Nappi, 1991). This assay also uses HPLC-ECD, and it measures the hydroxylation of tyrosine using *N*-acetyldopamine as cofactor and ascorbate as a reducing agent. The ascorbate maintains the level of L-dopa produced during tyrosine hydroxylation by reducing dopaquinone back to the *o*-diphenol nonenzymatically. The utility of this assay was demonstrated by Li and Nappi (1991) using microliter quantities of hemolymph from the larvae of *D. melanogaster.*

With the availability of these highly sensitive radiometric and HPLC-ECD assays for both monophenol and diphenol oxidase activity, researchers working with any insect system should be able to more accurately assess the role these enzymes might play in host–pathogen associations. Contrary to many other insect model systems where various means of phenol oxidase activation must be employed *in vitro* to measure activity by the spectrophotometric method, enzyme activity and its augmentation due to foreign insult can be assessed without activation using these new protocols. Using the radioenzymatic assay for monophenol oxidase, Beckage *et al.* (1990) were able to measure activity directly in *Manduca sexta* hemolymph samples without using an activator. Because of the limitations of the spectrophotometric phenol oxidase assay commonly used (Garcia-Canovas *et al.*, 1982; Garcia-Carmona *et al.*, 1982; Townsend *et al.*, 1984; Vachtenheim *et al.*, 1985), the use of more specific and sensitive HPLC-ECD assays should be encouraged for research in the field of arthropod immunity.

Although hemocytes seem to contribute significant phenol oxidase activity for the melanotic encapsulation of microfilariae within the hemocoel environment of mosquitoes (Li *et al.*, 1992), the encasement of ookinetes of *Plasmodium* on the surface of the midgut in *A. gambiae* likely does not involve the direct participation of circulating cells. Following the selection of a resistant strain of *A. gambiae* that melanizes the ookinete between the basal membrane labyrinth and basal lamina of the midgut (Collins *et al.*, 1986), Paskewitz *et al.* (1988, 1989) reported that resistance might be due to increased phenol oxidase activity within midgut cells. Incubation of susceptible and resistant midguts with L-Dopa or dopamine resulted in significant differences in the pattern of melanin deposition, with the refractory strain showing intense deposition of melanotic materials around encapsulated parasites and within midgut cells in the immediate proximity of the ookinete. There is no evidence to suggest, however, that differences in phenoloxidase activity between these strains reflect differences in the genes responsible for the expression of this enzyme.

Even though a key role for phenol oxidase in immune effector mechanisms of mosquitoes and other insects is apparent, the specific substrates involved, and the biochemical pathways by which they are converted to melanotic substances, have not been characterized for any insect species (Christensen and Tracy, 1989). Undoubtedly, other catecholamine-metabolizing enzymes in addition to phenol oxidases also are involved in melanotic encapsulation reactions. Studies of dopa decarboxylase conducted in the laboratory of Dr. A. J. Nappi (Loyola University of Chicago) were reported at the American Society of Parasitologists annual meeting in Madison, Wisconsin, in August, 1991. Using a temperature-sensitive dopa decarboxylase mutant strain of *D. melanogaster* and the parasitoid *Leptopilina,* this laboratory showed that the cellular and melanotic immune response mechanisms were severely compromised in hosts with reduced levels of this enzyme. Likewise, we have unpublished data from our studies with *A. aegypti* that suggest dopa decarboxylase might also be a factor participating in the melanotic encapsulation process.

Studies by Munkirs *et al.* (1990) identified tyrosine, dopamine, and *N*-β-alanyldopamine in hemolymph plasma from *A. aegypti* using HPLC-ECD. Following immune activation by the intrathoracic inoculation of heat-killed microfilariae, a single major peak (called peak I) was seen on the chromatogram that was not present in naive hemolymph samples. Wounding by the inoculation of saline without microfilariae resulted in hemolymph samples that contained only 5% of the peak I seen in immune-reactive plasma. This unknown catecholamine also was detected after treatment of hemolymph plasma with mild alkaline conditions, indicating that it is normally present as an electrochemically inert form. Peak I, however, did not cochromatograph with any of the catecholamines implicated in melanotic encapsulation reactions.

No other published accounts address changes in catecholamine substrates associated with melanotic encapsulation reactions in insects. Furthermore, there are no reports evident that evaluate catecholamine-metabolizing enzymes, such as decarboxylases, transferases, isomerases, or synthetases, regarding their potential role in the production of protein–polyphenol complexes required for parasite destruction. Several earlier reviews strongly suggest that more detailed studies of other enzymes and their substrates in immune-reactive insects are vital for our understanding of the biochemical pathways involved in the effector mechanisms of mosquitoes used to kill parasites in resistant hosts (Nappi and Christensen, 1987; Christensen and Nappi, 1988; Christensen and Tracy, 1989). Clearly, there remains a need to adequately address this critical area of research.

III. Genetic Control of Susceptibility/Refractoriness

A. Historical Perspective

It has been realized for some time that susceptibility to parasitic infection varies between different mosquito species and even among different geographical strains of the same species, and early studies with mosquito–*Plasmodium* models were the first to demonstrate that susceptibility of mosquitoes could be increased by selection (Huff, 1929; Trager, 1942; Micks, 1949). Likewise, Roubaud (1937) reported a greater susceptibility of certain strains of *A. aegypti* to infection with the filarial worm *Dirofilaria immitis* and suggested that susceptibility was an inherited character. Kartman (1953) compared susceptibilities of various strains of *A. aegypti, A. albopictus, Culex pipiens,* and *C. p. quinquefasciatus* to *D. immitis* and was able to obtain, by mass selection, strains of *A. aegypti* more refractory and more susceptible to this parasite.

In a classic series of studies, Macdonald (1962a,b, 1963) clarified the mode of inheritance of susceptibility of *A. aegypti* to semiperiodic *B. malayi*. From an initial population of mosquitoes showing a 17–31% susceptibility to this parasite, Macdonald obtained by familial selection through 15 generations a strain with a mean susceptibility of 84%. Through a series of crosses and backcrosses between this highly susceptible strain and uniformly refractory strains, he demonstrated that susceptibility was controlled by a sex-linked recessive gene that he designated f^m for filarial worm susceptibility, *B.*

malayi. Although f^m is the major controlling factor, Macdonald (1963) also demonstrated that this gene did not control susceptibility entirely and therefore concluded that modifying genes also must play a role. Macdonald and Ramachandran (1965) subsequently showed that f^m also controlled susceptibility of *A. aegypti* to periodic *B. malayi*, *B. pahangi*, *Wuchereria bancrofti*, and subperiodic *W. bancrofti*, but had no influence on susceptibility to *D. immitis* or *D. repens*. Because f^m homozygotes are susceptible to all muscle-inhabiting filarial worms tested, Barr (1975) suggested that selection for the f^m homozygote disrupted a basic defense mechanism of the mosquito.

Subsequent studies by Raghaven *et al.* (1967), Zielke (1973), and McGreevy *et al.* (1974) clearly showed that susceptibility of *A. aegypti* to *D. immitis* also was controlled by a sex-linked recessive gene, but not f^m. McGreevy *et al.* (1974) designated this gene f^t, and suggested that genes controlling the susceptibility of *A. aegypti* to filarial worm species directly affect the host tissue supporting parasite development rather than the parasite itself. Studies by Nayar *et al.* (1988) also suggest that genes influencing susceptibility of mosquitoes to *D. immitis* are expressed in the tissue site of parasite development. By transplanting *D. immitis*-infected Malpighian tubules from a refractory mosquito line to the hemocoel of uninfected susceptible lines, and by conducting the reciprocal transplants, they reported that the success or failure of parasite development was determined by the genotype of the Malpighian tubules. The genotype of the hemolymph bathing the Malpighian tubules had no influence on parasite development. However, Kobayashi *et al.* (1986) used parabiotic twinning with *B. malayi* susceptible and refractory strains of *A. aegypti* to demonstrate that development of larvae inhabiting the thoracic musculature was inhibited in susceptible mosquitoes when they were twinned with a refractory mosquito. They suggested that hemolymph factors might be responsible for the refractory condition of *A. aegypti* for this filarial worm species.

Although Huff (1929), Trager (1942), and Ward (1963) reported a genetic basis for variation in susceptibility of several different mosquito species for *Plasmodium* infections, the first clear genetic evidence was provided by Kilama and Craig (1969). They demonstrated that *P. gallinaceum* susceptibility in *A. aegypti* is determined primarily by a single autosomal dominant gene, *pls*. Their data, however, also suggest that other genetic factors are involved in *Plasmodium* susceptibility in this vector, because up to six developing oocytes were frequently isolated from the midguts of individual mosquitoes that they classified as completely refractory. They arbitrarily selected a count of ten developing oocytes as the point of demarcation between refractory and susceptible individuals. Likewise, Ward (1963) was unable to develop a 100% refractory strain of *A. aegypti* even after 26 generations of selection, suggesting a multigenic mode of inheritance. More recently, a strain of *A.*

gambiae, a principal vector of human malaria, has been selected for resistance to several species of *Plasmodium* by Collins *et al.* (1986). Genetic analysis of *Plasmodium* resistance in *A. gambiae* has suggested that two unlinked genes control the expression of susceptible and resistant phenotypes (Vernick *et al.*, 1989).

Even though this genetic basis for susceptibility and refractoriness of certain mosquitoes to both filarial worm and *Plasmodium* infections has been known for several decades, very little information has been generated concerning the genes or gene products controlling these conditions in any mosquito species.

B. Genes and Gene Products

Recent biotechnological advances have spurred a rapidly developing research effort aimed at advancing our knowledge in the area of vector–parasite relationships and vector genetics, as evidenced by a meeting in January, 1991, in Tucson, Arizona, on the "Prospects for Malaria Control by Genetic Manipulation of its Vectors" organized by the WHO Special Program for Research and Training in Tropical Diseases and the MacArthur Foundation [World Health Organization (WHO), 1991]. However, the genetic information presently available as a basis for this research effort is limited for the majority of mosquito vectors.

Aropheles gambiae has received considerable attention because of its importance in the transmission of human malaria and the selection of a resistant strain that successfully destroys ookinetes on the midgut surface (Collins *et al.*, 1986). In addition to the data suggesting that two unlinked genes control susceptibility in this species (Vernick *et al.*, 1989), Vernick and Collins (1989) provide evidence that one of these loci exhibits a tight linkage association with a polymorphic autosomal esterase locus. The effector mechanisms responsible for the killing of ookinetes involve the melanotic encasement of the parasite; consequently, studies by Paskewitz *et al.* (1989) have suggested that differences in phenol oxidase activity might account for differences in susceptibility between strains of *A. gambiae*. However, Romans *et al.* (1991) conducted backcrosses between two inbred lines of *A. gambiae* to determine linkage associations between a presumed diphenol oxidase gene (Dox-A2 derived from *Drosophila*) and the esterase locus linked with refractoriness. Results indicted that the two genes segregated independently. From data presently available, it is difficult to assess the roles that these enzymes actually play in influencing *Plasmodium* susceptibility in this vector. Because of the lack of identified genetic markers for *A. gambiae*, efforts to define the precise genetic basis for resistance are limited.

The most detailed information concerning the physical location of genes controlling susceptibility is known for *A. aegypti,* because this is the only species for which extensive genetic linkage data are available. A genetic linkage map, based on about 70 morphological mutant and isozyme markers (consisting of three linkage groups), has been developed for this species (Munstermann and Craig, 1979; Munstermann, 1990). Consequently, determining the genome position of genes of interest can be accomplished through linkage analysis with these identified marker loci. Kilama and Craig (1969) and Macdonald (1962b) used this technique to determine that *pls* is located on linkage group II and f^m is located on linkage group I. These vector–parasite model systems (*A. aegypti–P. gallinaceum* and *A. aegypti–Brugia* spp.) therefore are particularly important because immediate reference points for genes of interest are available for research efforts designed to isolate and characterize genes controlling susceptibility. Traditionally, however, a large number of segregating populations are required to develop linkage associations for genes of interest, because only a limited number of loci segregate in each cross. Because restriction fragment length polymorphism (RFLP) markers represent a new class of molecular markers that permit the development of detailed genetic maps from a limited number of crosses (Landry and Michelmore, 1987), we have taken this approach in our initial efforts to isolate the genes influencing susceptibility of *A. aegypti* to *P. gallinaceum* and *B. malayi.*

We have constructed a RFLP genetic linkage map consisting of 50 clones that identify 53 loci covering 134 map units across the *A. aegypti* genome (Severson *et al.,* 1993). The RFLP markers include 42 random cDNA clones, 3 random genomic DNA clones, and 5 cDNA clones of known genes. Our segregating populations consisted of F_2 progeny obtained from crosses between the Liverpool and RED or BZ strains of *A. aegypti.* The RED strain carries the red-eye (*re*) locus on linkage group I, the spot-abdomen (*s*) locus on linkage group II, and the black-tarsus (*blt*) locus on linkage group III. The BZ strain carries the *re* locus and the bronze-cuticle (*bz*) locus on linkage group I. Determination of linkage associations between our RFLP markers and several mutant marker loci allowed us to integrate them into the three previously identified linkage groups (Munstermann, 1990).

Although the RFLP markers are randomly distributed throughout the *A. aegypti* genome, a considerable proportion (30%) of them fitted a 1:1 segregation ratio in at least one of our mapping populations. This reflects progeny of a mating between a heterozygous individual and a homozygous individual for each of these loci. Therefore, despite many generations of maintenance as independent colonies, our segregation data indicate that laboratory strains of *A. aegypti* remain highly heterozygous. This has important implications for future mapping efforts, since crosses involving heterozygous individuals can limit the number of diagnostic loci identifiable with a given cross.

The availability of a saturated RFLP map provides a valuable mechanism to address aspects of vector biology at their most fundamental level. This linkage map could facilitate the isolation of genes of interest whose gene product is unknown (i.e., filarial worm or malaria susceptibility) by providing points for chromosome walking. In addition, many important phenotypic characters, such as filarial worm and *Plasmodium* susceptibility, are a consequence of the joint action of two or more individual genes. These characters are generally referred to as multigenic or quantitative traits, and expression of a particular phenotype depends upon a balance between individual gene effects and their interactive effects with each other. A saturated RFLP genetic linkage map provides a useful mechanism for the resolution of complex phenotypic traits into their individual genetic components (Nienhuis *et al.*, 1987; Osborn *et al.*, 1987; Paterson *et al.*, 1988; Tanksley and Hewitt, 1988; Lander and Botstein, 1989), by identifying all intervals between RFLP markers (referred to as quantitative trait loci, QTLs) that contain a gene(s) affecting a character throughout the genome. Each QTL can then be examined as a discrete genetic entity and its individual and interactive properties measured. These RFLP markers also can be used to examine chromosomal content and linear gene order among related species (Bonierbale *et al.*, 1988; Tanksley *et al.*, 1988). This will likely prove valuable for other mosquito species because an earlier examination of linkage relationships of phenotypic characters among higher dipteran species suggests that the major linkage groups are largely conserved (Foster *et al.*, 1981).

Since the basic chromosome number is conserved among mosquito species ($n = 3$), one can predict that chromosomal content and linear gene order are highly conserved among mosquito species. This should allow the utilization of the wealth of information concerning the genetics of *A. aegypti* to rapidly develop a detailed RFLP genetic linkage map for a genetically uncharacterized but related mosquito species. This scenario appears quite probable as we are successfully using our RFLP markers to examine linkage within segregating populations of *Aedes albopictus*. Preliminary results indicate that the linear order for three of our RFLP markers on linkage group I is conserved between *A. aegypti* and *A. albopictus* (Y. Mori, D. W. Severson, and B. M. Christensen, unpublished data). We have also successfully hybridized many of our RFLP markers to genome blots of other mosquito species including: *Aedes togoi*, *Anopheles gambiae*, *Armigeres subalbatus*, and several subspecies of the *Culex pipiens* complex (*C.p. molestus*, *C.p. pallans*, *C.p. pipiens*, *C.p. quinquefasciatus*).

No direct evidence is presently available, however, that identifies a gene product responsible for susceptibility or refractoriness of any mosquito species for a particular parasite, but recently we have identified several induced polypeptides that are strongly associated with *A. aegypti* strains that are refractory for the development of *B. malayi* (Wattam and Christensen, 1992). Polypeptide synthesis in thoracic tissue was assessed, using *in vivo* radiolabeling techniques (Beerntsen and Christensen, 1990), in susceptible and refractory strains of *A. aegypti* and a refractory strain newly isolated from a susceptible strain. Six polypeptide differences were recognized by SDS–PAGE and two-dimensional electrophoresis. These polypeptides were seen only in the refractory strains for 48 hr following a blood meal. A seventh polypeptide was present in those refractory mosquitoes that had ingested sucrose, but increased in intensity following bloodfeeding. The presence of microfilariae in the bloodmeal was not necessary to stimulate the synthesis of these molecules. Although these differences in bloodmeal-induced polypeptide production were strongly correlated with variation in susceptibility, direct evidence is not yet available regarding the role these polypeptides might play in mediating the genetically determined variation in susceptibility (Wattam and Christensen, 1992).

There are numerous physiological events that take place within vector mosquitoes that could influence the suitability of mosquitoes as hosts for parasites. It is not known, for instance, what role salivary gland components might play in influencing susceptibility and/or refractoriness. Earlier studies by Rosenberg (1985) clearly showed that sporozoites may not be capable of recognizing the appropriate membrane receptors on salivary glands of refractory *Anopheles* in certain mosquito–*Plasmodium* model systems. Because of the importance of salivary glands in the transmission of several pathogens by mosquitoes, researchers in James' laboratory at the University of California, Irvine, have undertaken a molecular analysis of the salivary glands of *A. aegypti* (James *et al.*, 1989, 1991). It is possible that the isolation and characterization of genes expressed specifically in the salivary glands could provide new insight regarding physiological mechanisms influencing susceptibility and/or refractoriness. In a similar manner, investigators in Raikhel's laboratory at Michigan State University, East Lansing, have been assessing the molecular biology of vitellogenesis with the ultimate goal of finding new and innovative means of disrupting reproduction in vector mosquitoes (Raikhel, 1992; Raikhel and Dhadialla, 1992). It would be extremely valuable if other research groups would make similar commitments to a molecular analysis of specific tissue sites of parasite development within select mosquito species, such as midgut, thoracic musculature, and Malpighian tubules.

C. Molecular Approaches in Vector Control: Future Directions

The genetic plasticity of mosquito vectors has resulted in the emergence of multiple-pesticide-resistant vector populations, thereby limiting the success of numerous vector control programs in areas endemic for malaria and filariasis. An alternate strategy for vector control would be to utilize elements of the observed genetic plasticity within vector populations to disrupt parasite transmission, by manipulating those aspects of vector competence under genetic control. The steps outlined (WHO, 1991) for the engineering and release of refractory mosquitoes involve (1) obtaining a thorough understanding of the molecular basis of vector competence, (2) the development of necessary molecular and genetic approaches required to engineer refractory mosquitoes, and (3) a more complete understanding of the population dynamics necessary to effectively apply these technologies in field situations. It is apparent from the previous discussion in this review that studies designed to elucidate the molecular basis of vector competence are in their infancy. Likewise, progress in the development of transgenic technology in mosquitoes has been limited.

Stable transformation of mosquitoes has been demonstrated in *A. gambiae* (Miller *et al.*, 1987), *Aedes triseriatus* (McGrane *et al.*, 1988), and *A. aegypti* (Morris *et al.*, 1989), although integration in each of these species occurred by recombination that was independent of P elements. For a thorough assessment of some of the problems associated with the production of transgenic mosquitoes, and the potential for using engineered mosquitoes in vector control strategies, several recent reviews are recommended (Crampton *et al.*, 1990; Eggleston, 1991; WHO, 1991). It is apparent, however, that techniques for efficient transformation systems for mosquitoes presently are not available.

Although the concept of releasing engineered mosquitoes with reduced vectorial capacity and/or increased resistance to pesticides might be an ultimate goal that is never realized, the development of this technology would have an enormous impact on our most fundamental understanding of gene regulation and expression in vectors. The definitive determination of gene function becomes possible if transformation systems for mosquitoes become as efficient as those developed for *Drosophila*.

Acknowledgments

This work was supported in part by National Institutes of Health Grants AI 19769, AI 28781, and AI 33127. We thank Dr. Jianyong Li for his critical review of this manuscript.

References

Ashida, M., and Yamazaki, H. I. (1990). Biochemistry of the phenoloxidase system in insects: With special reference to its activation. *In* "Molting and Metamorphosis" (E. Ohnishi and H. Ishizaki, eds.), pp. 239–265. Springer-Verlag, Berlin.

Barr, A. R. (1975). Evidence for genetic control of invertebrate immunity and its field significance. *In* "Invertebrate Immunity" (K. Maramorosch and R. E. Shope, eds.), pp. 129–135. Academic Press, New York.

Bayne, C. J. (1990). Phagocytosis and non-self recognition in invertebrates. *BioScience* **40:**723–731.

Beckage, N. E., Metcalf, J. S., Nesbit, D. J., Schleifer, K. W., Zetlan, S. R., and Buron, I. D. (1990). Host hemolymph monophenoloxidase activity in parasitized *Manduca sexta* larvae and evidence for inhibition by wasp polydnavirus. *Insect Biochem.* **20:**285–294.

Beckmann, J. S., and Soller, M. (1983). Restriction fragment length polymorphisms in genetic improvement: Methodologies, mapping and costs. *Theor. Appl. Genet.* **67:**35–43.

Beerntsen, B. T., and Christensen, B. M. (1990). *Dirofilaria immitis:* Effect on hemolymph protein synthesis in *Aedes aegypti* during melanotic encapsulation reactions against inoculated microfilariae. *Exp. Parasitol.* **71:**406–414.

Beerntsen, B. T., Luckhart, S., and Christensen, B. M. (1989). *Brugia malayi* and *B. pahangi:* Inherent differences in immune activation in the mosquitoes *Armigeres subalbatus* and *Aedes aegypti. J. Parasitol.* **75:**76–81.

Bishop, D.H.L., and Beaty, B. J. (1986). Interference-immunity of mosquitoes to bunyavirus superinfection. *In* "Immune Mechanisms in Invertebrate Vectors" (A. M. Lackie, ed.), pp. 95–115. Clarendon Press, Oxford.

Boman, H. G., and Hultmark, D. (1987). Cell-free immunity in insects. *Annu. Rev. Microbiol.* **41:**103–126.

Bonierbale, M. W., Plaisted, R. L., and Tanksley, S. D. (1988). RFLP maps based on a common set of clones reveal modes of chromosomal evolution in potato and tomato. *Genetics* **120:**1095–1103.

Brehélin, M. (1986). "Immunity in Invertebrates." Springer-Verlag, Berlin.

Chen, C. C., and Laurence, B. H. (1985). An ultrastructural study on the encapsulation of microfilariae of *Brugia pahangi* in the haemocoel of *Anopheles quadrimaculatus. Int. J. Parasitol.* **15:**421–428.

Chen, C. C., and Laurence, B. H. (1987). In vitro study on humoral encapsulation of microfilariae: Establishment of technique and description of reaction. *Int. J. Parasitol.* **17:**781–787.

Christensen, B. M. (1986). Immune mechanisms and mosquito–filarial worm relationships. *In* "Immune Mechanisms in Invertebrate Vectors" (A. M. Lackie, ed.), pp. 145–160. Clarendon Press, Oxford.

Christensen, B. M., and Forton, K. F. (1986). Hemocyte-mediated melanization of microfilariae in *Aedes aegypti. J. Parasitol.* **72:**220–225.

Christensen, B. M., and Nappi, A. J. (1988). Immune responses of arthropods. *ISI Atlas Sci.: Plant Anim. Sci.* **1:**15–19.

Christensen, B. M., and Tracy, J. W. (1989). Arthropod-transmitted parasites: Mechanisms of immune interaction. *Am. Zool.* **29:**387–398.

Christensen, B. M., Huff, B. M., Miranpuri, G. S., Harris, K. L., and Christensen, L. A. (1989). Hemocyte population changes during the immune response of *Aedes aegypti* to inoculated microfilariae of *Dirofilaria immitis. J. Parasitol.* **75:**119–123.

Collins, F. H., Sakai, R. K., Vernick, K. D., Paskewitz, S., Seeley, D. C., Miller, L. H., Collins, W. E., Campbell, C. C., and Gwadz, R. W. (1986). Genetic selection of a *Plasmodium* refractory strain of the malaria vector *Anopheles gambiae. Science* **234:**607–610.

Crampton, J., Morris, A., Lycett, G., Warren, A., and Eggleston, P. (1990). Transgenic mosquitoes: A future vector control strategy? *Parasitol. Today* **6:**31–36.

Dunn, P. E. (1986). Biochemical aspects of insect immunology. *Annu. Rev. Entomol.* **31:**321–339.

Dunn, P. E. (1990). Humoral immunity in insects. *BioScience* **40:**738–744.

Eggleston, P. (1991). The control of insect-borne disease through recombinant DNA technology. *Heredity* **66:**161–172.

Forton, K. F., Christensen, B. M., and Sutherland, D. R. (1985). Ultrastructure of the melanization response of *Aedes trivittatus* against inoculated *Dirofilaria immitis* microfilariae. *J. Parasitol.* **71:**331–341.

Foster, G. G., Whitten, M. J., Konovalov, C., Arnold, J.T.A., and Maffi, G. (1981). Autosomal genetic maps of the Australian sheep blowfly, *Lucilia cuprina dorsalis* R.-D. (Diptera: Calliphoridae), and possible correlations with the linkage maps of *Musca domestica* L. and *Drosophila melanogaster* (Mg.). *Genet. Res.* **37:**55–69.

Garcia-Canovas, F., Garcia-Carmona, F., Sanchez, J. V., Pastor, J.L.I., and Teruel, J.A.L. (1982). The role of pH in the melanin biosynthesis pathway. *J. Biol. Chem.* **257:**8738–8744.

Garcia-Carmona, F., Garcia-Canovas, F., Iborra, J. L., and Lozano, J. A. (1982). Kinetic study of the pathway of melanization between L-dopa and dopachrome. *Biochim. Biophys. Acta* **717:**124–131.

Götz, P. (1986). Mechanisms of encapsulation in dipteran hosts. *In* "Immune Mechanisms in Invertebrate Vectors" (A. M. Lackie, ed.), pp. 1–19. Clarendon Press, Oxford.

Gupta, A. P. (1986). "Hemocytic and Humoral Immunity in Arthropods." Wiley, New York.

Gwadz, R. W., Kaslow, D., Lee, J.-Y., Malowy, W. L., Zasloff, M., and Miller, L. H. (1989). Effects of magainins and cecropins on the sporogonic development of malaria parasites in mosquitoes. *Infect. Immun.* **57:**2628–2633.

James, A. A., Blackmer, K., and Racioppi, J. V. (1989). A salivary gland-specific, maltase-like gene of the vector mosquito, *Aedes aegypti*. *Gene* **75:**73–83.

James, A. A., Blackmer, K., Marinotti, O., Ghosn, C. R., and Racioppi, J. V. (1991). Isolation and characterization of the gene expressing the major salivary gland protein of the female mosquito, *Aedes aegypti*. *Mol. Biochem. Parasitol.* **44:**245–254.

Huff, C. J. (1929). The effects of selection upon susceptibility to bird malaria in *Culex pipiens* Linn. *Ann. Trop. Med. Parasitol.* **23:**247.

Karp, R. D. (1990). Cell-mediated immunity in invertebrates. *BioScience* **40:**732–737.

Kartman, L. (1953). Factors influencing infection of the mosquito with *Dirofilaria immitis* (Leidy, 1856). *Exp. Parasitol.* **2:**27–78.

Kilama, W. L., and Craig, G. B., Jr. (1969). Monofactorial inheritance of susceptibility to *Plasmodium gallinaceum* in *Aedes aegypti*. *Ann. Trop. Med. Parasitol.* **63:**419–432.

Kobayashi, M., Ogura, N., and Yamamoto, H. (1986). Studies of filariasis X: A trial to analyse refractory mechanisms of the mosquito *Aedes aegypti* to the filarial larvae *Brugia malayi* by means of parabiotic twinning. *Dokkyo J. Med. Sci.* **13:**61–67.

Korner, A., and Pawelek, J. (1982). Mammalian tyrosinase catalyzes three reactions in the biosynthesis of melanin. *Science* **217:**1163–1165.

Lackie, A. M., ed. (1986). "Immune Mechanisms in Invertebrate Vectors," Zool. Soc. London Symp. 56. Clarendon Press, Oxford.

Lackie, A. M. (1988). Immune mechanisms in insects. *Parasitol. Today* **4:**98–105.

Lander, E. S., and Botstein, D. (1989). Mapping Mendelian factors underlying quantitative traits using RFLP linkage maps. *Genetics* **121:**185–199.

Landry, B. S., and Michelmore, R. W. (1987). Methods and applications of restriction fragment length polymorphism analysis to plants. *In* "Tailoring Genes for Crop Improvement: An Agricultural Perspective" (G. Bruening, J. Harada, and A. Hollaender, eds.), pp. 25–44. Plenum Press, New York.

Law, J. H., and Wells, M. A. (1989). Insects as biochemical models. *J. Biol. Chem.* **264:**16335–16338.

Lee, C. C., Wu, Z., Gibbs, R. A., Cook, R. G., Muzny, D. M., and Caskey, C. T. (1988).

Generation of cDNA probes directed by amino acid sequence: Cloning of urate oxidase. *Science* **239:**1288–1291.

Li, J., and Christensen, B. M. (1990). Immune competence of *Aedes trivittatus* hemocytes as assessed by lectin binding. *J. Parasitol.* **76:**276–278.

Li, J., and Nappi, A. J. (1991). Analysis of monophenol oxidase activity using high pressure liquid chromatography with electrochemical detection. *J. Liq. Chromatogr.* **14:**2089–2108.

Li, J., Tracy, J. W., and Christensen, B. M. (1989). Hemocyte monophenol oxidase activity in mosquitoes exposed to microfilariae. *J. Parasitol.* **75:**1–5.

Li, J., Christensen, B. M., and Tracy, J. W. (1990). Electrochemical determination of diphenol oxidase activity using high pressure liquid chromatography. *Anal. Biochem.* **190:**354–359.

Li, J. Y., Tracy, J. W., and Christensen, B. M. (1992). Phenol oxidase activity in hemolymph compartments of *Aedes aegypti* during melanotic encapsulation reactions against microfilariae. *Dev. Comp. Immunol.* **16:**41–48.

Macdonald, W. W. (1962a). The selection of a strain of *Aedes aegypti* susceptible to infection with semi-periodic *Brugia malayi*. *Ann. Trop. Med. Parasitol.* **56:**368–372.

Macdonald, W. W. (1962b). The genetic basis of susceptibility of infection with semi-periodic *Brugia malayi* in *Aedes aegypti*. *Ann. Trop. Med. Parasitol.* **56:**373–382.

Macdonald, W. W. (1963). Further studies on a strain of *Aedes aegypti* susceptible to infection with semi-periodic *Brugia malayi*. *Ann. Trop. Med. Parasitol.* **57:**452–460.

Macdonald, W. W., and Ramachandran, C. P. (1965). The influence of the gene f^m (filarial susceptibility, *Brugia malayi*) on the susceptibility of *Aedes aegypti* to seven strains of *Brugia, Wuchereria* and *Dirofilaria*. *Ann. Trop. Med. Parasitol.* **59:**64–73.

McGrane, V., Carlson, J. O., Miller, B. R., and Beaty, B. J. (1988). Microinjection of DNA into *Aedes triseriatus* ova and detection of integration. *Am. J. Trop. Med. Hyg.* **39:**502–510.

McGreevy, P. B., McClelland, G.A.H., and Lavoipierre, M.M.J. (1974). Inheritance of susceptibility to *Dirofilaria immitis* infection in *Aedes aegypti*. *Ann. Trop. Med. Parasitol.* **68:**97–109.

Micks, D. W. (1949). Investigations on the mosquito transmission of *Plasmodium elongatum* Huff, 1930. *J. Natl. Malaria Soc.* **8:**206.

Miller, L. H., Sakai, R. K., Rownans, P., Gwadz, R. W., Kantoff, P., and Coon, H. G. (1987). Stable integration and expression of a bacterial gene in the mosquito *Anopheles gambiae*. *Science* **237:**779–781.

Morris, A. C., Eggleston, P., and Crampton, J. M. (1989). Genetic transformation of the mosquito *Aedes aegypti* by microinjection of DNA. *Med. Vet. Entomol.* **3:**1–7.

Munkirs, D. D., Christensen, B. M., and Tracy, J. W. (1990). Analysis of hemolymph plasma substrates in immune reactive *Aedes aegypti*. *J. Invertebr. Pathol.* **56:**267–279.

Munstermann, L. E. (1990). Linkage map of the yellow fever mosquito, *Aedes aegypti*. *In* "Genetic Maps: Locus Maps of Complex Genomes" (S. J. O'Brian, ed.), Vol. 5, pp. 3.179–3.183. Cold Spring Harbor Lab., Cold Spring Harbor, New York.

Munstermann, L. E., and Craig, G. B., Jr. (1979). Genetics of *Aedes aegypti*. *J. Hered.* **70:**291–296.

Nappi, A. J., and Christensen, B. M. (1986). Hemocyte cell surface changes in *Aedes aegypti* in response to microfilariae of *Dirofilaria immitis*. *J. Parasitol.* **72:**875–879.

Nappi, A. J., and Christensen, B. M. (1987). Insect immunity and mechanisms of resistance by nematodes. *In* "Vistas on Nematology" (J. A. Veech and D. W. Dickson, eds.), pp. 285–291. Society of Nematologists, Hyattsville, MD.

Nappi, A. J., and Silvers, M. (1984). Cell surface changes associated with cellular immune reactions in *Drosophila*. *Science* **225:**1166–1168.

Nappi, A. J., Christensen, B. M., and Tracy, J. W. (1987). Quantitative analysis of hemolymph monophenol oxidase activity in immune reactive *Aedes aegypti*. *Insect Biochem.* **17:**685–688.

Nayar, J. K., Knight, J. W., and Bradley, T. J. (1988). Further characterization of refractoriness in *Aedes aegypti* (L.) to infection by *Dirofilaria immitis* (Leidy). *Exp. Parasitol.* **66:**124–131.

Nienhuis, J., Helentjaris, T., Slocum, M., Ruggero, B., and Schaefer, A. (1987). Restriction fragment length polymorphism analysis of loci associated with insect resistance in tomato. *Crop Sci.* **27:**797–803.

Ogura, N. (1986). Haemagglutinating activity and melanin deposition on microfiliariae of *Brugia pahangi* and *B. malayi* in the mosquito, *Armigeres subalbatus. Jpn. J. Parasitol.* **35:**542–549.

Ogura, N. (1987). The effect of exogenous haemagglutinin on *in vitro* melanin deposition on microfilariae of *Brugia pahangi* in haemolymph of the mosquito, *Armigeres subalbatus. Jpn. J. Parasitol.* **36:**291–297.

Ogura, N., Kobayashi, M., and Yamamoto, H. (1985). Haemagglutinating activity in fluid collected from the mosquito, *Armigeres subalbatus,* by centrifugation method. *Dokkyo J. Med. Sci.* **12:**217–221.

Osborn, T. C., Alexander, D. C., and Fobes, J. F. (1987). Identification of restriction fragment length polymorphisms linked to genes controlling soluble solids content in tomato fruit. *Theor. Appl. Genet.* **73:**350–356.

Paskewitz, S. M., Brown, M. R., Lea, A. O., and Collins, F. H. (1988). Ultrastructure of the encapsulation of *Plasmodium cynomolgi* (B strain) on the midgut of a refractory strain of *Anopheles gambiae. J. Parasitol.* **74:**432–439.

Paskewitz, S. M., Brown, M. R., Collins, F. H., and Lea, A. O. (1989). Ultrastructural localization of phenoloxidase in the midgut of refractory *Anopheles gambiae* and association of the enzyme with encapsulated *Plasmodium cynomolgi. J. Parasitol.* **75:**594–600.

Paterson, A. H., Lander, E. S., Hewitt, J. D., Peterson, S., Lincoln, S. E., and Tanksley, S. D. (1988). Resolution of quantitative traits into Mendelian factors by using a complete linkage map of restriction fragment length polymorphisms. *Nature* (*London*) **335:**721–726.

Pentz, E. S., Black, B. C., and Wright, T.R.F. (1986). A diphenol oxidase gene is part of a cluster of genes involved in catecholamine metabolism and sclerotization in *Drosophila.* I. Identification of the biochemical defect in Dox-A2 [1(2)37Bf] mutants. *Genetics* **112:**823–841.

Raghaven, N.G.S., Das, M., Mammen, M. L., Singh, N. N., and Wattal, B. L. (1967). Genetic basis of differential susceptibility of *Aedes aegypti* to dirofilarial infection. Part I: Preliminary observations on selection of *Aedes aegypti* strains susceptible and refractory to *Dirofilaria immitis* infections. *Bull. Indian Soc. Mal. Communicable Dis.* **4:**318–323.

Raikhel, A. S. (1992). Vitellogenesis in mosquitoes. *Adv. Dis. Vector Res.* **9:**1–58.

Raikhel, A. S., and Dhadialla, T. S. (1992). Accumulation of yolk proteins in insect oocytes. *Annu. Rev. Entomol.* **37:**217–251.

Ratcliffe, N. A., Leonard, C. M., and Rowley, A. F. (1984). Prophenoloxidase activation: Nonself recognition and cell co-operation in insect immunity. *Science* **226:**557–559.

Renwrantz, L. (1986). Lectins in molluscs and arthropods: Their occurrence, origin and roles in immunity. *In* "Immune Mechanisms in Invertebrate Vectors" (A. M. Lackie, ed.), pp. 81–93. Clarendon Press, Oxford.

Rizki, R. M., and Rizki, T. M. (1990). Encapsulation of parasitoid eggs in phenoloxidase-deficient mutants of *Drosophila melanogaster. J. Insect Physiol.* **36:**523–529.

Rizki, T. M., and Rizki, R. M. (1983). Blood cell surface changes in *Drosophila* mutants with melanotic tumors. *Science* **220:**73–75.

Romans, P., Seeley, D. C., Kew, Y., and Gwadz, R. W. (1991). Use of a restriction fragment length polymorphism (RFLP) as a genetic marker in crosses of *Anopheles gambiae* (Diptera: Culicidae): Independent assortment of a diphenol oxidase RFLP and an esterase locus. *J. Med. Entomol.* **28:**147–151.

Rosenberg, R. (1985). Inability of *Plasmodium knowlesi* sporozoites to invade *Anopheles freeborni* salivary glands. *Am. J. Trop. Med. Hyg.* **34:**687–691.

Roubaud, E. (1937). Nouvelles recherches sur l'infection du moustique de la fievre jaune par *Dirofilaria immitis* chez quelques culcides indégènes. *Bull. Soc. Pathol. Exot.* **30:**480–484.

Severson, D. W., Mori, A., Zhang, Y., and Christensen, B. M. (1993). Linkage map for *Aedes aegypti* using restriction fragment length polymorphisms. *J. Heredity* (in press).

Söderhäll, K., and Smith, V. J. (1986). The prophenoloxidase system: The biochemistry of its activation and role in arthropod cellular immunity with special references to crustaceans. *In* "Immunity in Invertebrates" (M. Brehélin, ed.), pp. 208–223. Springer-Verlag, Berlin.

Spray, F. J., and Christensen, B. M. (1991). *Aedes aegypti:* Characterization of hemocyte polypeptide synthesis during wound healing and immune response to inoculated microfilariae. *Exp. Parasitol.* **73:**481–488.

Stanley-Samuelson, D. W., Jensen, E., Nickerson, K. W., Tiebel, K., Ogg, C. L., and Howard, R. W. (1991). Insect immune response to bacterial infection is mediated by eicosanoids. *Proc. Natl. Acad. Sci. U.S.A.* **88:**1064–1068.

Sugumaran, M. (1986). Tyrosinase catalyzes an unusual oxidative decarboxylation of 3,4-dihydroxymandelate. *Biochemistry* **25:**4489–4492.

Sugumaran, M. (1990). Prophenoloxidase activation and insect immunity. *UCLA Symp. Mol. Biol.* **121:**47–62.

Sun, S., Lindstrom, I., Boman, H. G., Faye, I., and Schmidt, O. (1990). Hemolin: An insect immune protein belonging to the immunoglobulin superfamily. *Science* **250:**1729–1732.

Tanksley, S. D., and Hewitt, J. (1988). Use of molecular markers in breeding for soluble solids content in tomato; A re-examination. *Theor. Appl. Genet.* **75:**811–823.

Tanksley, S. D., Bernatzky, R., Lapitan, N. L., and Prince, J. P. (1988). Conservation of gene repertoire but not gene order in pepper and tomato. *Proc. Natl. Acad. Sci. U.S.A.* **85:**6419–6423.

Townsend, D., Guillery, P., and King, R. A. (1984). Optimized assay for mammalian tyrosinase (polyhydroxyl phenyloxidase). *Anal. Biochem.* **139:**345–352.

Townson, H., and Chaithong, U. (1991). Mosquito host influences on development of filariae. *Ann. Trop. Med. Parasitol.* **85:**149–163.

Trager, W. (1942). A strain of mosquito *Aedes aegypti* selected for susceptibility to the avian malaria parasite *Plasmodium lophurae*. *J. Parasitol.* **28:**457.

Vachtenheim, J., Duchon, J., and Matous, B. (1985). A spectrophotometric assay for mammalian tyrosinase utilizing the formation of melanochrome from L-dopa. *Anal. Biochem.* **146:**405–410.

Vernick, K. D., and Collins, F. H. (1989). Association of a *Plasmodium*-refractory phenotype with an esterase locus in *Anopheles gambiae*. *Am. J. Trop. Med. Hyg.* **40:**593–597.

Vernick, K. D., Collins, F. H., and Gwadz, R. W. (1989). A general system of resistance to malaria infection in *Anopheles gambiae* controlled by two main genetic loci. *Am. J. Trop. Med. Hyg.* **40:**585–592.

Ward, R. A. (1963). Genetic aspects of the susceptibility of mosquitoes to malarial infection. *Exp. Parasitol.* **13:**328–341.

Wattam, A. R., and Christensen, B. M. (1992). Induced polypeptides associated with filarial worm refractoriness in *Aedes aegypti*. *Proc. Natl. Acad. Sci. U.S.A.* **89:**6502–6503.

World Health Organization (WHO) (1991). "Prospects for Malaria Control by Genetic Manipulation of its Vectors," TDR/BCV/MAL-ENT/91.3. UNDP/World Bank/WHO Special Programme for Research and Training in Tropical Diseases.

Yamamoto, H., Kobayashi, M., Ogura, N., Tsuruoka, H., and Chigusa, Y. (1985). Studies on filariasis. VI. The encapsulation of *Brugia malayi* and *B. pahangi* larva in the mosquito, *Armigeres subalbatus*. *Jpn. J. Sanit. Zool.* **36:**1–6.

Zielke, E. (1973). Untersuchungen zur Vererbung der Empfänglichkeit gegenüber der Hundefilarie *Dirofilaria immitis* bei *Culex pipiens fatigans* und *Aedes aegypti*. *Z. Trop. Parasitol.* **24:**36–44.

Chapter 12

Cellular Defense Responses of Insects: Unresolved Problems

Norman A. Ratcliffe
Biomedical and Physiological
Research Group
School of Biological Sciences
University College of Swansea
Swansea, Wales, United Kingdom

I. Introduction

In recent years, there have been a number of important developments in our understanding of insect immune defense reactions. For example, as far as humoral factors are concerned, inducible immune proteins have been sequenced and characterized and some of the genes for their production cloned (e.g., Boman and Hultmark, 1987; Hoffmann and Hoffmann, 1990; Faye and Hultmark, Volume 2, Chapter 2), the role of the prophenoloxidase system (e.g., Ratcliffe *et al.*, 1984; Sugumaran and Kanost, Volume 1, Chapter 14) and the hemolymph lectins (Pendland *et al.*, 1988) in the recognition of foreign antigens have been confirmed, and the hemolymph lectins too have been isolated, sequenced, and cDNA clones produced for ontogenetic studies (e.g., Takahashi *et al.*, 1985, 1986). As regards cellular immunity, important advances have also been made, such as the detailed characterization of cellular reactivity (reviewed in Ratcliffe and Rowley, 1979; Götz and Boman,

1985), including allogenic recognition of grafts (George *et al.*, 1987) and the demonstration of cell–cell cooperative events (Schmit and Ratcliffe, 1977) utilizing, more recently, purified populations of blood cells (Anggraeni and Ratcliffe, 1991).

Despite this progress in our knowledge of insect immunity, many problems remain unanswered. The purpose of this brief overview is to identify at least some of these unresolved questions and to indicate possible directions for future endeavors. Attention will be focused particularly on the insect cellular defenses but, since evidence is now accumulating for a complex interplay between the cellular and humoral components in the production of the overall immune reactivity of insects, such divisions are purely arbitrary.

II. Main Events in Insect Immune Reactivity

Figure 1 (modified from Ratcliffe and Rowley, 1987) summarizes the main phases and interactions occurring during the insect response to microbial and macrobial invasion. The question marks in this figure indicate some, but not all, of the unresolved problems in our understanding of the processes involved. For convenience, the insect immune defenses can be divided into three interacting phases (see also Chadwick and Aston, 1978):

(a) An immediate and inductive stage
(b) A mainly cellular and synthetic stage
(c) A final humoral/cellular stage

A. Immediate and Inductive Stage

Insect immunity is induced by wounding or by parasitic invasions through the gut, integument, or tracheae. Hemolymph extruded through the injury upon exposure to the air rapidly coagulates to seal the wound. The coagulation process involves the degranulation of blood cells (hemocytes) termed granular cells, cystocytes, and/or plasmatocytes, which interact with plasma components to form an insoluble gel (Bohn, 1986). This initiation of immune reactivity poses a number of unresolved problems in insect immunity, and it is very noticeable that in Fig. 1 most of the question marks occur during this first phase.

The site of origin and mode of development of the blood cells involved in immunity in most insect species are thus still unknown. A similar problem relates to the identification of the immune reactive cells with a multitude of names published in the literature for only six to eight basic hemocyte types.

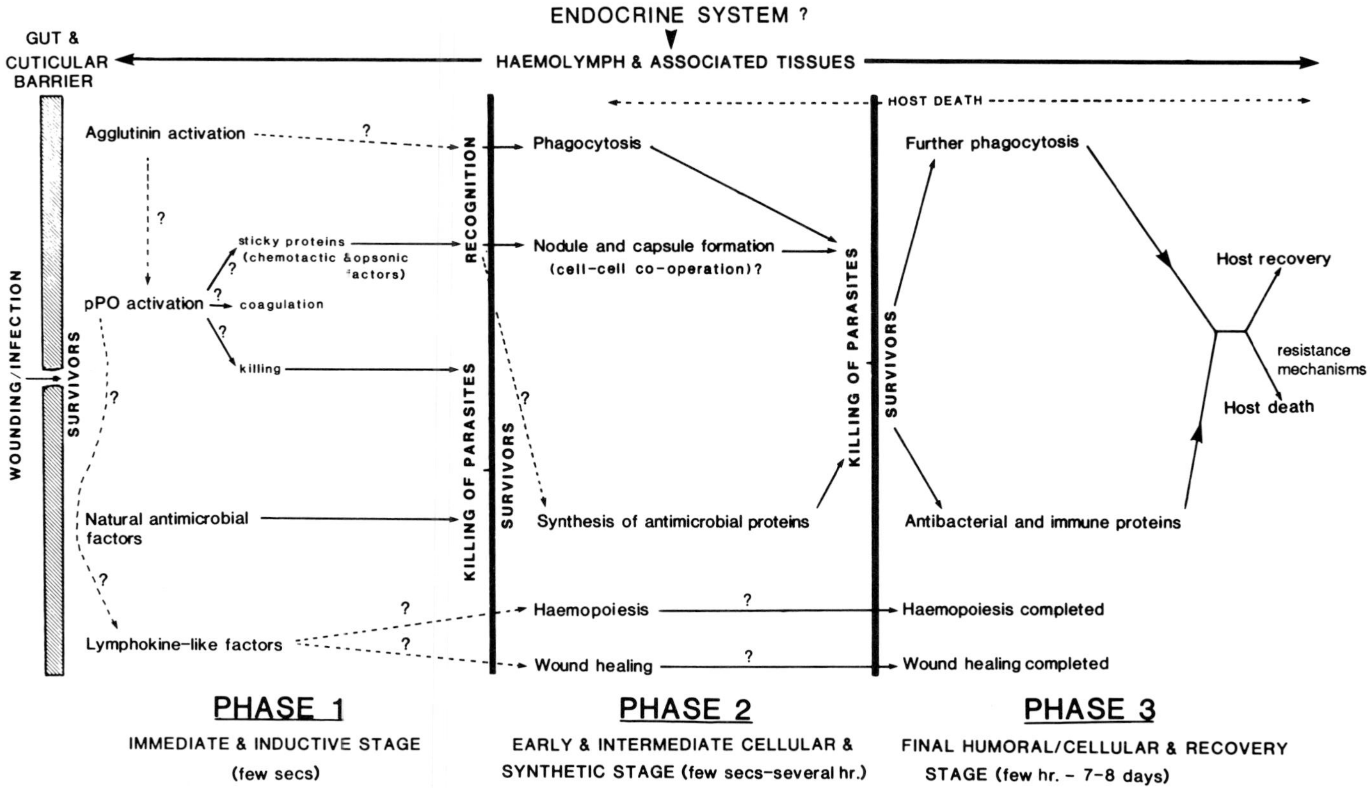

Figure 1 Proposed scheme for stages in insect host defense reactions. Modified from Ratcliffe and Rowley (1987) with permission of CRC Press, Inc.

In addition, a vital immune process occurring early on in the insect's response to exposure to non-self antigens is the recognition of foreignness about which much has yet to be learned. For example, is this process cell and/or plasma mediated? Which cell types and molecules are involved? How is it controlled and is it broad spectrum or highly specific? How does the recognition process lead to later events in the immune response? The interactive nature of the cellular and humoral components should also be considered at this stage. These are just a few questions that have been posed. Other unresolved problems in this early phase relate to our poor understanding of the coagulation process itself in insects and whether, as in crustaceans, the prophenoloxidase (pPO) complex is involved.

The end result of the first phase then is the sealing of the wound site or portal of invasion, recognition of the invader, and activation of the immune system. In many cases, invasion is halted at this stage by an array of antiparasitic substances such as naturally occurring antimicrobial factors in the hemolymph (e.g., Croft *et al.*, 1982; Jones *et al.*, 1989), lysozyme and lysosomal enzymes released from the degranulating blood cells (Zachary and Hoffmann, 1984), and antimicrobial substances generated by activation of the pPO complex (Rowley *et al.*, 1990). Parasites and pathogens, however, have evolved many strategies for avoiding the insect immune defenses so that progression to the second phase, which is characterized by cell-mediated reactions and the synthesis of the immune proteins, may be curtailed. Only recently have we learned anything about the avoidance strategies of invading microbial and macrobial parasites (e.g., Brehélin, 1990).

B. Mainly Cellular and Synthetic Stage

If the invading organism survives the initial onslaught by the insect immune defenses but is recognized as foreign, then it may rapidly be phagocytosed and/or encapsulated by the circulating blood cells. The type of reaction elicited depends upon the numbers and size as well as the pathogenicity of the invading organism. In *Galleria mellonella,* phagocytosis alone can deal with 10^2–10^3 bacteria per microliter hemolymph while larger numbers are removed by nodule formation (Walters and Ratcliffe, 1983).

There is considerable disagreement as to the cell type(s) mainly responsible for phagocytosis in insects. Ratcliffe and Rowley (1979) reported that the plasmatocytes are the most important phagocytic cells, whereas Brehélin and Zachary (1986) believe that the granular hemocytes may be the main effector cells. These differences probably result from problems in classifying insect blood cells and from the large variations occurring in hemocyte types even between closely related species. How the relevant phagocytic blood cell recognizes the foreign invader is also a problem mentioned previously. In insects, and in many other arthropods too, at least two types of molecules seem to be responsible for non-self recognition by the blood cells, namely, the hemolymph agglutinins and factors generated during activation of the pPO cascade. Since both types of molecules may be produced or stored by the same hemocyte type (Ratcliffe, 1991), then the interactive nature of these factors seems to be highly probable and evidence for this possibility is presented in Section III.B. Details of molecular events occurring at the hemocyte membrane and subsequently in the cytoplasm of these cells following binding of the invader or non-self molecules are also virtually unknown.

What is becoming increasingly evident, with the development of cell separation techniques and the purification of reactive molecules associated with the pPO system, is that cellular immunity in arthropods involves cell-to-cell cooperation that is carefully orchestrated by products of the activated cells themselves (Johansson and Söderhäll, 1989a; Ratcliffe, 1991). Thus, interactive immune competent cells are not confined to the vertebrate immune system.

This second stage of insect immunity is also characterized by the synthesis of over 15 immune proteins (Boman and Hultmark, 1987; Faye and Hultmark, Volume 2, Chapter 2). The question arises as to when and how the synthetic machinery in the fat body, hemocytes, and other tissues is turned on by the invading organisms? There is now some tentative evidence that the non-self recognition events in insects activate the relevant immune protein genes (Faye, 1990). In addition, it has recently been shown that endotoxin (LPS) and laminarin (a β-1,3-glucan), which are both components of microbial cell walls and activators of the pPO recognition system, also induce the expression of cecropins (one group of the induced antibacterial proteins in insects) in an insect cell line (Samakovlis *et al.*, 1990; see Section III.D). Finally, if these immune proteins are such important components of the insect defenses, then why are some insects apparently incapable of producing them?

C. Final Humoral/Cellular Stage

Due to the efficiency of the first two stages in insect immunity, probably only a minority of invading organisms survive until the final phase, which results in either host recovery or death. This stage occurs a few hours up to several days after invasion of the insect and is mainly mediated by the newly synthesized immune proteins. Parasites and pathogens resistant to the killing mechanisms within the blood cells and capsules will, however, break out of these cellular structures during this phase and elicit a further round of phagocytosis and encapsulation by the hemocytes.

To date, the insect immune proteins, which include the cecropins, attacins, diptericins, defensins, sarcotoxins, apidaecins, and lysozyme (Boman and Hultmark, 1987; Casteels *et al.*, 1990; Hoffmann and Hoffmann, 1990), have all been shown to be antibacterial. Does this mean that no such final humoral stage is mounted against other microbial and macrobial parasites? This seems unlikely especially following the report by Ham *et al.* (1991) that anti-immune protein antibodies block the normally strong microfilaricidal activity of the hemolymph of *Aedes aegypti* and *Simulium ornatum*. In addition, attention should be drawn to the "antibodylike" response of *Periplaneta americana* during which a large 700-kDa molecule is induced (George *et al.*, 1987; Karp, 1990, and Volume 1, Chapter 13), which may well be able to neutralize a range of microbial invaders.

Another phenomenon that characterizes this final stage is the completion of wound healing probably as a result of the synthesis and repair of the basement membrane by the hemocytes (Knibiehler *et al.*, 1987). The recruitment of hemocytes around capsules and the replacement by hemopoiesis of blood cells utilized in the cellular defenses cease too. It has been suggested that the hemocytes in capsules may secrete both "recruiting" and "cessation factors" and that the recruitment of blood cells may depend upon the relative amounts of these substances (Lackie, 1988). In early capsules, the recruiting signal would be strong but would weaken with the addition of more cell layers. As recruitment slows, the hemocytes would have more opportunity to secrete the cessation factors, which may be the homogeneous, glycosaminoglycans reported coating many mature capsules and possibly related to basement membrane material (Lackie, 1988).

Regarding the replacement of cells utilized in the cellular defenses, these may originate from specific hemopoietic tissue and/or by release of sessile populations attached to the internal organs. No doubt humoral factors, such as the "plasmatocyte depletion factor" of *Galleria mellonella* (Chain and Anderson, 1983), are involved, but the whole subject of hemopoiesis and the control of hemocyte numbers in insects is not only neglected but somewhat confused. A fascinating article by Pathak (1986) indicates that the endocrine glands of insects probably play a key role in the control of blood cell numbers and cell types as well as in mediation of the immune system itself. The subject of the interaction of the insect endocrine and immune systems is dealt with briefly in Section III.A.

This final phase results either in the recovery of the insect host due to the efficacy of the immune system or else the invader prevails and the host is killed. The killing of bacteria by the immune proteins is well documented and has been mentioned previously (see Section II) but little is known of the killing mechanism of the hemocytes themselves that have ingested or surrounded invaders. Anderson *et al.* (1973) described the killing of bacteria by *Blaberus craniifer* hemocytes and Walters and Ratcliffe (1983) showed that when *E. coli* were sequestered in nodules they were rapidly killed (Walters and Ratcliffe, 1983) (Table 1). *Blaberus* hemocytes too have been shown to contain endogenous peroxidase (H. Mullett, unpublished observations). In addition, insect hemocytes have been reported to contain acid phosphatase (e.g., Crossley, 1968; Rowley and Ratcliffe, 1979), β-glucuronidase and β-glucosaminidase (Walters and Ratcliffe, 1981), lysozyme (e.g., Zachary and Hoffmann, 1984), cecropins and attacins (Trenczek, 1986), and the pPO system, which has recently been shown to have limited antimicrobial activity (Rowley *et al.,* 1990).

Finally, many microbial and macrobial parasites have evolved mechanisms for avoidance or depression of the insect host responses (Table 2). This subject has been reviewed extensively (e.g., Ratcliffe *et al.,* 1985; Götz and Boman, 1985; Ratcliffe and Rowley, 1987; Vinson, 1988; Brehélin, 1990; Schmidt and Theopold, 1991) and is of extreme importance as it not only reveals clues as to the mechanisms controlling insect immunity, but also allows parasites to survive for long periods in their hosts, which then act as the vectors of the most devastating diseases (see Section III.E).

III. Specific Problems in Insect Immunity

The aforementioned brief scenario has identified a large number of unresolved problems in our understanding of insect immunity. In the remainder of this chapter, some of the more important of these unanswered questions are considered in more detail. The subjects selected are those that form the basis of control and coordination of the insect immune system and include the following:

(a) The classification, origins, and manipulation of insect blood cells
(b) The recognition of non-self invaders
(c) Cell–cell communication in immunity
(d) The interrelationship of cellular and humoral immunity
(e) Future areas for study

The final topic is included since recently there has been additional interest in the immune systems of invertebrates from more applied aspects, including the possible importance of the host defenses as determinants of infectivity in vector species (Christensen and Severson, Volume 1, Chapter 11) and the use of changes in immune reactivity as monitors of environmental pollution.

Table 1

Percentage Relative Viability[a] (% R.V.) of *E. coli* and *B. cereus* in Nodules from *Galleria mellonella* and in Corresponding Haemolymph or Grace's Insect Medium Incubated Controls

Time (hr postinjection)	*B. cereus*			*E. coli*		
	Experimental (in nodules)	Hemolymph	Grace's	Experimental (in nodules)	Hemolymph	Grace's
4	10.11 ± 4.65[b]	558.7 ± 273.5	132.9 ± 40.5	16.14 ± 8.12	513.0 ± 291	327.9 ± 122
6	65.66 ± 41.6	2087.6 ± 451	1295.0 ± 648	4.10 ± 3.07	1152.2 ± 319	317.8 ± 112
9	210.0 ± 86.8	2334.2 ± 489	1409.9 ± 511	3.99 ± 2.05	1207.0 ± 436	475.0 ± 174
12	510.1 ± 153	2874.0 ± 586	2065.1 ± 715	3.29 ± 2.5	1661.2 ± 730	502.9 ± 176
15	1270.0 ± 283	2598.5 ± 520	1895.2 ± 500	3.10 ± 2.1	1147.4 ± 378	538.2 ± 218
18	Death	—	—	1.74 ± 1.38	1156.5 ± 520	377.3 ± 153
21	—	—	—	1.64 ± 1.27	1216.0 ± 455	572.7 ± 215
24	—	—	—	1.75 ± 1.28	1367.1 ± 419	573.0 ± 181

[a]Percentage relative viability at time "0" = 100%.
[b]Standard error.

Modified from Walters and Ratcliffe (1983) with permission of Pergamon Press.

Table 2

Some Avoidance and Resistance Strategies Adopted by Parasites and Pathogens against Invertebrate Host Defense Mechanisms

A. Passive avoidance	
	1. Organs colonized out of reach of blood cells
	2. Low-reactivity regions of host occupied
	3. Immature hosts invaded
B. Active mechanisms	
	1. Refractory envelopes/cell walls, etc.
	2. Mimicry of host tissues
	3. Acquisition of host antigens
	4. Lysis and abrogation of leukocytes
	5. Inhibition of immune recognition factors
	6. Inhibition of chemotaxis
	7. Inhibition of phagocytosis
	8. Inhibition of nodule formation/ encapsulation
	9. Provision of alternative targets
	10. Utilization of host response
	11. Abrogation of immune proteins
C. No obvious mechanism	
	1. Host swamped by numbers or rate of growth of parasite/pathogen?

Modified from Ratcliffe et al. *(1985).*

A. Classification, Origins, and Manipulation of Insect Blood Cells

The characterization of insect blood cell types is a controversial topic that has been reviewed extensively in the last twenty years (e.g., Jones, 1962; Price and Ratcliffe, 1974; Gupta, 1979; Rowley and Ratcliffe, 1981; Brehélin and Zachary, 1986). The problems in the classification of insect blood cells are due mainly to the sheer diversity in insect forms, which is reflected in the great variation in the structure of the blood cells present. The situation is exacerbated by a vast range of different names in the literature for six to eight basic types of cells. Table 3, from Rowley and Ratcliffe (1981), lists some of the synonyms used in publications for the six basic types of insect hemocytes identified in that study. In addition, inconsistencies in nomenclature have arisen due to the use of insects in different developmental stages and as a result of the utilization of a range of experimental techniques. Attempts have

Table 3

List of Various Synonyms for Insect Hemocytes

Classification scheme[a]	Author's equivalent classes to:					
	Prohemocytes	Plasmatocytes	Granular cells	Cystocytes	Spherule cells	Oenocytoids
Hollande (1911)	proleucocytes	phagocytes	granular leucocytes adipoleucocytes[b]		spherule cells	oenocytoids
Metalnikov (1924)	lymphocytes proleucocytes	← leucocytes	→		cell spheruleuses	
Cameron (1934)	lymphocytes	leucocytes	leucocytes?		spherule cells	oenocytes
Yeager (1945)	proleucocytoids	plasmatocytes podocytes[c] vermiform cells[c]	chromophilic cells?	cystocytes	spheroidocytes	oenocytelike cells
Wigglesworth (1959)	proleucocytes	phagocytic amoebocytes			phagocytic amoebocytes?	oenocytoids
Jones (1962); Arnold (1974)	prohemocytes	plasmatocytes podocytes[c] vermiform cells[c]	granular hemocytes adipohemocytes[b]	cystocytes	spherule cells	oenocytoids
Brehélin *et al.* (1978)	prohemocytes	plasmatocytes granulocytes? thrombocytoids[c] podocytes[c]	granulocytes	coagulocytes	spherule cells	oenocytoids
Gupta (1979)	prohemocytes	plasmatocytes	granulocytes	coagulocytes	spherulocytes	oenocytoids

[a]References cited in Rowley and Ratcliffe (1981).
[b]Some workers class adipohemocytes as granular cells.
[c]In some instances these may be classfied with the plasmatocytes.

even been made also to classify the blood cells just on the ultrastructural appearance of their granules. Some variation in granule structure is to be expected in a group containing 1–3 million species! A common error made by workers new to this field is to adopt vertebrate methods in the preparation of insect blood cells for microscopical examination. Nothing is guaranteed more to distort and lyse the fragile hemocytes of insects than smearing them on a microscope slide. A final problem certain to confuse inexperienced workers is the retention of vertebrate terms such as "granulocyte" for describing insect blood cells when there is little evidence for homology of any cell types in these two groups of animals.

Brehélin and Zachary (1986) have also discussed in detail some of the problems in categorizing insect blood cells and stressed the point that designations based purely on hemocyte functions are also unreliable. Thus, in *Locusta migratoria* and *Melolontha melolontha,* capsules are formed by granular cells in the former and by plasmatocytes in the latter. Functionally these two cell types are identical but morphologically they are quite distinct. They proposed that studies of insect blood cells should include ultrastructural, functional, and enzymatic aspects in order for a clear definition to be derived (Brehélin and Zachary, 1986). To this list should also be added phase-contrast observations of unfixed cells. This technique is particularly useful as it reveals not only morphological characteristics but also other properties of the cells such as their stability and behavior *in vitro.*

Recently, cell types have been identified by the nature of their cell-surface determinants as detected by staining with lectins and monoclonal antibodies (mAb). In *Drosophila,* for example, the hemocytes responsible for the encapsulation of aberrant tissues and parasites, as well as for the rejection of foreign tissue implants, have been shown to bind fluorescein-labeled wheat germ agglutinin (Rizki and Rizki, 1982; Nappi and Silvers, 1984). Similarly, in *Aedes aegypti* there is a fivefold increase in the percentage of cells binding wheat germ agglutinin following inoculation with *Dirofilaria immitis* microfilaria (Nappi and Christensen, 1986). Lectin-binding can also distinguish between certain hemocyte types in the phasmid *Extatosoma tiaratum,* although there is considerable staining cross-reactivity with the different cell types (Richards *et al.,* 1989), which limits its usefulness for functional work. An alternative approach by Trenczek and Bennich (1992) has been to raise mAb against specific subpopulations of plasmatocytes and granular cells from cecropia moths. Mullett *et al.* (1993) have likewise produced mAb against *Blaberus discoidalis* and *Galleria mellonella* hemocytes so that these workers now have powerful tools with which to probe blood cell interrelationships and interactions. One drawback of mAb is that they may be genus-specific (Table 4) and therefore unlikely to provide a means of deriving a unified classification of insect blood cells across a range of insect orders.

Table 4

Cross-reactivity of two *Blaberus discoidalis* Monoclonal Antibodies against the Hemocytes of a Number of Insect Species[a]

Species	Monoclonal	
	B5	1Hb
Blaberus discoidalis	+++	+++
Blaberus craniifer	++	++
Gromphadorhina sp.	+	+
Periplaneta americana	−	+
Galleria mellonella	−	−

[a]+++/++/+/− = intensity of staining by FITC labeling.

From Mullett et al. *(1993).*

The message is clear for insect hematologists and other scientists in this field. Very useful classification schemes for insect hemocytes are available in the literature (e.g., Jones, 1962; Price and Ratcliffe, 1974; Brehélin and Zachary, 1986). These schemes provide the basis for categorizing the hemocytes of most insect species but be prepared for the unusual (e.g., Fig. 2). In addition, as a first step, examine the cells carefully under phase-contrast optics but beware of artifacts that readily occur with this technique (see Azambuja *et al.,* 1991).

The apparent diversity of blood cells in insects is likewise reflected in the variation in location and structure of the hemopoietic tissues in these animals (e.g., Jones, 1970; Gupta, 1979; Rowley and Ratcliffe, 1981). In some dipterans such as *Musca domestica,* they may be formed by diffuse collections of cells, whereas in the orthopterans *Gryllus bimaculatus* and *Locusta migratoria,* they may be highly developed with some signs in the former species of differentiation into various zones, as in higher animal groups (Hoffmann *et al.,* 1979). Hemocytes are also probably derived by the mitotic division of the circulating cells. Jones and Liu (1968) showed that in the wax moth *Galleria mellonella,* although the mitotic rate was only 1%, this is sufficient to maintain larval hemocyte numbers and preclude the need for hemopoietic tissue. Another source of circulating hemocytes is by the release of sessile cells loosely attached to the internal organs. Recently, for example, it has been shown that the treatment of *Tipula paludosa* larvae with ether and acetic acid vapors increased the total hemocyte counts by 2.8-fold due to the release of sessile cells (Green and Carter, 1991).

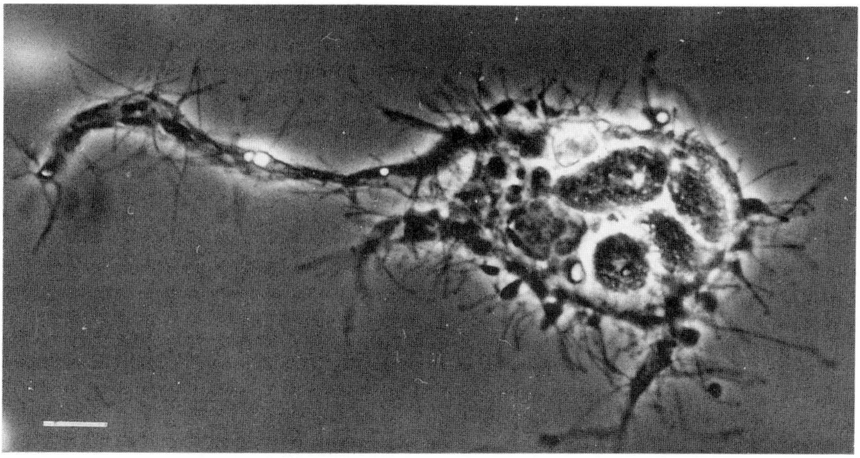

Figure 2 Giant cell of *Rhodnius robustus* with ingested granular cells. Scale bar is 10 μm. From Azambuja *et al.* (1991) with permission of Memorias Do Instituto Oswaldo Crus.

Despite these observations, we still have only limited knowledge about the interrelationships of the various hemocyte types and the mechanisms involved in the control of the development and liberation of the blood cells into the circulation (Lackie, 1988). Unfortunately, few scientists are attracted to this area of comparative hematology.

Most workers recognize the basic blood cell type, the prohemocyte, from which the other hemocyte types may be derived (Lackie, 1988). Many insect hematologists believe that the commonly described blood cells, that is, the plasmatocytes, granular cells, cystocytes, spherule cells, and oenocytoids, may be stages in development from one or two basic cell types (Rowley and Ratcliffe, 1981). Shrivastava and Richards (1965) showed in an autoradiographic study of *G. mellonella* that a developmental series exists passing from prohemocytes to plasmatocytes to granular cells, whereas the lepidopteran oenocytoid may have a separate lineage (Beeman *et al.*, 1983). One possible, but speculative, developmental scheme is shown in Fig. 3 for the blood cells of the stick insect *Clitumnus extradentatus*, with the main ontogenetic pathway occurring from the prohemocyte (stem cell) to the cystocyte (Ratcliffe and Rowley, 1987).

Regarding the control of the development and liberation of the hemocytes into the circulation, there is evidence that the hemogram, and hence the insect immune system, is under endocrine control (reviewed in Pathak, 1986). Ligature experiments to isolate the anterior endocrine glands show

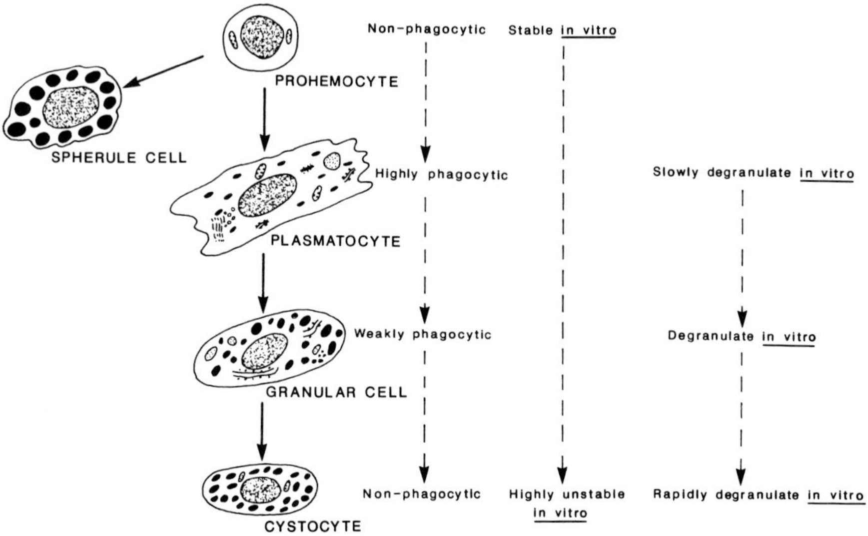

Figure 3 Diagrammatic, but speculative, scheme for hemocyte differentiation in the stick insect *Clitumnus extradentatus*. Modified from Ratcliffe and Rowley (1987) with permission of CRC Press, Inc.

that these structures greatly influence the interconversion of hemocyte types, increase the mitotic indices, release the hemocytes from the hemopoietic organs, and mobilize the sessile cells (e.g., Hinks and Arnold, 1977). Studies involving injuring, bleeding, and cellular reactions such as encapsulation also suggest that wounds and/or parasites produce some factor(s) causing the neuroendocrine system to produce more hormones that then influence the mitotic indices and release of hemocytes (Pathak, 1986). The fact that many parasites, such as the parasitoids, induce radical host endocrine changes (Beckage, 1990) must have a profound effect on the immune potential of the host insect. Finally, ablation and implantation studies of endocrine glands indicate too that these organs greatly affect hemocyte numbers (Pathak, 1983).

In order to study the interrelationships of insect blood cells, as well as their functions in immune reactivity, techniques must be developed for manipulating the cells and studying them *in vitro*. Insect hemocytes are notoriously difficult to manipulate and this is reflected in the fact that it was not until 1974 that phagocytosis was first studied *in vitro* (Ratcliffe and Rowley, 1974) and only in 1991 did it become possible to separate the main immune reactive cells and obtain functional monolayers *in vitro* (Anggraeni and Ratcliffe, 1991). The extent of this problem is illustrated by attempts to separate the different hemocyte types of two *Blaberus* species. We reported previously the successful separation of the blood cells of *B. craniifer* using continuous gradients of Percoll (Mead *et al.*, 1986). However, attempts to isolate the hemocytes of *B. discoidalis* have failed even though the cells appear identical to those of *B. craniifer* under the light microscope and the protocol adopted was identical to that used for the *B. craniifer* cells (H. Mullett, unpublished observations). Possibly, as in *Schistocerca gregaria* (Huxham and Lackie, 1988), the *B. discoidalis* cells may comprise a more finely graded range of granule-containing cells than those of *B. craniifer,* which makes separation on continuous gradients impossible.

Thus, to date, only in one insect species, *G. mellonella,* have separated and immune reactive hemocytes been obtained *in vitro* (Table 5) (Anggraeni and Ratcliffe, 1991). One of the main problems is the almost instantaneous coagulation and gelation of the hemolymph following bleeding. This coagulation is due to the fragile nature of the granular cells or cystocytes (designation depends upon species). These cells are primed to discharge their contents upon contact with minute quantities of microbial products such as β-1,3-glucans and endotoxin (Leonard *et al.*, 1985b; Ratcliffe *et al.*, 1991), which are present in abundance on all glassware and solutions unless specific precautions are taken. Now, fortunately, this problem can be partially overcome by using an ice-cold isotonic, anticoagulant solution containing EDTA and at low pH to stabilize the cells (Mead *et al.*,1986). Even so, however, the

Table 5

Percentage of *Galleria mellonella* Plasmatocytes and Granular Cells Alone or Mixed, from Washed Cell Preparations Phagocytosing *Bacillus cereus* after 1 hr Incubation of Monolayers at 25°C

Type of monolayer	Cell type counted	Percentage of cells phagocytic[a]	
		Control[b]	Experimental[c]
Plasmatocytes alone	Plasmatocytes	5.3 ± 1.3	13.4 ± 2.1*
Granular cells alone	Granular cells	0	0
Mixed	Plasmatocytes	8.7 ± 2.7	30.1 ± 3.4*/**
	Granular cells	0	0

[a]Mean value of 18 monolayers from six gradients ± SD.
[b]Monolayers were overlaid with *B. cereus* suspended in Tris overlay buffer.
[c]Monolayers were overlaid with *B. cereus* suspended in Tris overlay buffer containing 0.1 mg ml^{-1} laminarin.

*Significantly different compared with control ($P < 0.05$).
**Significantly different compared with plasmatocytes alone ($P < 0.005$).

Modified from Anggraeni and Ratcliffe (1991) with permission of Pergamon Press.

amount of manipulation possible in terms of centrifugation and pipetting is strictly limited and depends very much upon the species under study. Thus, *G. mellonella* blood cells are easier to handle than those of *Locusta migratoria* or *Schistocerca gregaria,* which in turn are more stable than those of some cockroaches and phasmids such as *B. discoidalis* and *Extatosoma tiaratum.* Much additional work is required into the stabilization and separation of the blood cells of a range of insect species in order for additional *in vitro* models to be available for study.

B. Recognition of Non-Self Invaders

The basis of any immune response is the ability to recognize as foreign any organisms invading the body. Insects have been shown to be very effective at clearing foreign invaders of the hemocoel by phagocytosis and capsule formation (Ratcliffe and Rowley, 1987), so they can clearly discriminate self from non-self. The level of this discrimination may also be finely tuned since Karp and his colleagues (reviewed in Karp, 1990) have reported the presence of allograft rejection of both integumental grafts and implants by *Periplaneta americana.* As mentioned previously (Section II.B), two possible factors may act as non-self recognition molecules in insects, that is, the hemolymph agglutinins (often termed lectins by many authors although not totally synonymous) and components of the prophenoloxidase cascade. To these recognition factors may also be added the immune protein haemolin (p4), which binds to bacteria and may initiate the synthesis of the immune proteins (Faye, 1990) (see Section III.D).

The agglutinins are present in the hemolymph and in tissues, such as the gut, of most insects. The hemolymph lectins are synthesized by both the hemocytes (Fig. 4) and the fat body (reviewed by Rowley *et al.,* 1986). The occurrence and functioning of these molecules are, however, complex. Thus in some species, such as the flesh fly, *Sarcophaga peregrina* (Komano *et al.,* 1980), *P. americana,* and *S. gregaria* (Ingram *et al.,* 1984), they are inducible, whereas in others, including *Spodoptera exigua* (Pendland and Boucias, 1986), they are apparently not. In addition, the hemolymph of the cockroach *B. discoidalis* contains not one but three agglutinins each with its own characteristic sugar-binding specificity (Chen *et al.,* 1993).

There is also disagreement regarding the role of the agglutinins in immune recognition. Anderson *et al.* (1973) and Rowley and Ratcliffe (1980) reported that bacteria or erythrocytes presensitized with whole hemolymph of *B. craniifer, P. americana,* or *Clitumnus extradentatus* were not internalized any more effectively by the hemocytes than were untreated test particles. Likewise, Bradley *et al.* (1989) showed that the purified lectin from the grasshopper *Melanoplus differentialis* failed to enhance the *in vitro* association of *Bacillus*

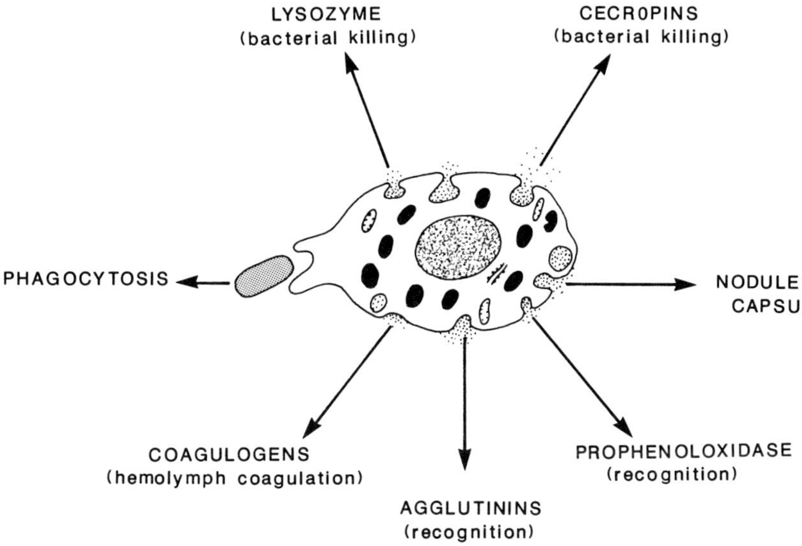

Figure 4 Diagram illustrating the multifunctional role of insect granule-containing cells (= granular cells or cystocytes). Modified from Ratcliffe and Rowley (1987) with permission of CRC Press, Inc.

thuringiensis, Nosema locustae spores, or erythrocytes with the hemocytes of this insect. In contrast to these studies, Pendland *et al.* (1988), working with *S. exigua,* isolated a D-galactose-specific hemolymph lectin that enhanced both the clearance of fungal spores from the body and their association with the hemocytes, whereas Richards and Ratcliffe (1990) showed that the purified agglutinin from the phasmid *Extatosoma tiaratum* in the presence of Ca^{2+} significantly increased the rosetting of test erythrocytes around the hemocytes of this species (Table 6). Drif and Brehélin (1989) reported, too, that the agglutinin in the hemolymph of *Locusta migratoria* acts as an opsonin to enhance erythrocyte uptake by the hemocytes. These apparent contradictions for a role for insect agglutinins in immune recognition are confusing. This situation is confounded by the recent studies of Kubo *et al.* (1990) and Kawaguchi *et al.* (1991), who have shown that, in *P. americana* and *S. peregrina,* the hemolymph lectins may be involved not only in defense but also in development and regeneration processes.

No doubt some, but not all, of the functional differences reported for insect agglutinins can be explained by suboptimal experimental conditions. For example, great care must be exercised in using vertebrate erythrocytes as test particles as they frequently clump following sensitization with lectin so impeding their uptake by the hemocytes (Rowley and Ratcliffe, 1980). Other

Table 6

Influence of Ca^{2+} and Purified *Extatosoma tiaratum* (Et) Lectin on Spontaneous Rosetting of Rabbit Erythrocytes (RE) around Live and Fixed Et Hemocyte Monolayers

		Live hemocytes[a]			Fixed hemocytes[a]		
Treatment 1	Treatment 2	PL	SG cells	Overall	PL	SG cells	Overall
1. TBS/CA^{2+}	RE/Ca^{2+}	11.3 ± 2.7[b]	57.9 ± 4.0	18.2 ± 4.1	9.4 ± 2.3[b]	48.5 ± 3.7	14.8 ± 2.6
2. TBS/No Ca^{2+}	RE/No Ca^{2+}	3.8 ± 1.2	24.6 ± 2.7	6.8 ± 2.7	2.0 ± 0.6	12.0 ± 1.3	4.2 ± 1.3
3. Et lectin/Ca^{2+}	RE/Ca^{2+}	43.7 ± 5.4	74.2 ± 5.7	45.0 ± 5.3	40.9 ± 5.9	73.3 ± 5.1	42.6 ± 5.5
4. Et lectin/No Ca^{2+}	RE/Ca^{2+}	16.1 ± 4.6	61.0 ± 5.8	24.0 ± 5.6	15.1 ± 4.2	60.8 ± 5.6	22.4 ± 4.1
5. Et lectin/Ca^{2+} 200 m*M* D-galactose	RE/Ca^{2+}	ND[c]	ND	ND	10.4 ± 3.6	48.7 ± 5.2	15.4 ± 4.7

[a]Categories counted: plasmatocytes (PL), spreading granular (SG) cells, all cell types (Overall).
[b]Mean of 8 values ± SD.
[c]ND = not done.

Modified from Richards and Ratcliffe (1990) with permission of Pergamon Press.

factors that greatly influence test particle–hemocyte association include whether the test particle or the hemocyte monolayers are presensitized with lectin and the extent to which the hemocytes are flattened in the monolayers. Richards and Ratcliffe (1990) found that there is a greater opsonic effect when the hemocytes rather than the test particles are sensitized with the lectin, while excessively flattened hemocytes may have too few receptors per unit surface area to bind test particles effectively.

If, as seems likely, insect hemolymph agglutinins do function as recognition molecules, the question of how they interact with the hemocyte surface to form bridging molecules attaching the foreign particles to the blood cells needs to be addressed. Naive insects contain hemolymph agglutinins and yet these molecules remain in the circulation rather than attaching to the blood cells. Subsequent exposure and attachment to foreign antigens must induce some sort of modification in the lectin so that it can interact with carbohydrate moieties on the hemocyte surface. Alternatively, the modified lectin may expose a peptide sequence for which the cell has a receptor. The work of Komano *et al.* (1981) on the lectin induced in injured *S. peregrina* larvae may be relevant here. They reported that the lectin consists of α and β subunits and postulated that some of the α subunits are converted to the β subunits by partial proteolysis to form the active lectin with the structure of $\alpha_4\beta_2$. This active molecule may possibly be able to bind to the outside of the hemocytes and the proteolysis involved in producing the active molecule may result from the release of serine proteases involved in activation of the pPO cascade (reviewed in Ashida and Yamazaki, 1990; Ratcliffe, 1991).

Regarding the role of the pPO cascade in immune recognition and other aspects of immune regulation, it is much more difficult to obtain definitive proof for the functioning of pPO in insect immunity because of the complexity of the system. The pPO system consists of a cascade of enzymes and associated factors such as various plasma receptors, serine proteases, and inhibitors (Fig. 5) (Johansson and Söderhäll, 1989a; Ashida and Yamazaki, 1990; Ratcliffe, 1991), which are often difficult to purify in a stable form.

The pPO system has been reported from both the blood cells (Ashida *et al.*, 1988) and plasma (Saul *et al.*, 1987). This difference probably results from the instability of the hemocytes of many insects that readily discharge their contents following bleeding. It is also significant that the granular cell type (= cystocyte in some species) may contain not only agglutinins but also components of the pPO system (Fig. 4) (reviewed in Rowley *et al.*, 1986). This cell type rapidly degranulates in contact with non-self materials (Ratcliffe and Rowley, 1979) to release the non-self recognition molecules. In *Bombyx mori*, however, the oenocytoids and plasmatocytes contain the majority of the pPO and the granular cells do not have appreciable amounts of this enzyme (Ashida *et al.*, 1988). These differences are difficult to reconcile,

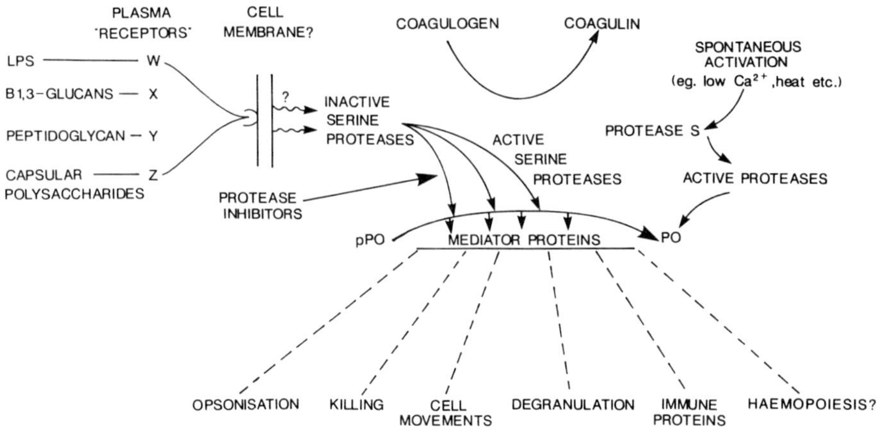

Figure 5 Hypothetical scheme for activation of arthropod prophenoloxidase (pPO). LPS = lipopolysaccharide; PO = phenoloxidase. Based largely on the work of Söderhäll, Ashida, and colleagues. See text for references. Modified from Ratcliffe (1991) with permission of CRC Press Inc.

although the majority of insect species *do not* contain oenocytoids and would require an alternative site for synthesis/storage of pPO.

Evidence for participation of pPO in insect immunity is reviewed elsewhere (Ratcliffe, 1991). The most convincing proof is provided by a series of *in vitro* phagocytosis experiments similar to those described by Smith and Söderhäll (1983) for crustacean hemocyte monolayers overlaid with a marine bacterial isolate, *Moraxella*. Hemocyte monolayers of *Galleria mellonella, Leucophaea maderae,* and *Blaberus craniifer* were set up and overlaid with the bacterium *Bacillus cereus,* plus β-1,3-glucan, dextran, lipopolysaccharide (LPS), and a potent inhibitor of serine protease (p-NPGB). The β-1,3-glucan and LPS, which are both microbial cell-wall components and activators of pPO, enhanced phagocytosis of *B. cereus* five- to sixfold (Table 7) (Ratcliffe *et al.*, 1984; Leonard *et al.*, 1985a). These experiments seem to indicate that since pPO generates sticky proteins that adhere to foreign invaders (Söderhäll *et al.*, 1979), then the test bacteria become coated in sticky opsonic material that is recognized as foreign by the phagocytic cells. These results, however, contrast with the work of Dularay and Lackie (1985), who failed to induce an encapsulation response in the locust *Schistocerea gregaria,* in which Sepharose beads were precoated with phenoloxidase and at least four additional proteins generated by the pPO cascade. Again, there is no ready explanation for this contradiction in experimental results except that the putative opsonin

Table 7

Effect of Laminarin on Phagocytosis of *Bacillus cereus* by Hemocytes from Three Different Insect Species

Treatment	% Hemocytes phagocytic	% Plasmatocytes phagocytic	% Cystocytes/granular cells phagocytic	Number of bacteria/100 hemocytes
G. mellonella				
Laminarin (1 mg ml^{-1})	8.0 ± 1.8[a,b]	14.1 ± 3.7[b]	4.3 ± 1.4[b]	17.4 ± 3.1[b]
GIM control[e]	1.2 ± 0.7	2.0 ± 0.6	0.7 ± 0.5	1.8 ± 0.9
0.01 m*M* *p*NPGB[f]	1.3 ± 0.5	2.4 ± 1.0	0.7 ± 0.4	2.2 ± 1.1
Laminarin + *p*NPGB	1.5 ± 0.6	2.8 ± 1.6	0.5 ± 0.4	2.2 ± 0.8
Dextran (1 mg ml^{-1})	1.9 ± 0.6	2.7 ± 1.1	1.0 ± 0.5	2.6 ± 0.8
L. maderae				
Laminarin (1 mg ml^{-1})	9.8 ± 1.6[b,c]	15.9 ± 4.3[b]	4.7 ± 1.6[b]	26.8 ± 6.8[b]
L-J control[g]	1.5 ± 0.7	2.2 ± 1.1	0.9 ± 0.8	2.2 ± 1.9
0.01 m*M* *p*NPGB	0.9 ± 0.5	1.2 ± 0.6	0.6 ± 0.4	1.2 ± 0.9
Laminarin + *p*NPGB	1.4 ± 0.7	0.7 ± 0.6	1.2 ± 0.8	2.3 ± 1.7
B. craniifer				
Laminarin (1 mg ml^{-1})	4.6 ± 0.6[b,d]	9.6 ± 1.4[b]	1.2 ± 0.4[b]	10.7 ± 1.8[b]
Carlson's control	0.7 ± 0.3	1.2 ± 0.4	0.6 ± 0.4	1.5 ± 0.6
0.01 m*M* *p*NPGB	0.9 ± 0.2	1.2 ± 0.4	0.8 ± 0.4	1.3 ± 0.5
Laminarin + *p*NPGB	1.1 ± 0.2	1.5 ± 0.6	0.8 ± 0.3	1.6 ± 0.5

[a]Mean value of 20 monolayers (from a total of 28 insects in four experiments) ± SD.
[b]$P \leq 0.05$ compared with controls.
[c]Mean value of 10 monolayers (from a total of 14 insects in two experiments) ± SD.
[d]Mean value of 10 monolayers (from a total of 8 insects in two experiments) ± SD.
[e]GIM = Grace's insect medium.
[f]*p*NPGB = *p*-nitrophenyl-*p*′-guanidinobenzoate.
[g]L-J = Landureau and Jollés medium.

From Leonard et al. *(1985a) with permission of Pergamon Press.*

may be very short-lived and lost or inactivated by the experimental procedure adopted by Dularay and Lackie (1985).

The multifunctional role of the insect granular hemocytes (= cystocytes) must also be considered since these cells contain not only factors associated with the pPO system but also agglutinins (Fig. 4) (Rowley *et al.*, 1986). Degranulation of the hemocytes in response to invading organisms will thus release components of the pPO cascade as well as the agglutinins that have been shown to act as recognition molecules (see the foregoing). Even the experiments described previously utilizing presensitization of hemocytes or test particles with purified lectin (Richards and Ratcliffe, 1990) do not provide definitive proof for an opsonic role for hemolymph agglutinins since they could be interpreted in terms of the lectin stimulating the release of the pPO to enhance the uptake of the foreign particles. Additional work is thus urgently required, preferably using purified molecules, to determine the relative roles of the pPO and agglutinins as well as surface charge (reviewed in Lackie, 1988) in the immune reactivity of insects.

Recently, clues to the relative roles of pPO and hemolymph agglutinins have become available. First, Richards and Ratcliffe (1990) showed in the phasmid *Extatosoma tiaratum* that similar percentages of erythrocyte rosetting around the hemocytes were induced by the purified phasmid lectin in both live and glutaraldehyde-fixed monolayers (Table 6). In other words, attachment of foreign particles to the hemocytes was not dependent upon the release of sticky pPO proteins from these cells. Table 6 also shows the importance of Ca^{2+} in this initial binding or recognition process (for a detailed discussion of possible influence of Ca^{2+} in lectin-mediated recognition in invertebrates, see Richards and Renwrantz, 1991).

Second, H. Durrant, N. A. Ratcliffe, and C. C. Chen (unpublished observations) reported that lectins isolated from the hemolymph of *Blaberus discoidalis* significantly enhanced the phenoloxidase activation of hemocyte lysate preparations (HLS) of this cockroach by the microbial activator laminarin (a β-1,3-glucan) (Table 8). One of the lectins, the so-called plasma factor (PF), was β-1,3-glucan-specific, whereas the other was galactose/glucose-specific. The PF was reported previously to enhance laminarin activation of *Blaberus* phenoloxidase (Söderhäll *et al.*, 1988), but only by modification of the purification procedure was its agglutinin activity revealed (C. C. Chen, N. A. Ratcliffe, and A. F. Rowley, unpublished results). Interestingly, preincubation of the two *Blaberus* lectins with their specific ligand sugars failed to abrogate their phenoloxidase-activating properties. Possibly, these lectins have multiple domains, as described for the *Limulus* factor C (Muta *et al.*, 1991), with the sugar-binding and phenoloxidase-activating domains being separate. In addition, various heterologous plant lectins such as concanavalin A, *Tetragonolobus purpureas*, and *Bandeiraea simplicifolia* also enhanced laminarin activation of the *Blaberus* phenoloxidase, whereas several control proteins such as thyroglobulin, aldolase, lysozyme, and bovine serum albumin did not (H. Durrant, N. A. Ratcliffe, and C. C. Chen, unpublished results) (Table 8).

The latter experiments, although only preliminary, are interesting as they help to explain the apparent presence of two immune recognition systems in insects and other arthropods. In fact, there may be only one non-self recognition scenario with early events controlled by the hemolymph agglutinins, as suggested originally in Fig. 1, and later events involving the pPO complex.

Table 8
Enhancement of Activation of *Blaberus discoidalis* Prophenoloxidase by Laminarin in the Presence of Various Lectins

Activation mixture	Maximum phenoloxidase activity[a] at optimal lectin/protein concentration ΔA490 10 min^{-1}
HLS[b] + cac. buffer	0.040 ± 0.012
HLS + lam[c]	0.111 ± 0.022
HLS + L2[d]	0.100 ± 0.015
HLS + L2 + lam	0.184 ± 0.029
HLS + pf[e]	0.119 ± 0.016
HLS + pf + lam	0.206 ± 0.033
HLS + con A	0.087 ± 0.017
HLS + con A + lam	0.157 ± 0.022
HLS + lysozyme[f]	0.033 ± 0.015
HLS + lysozyme + lam	0.115 ± 0.020

[a]Assay carried out in multiwell trays and read on a Biorad EIA reader.
[b]Hemocyte lysate supernatant at protein concentration of 2.0 mg ml^{-1}.
[c]Laminarin 1.0 mg ml^{-1}.
[d]L2 = purified *B. discoidalis* serum lectin.
[e]pf = purified *B. discoidalis* plasma factor (a β-1,3-glucan-specific lectin).
[f]Protein control.

From H. Durrant, N. A. Ratcliffe, and C. C. Chen (unpublished results).

The details of the interaction of these two sets of molecules await further investigation.

Third, additional evidence for the interaction of the hemolymph lectins and the pPO system is present in the literature. Thus Minnick *et al.* (1986) purified a glucose specific lectin from the tobacco hornworm *Manduca sexta*, termed M13, and showed that when added to hemolymph of the same species it provoked a coagulation response. The pPO system may well be involved in coagulation responses in insects as in some other arthropods. Also significant was the work of Dunphy and Chadwick (1989) with *G. mellonella* hemocytes *in vitro*. They studied the influence of various carbohydrates on the adhesion of the bacteria *Pseudomonas aeruginosa* and *Proteus mirabilis* to the hemocytes and noted that certain sugars enhanced both bacterial adhesion and pPO activation. As a result of these experiments, Dunphy and Chadwick (1989) suggested that both the lectins and the pPO system are opsonic and enhance the binding of bacteria to the blood cells.

C. Cell–Cell Communication in Immunity

Another area of insect immunity in need of further clarification is the determination of the relative roles and interactions of the immune reactive cells. The wax moth, *Galleria mellonella,* has proved to be an ideal subject for this work because of the ease of identification and relative stability of the blood cells. In *Galleria,* and probably other insects too, the two main blood cell types responsible for the cellular defenses are the granular cells (= cystocytes of some species) and the plasmatocytes, although the oenocytoids are also a site of storage/synthesis of pPO (Schmit *et al.,* 1977). The granular cells (see Fig. 4) contain both pPO and also probably agglutinins (Ratcliffe and Rowley, 1987), are nonphagocytic, unstable *in vitro,* and clearly distinguishable from the plasmatocytes, which are mainly agranular, phagocytic, and highly stable *in vitro.* The fact that the *Galleria* plasmatocytes are usually a granular contrasts with many other insect species (e.g., cockroaches, locusts, other lepidopterans, etc.) in which the presence of numerous granules often makes it difficult to discern them from granular cells and cystocytes. In addition, granule discharge from the plasmatocytes enhances the difficulty of understanding the events involved during cellular immunity in insects.

Earlier studies by Ratcliffe and Gagen (1977) and Schmit and Ratcliffe (1977) noted that in *Galleria* larvae, during nodule formation and encapsulation in response to bacteria and foreign implants, respectively, the first and immediate (within a few seconds usually) response to foreign antigens is the degranulation of the fragile granular cells. Plasma agglutinins too may well be involved in this initial foreign body–granular cell recognition event (see Section III.B). The granular cells thus release their putative non-self recognition molecules, that is, the pPO factors and agglutinins, onto the surface of the foreign material, which often then begins to melanize. Within 10–15 min the second immune reactive cell type in *Galleria,* the plasmatocytes, arrives at the scene and forms a multicellular sheath around the central melanotic core of degranulated and lysed granular cells. Events similar to these have also been observed in the stick insect *Clitumnus extradentatus,* the locust *Schistocerca gregaria,* and the beetle *Tenebrio molitor* (Ratcliffe and Rowley, 1979), so that they may be widespread in insects. Thus, cell–cell cooperation is involved during cellular reactivity and probably occurs during both capsule formation and phagocytosis in insects (Fig. 6) (see the following).

More recently, additional studies on cell–cell cooperation have been undertaken with purified populations of *Galleria* granular cells and plasmatocytes (Anggraeni and Ratcliffe, 1991). The hemocytes were purified on continuous Percoll gradients and monolayers prepared from the separated plasmatocytes or granular cells or from a mixture of the two. The monolayers were then overlaid with heat-killed *Bacillus cereus* either suspended in buffer alone (controls) or with laminarin (a β-1,3-glucan activator of pPO = experimentals) and incubated for 1 hr to allow phagocytosis to occur. The results confirmed plasmatocytes as the main phagocytic cell type in *Galleria;* no ingestion at all was recorded for the granular cells. Phagocytosis by the plasmatocytes increased from 5.3 to 8.7% for controls (i.e., without laminarin) and from more than 13.4 to 30% for experimentals (i.e., with laminarin) upon addition of granular cells back to the isolated plasmatocytes (Table 5). An increase in phagocytosis from 5.3 to 13.4% by plasmatocytes upon addition of laminarin was probably due to a small contaminant population of granular cells. These results clearly show that the granular cells, although not phagocytic themselves, contain some factor(s) (probably pPO and/or agglutinin) essential for optimal functioning of the plasmatocytes during cellular reactivity.

Similar cell–cell cooperative events have been described during cellular immune events with crustacean blood cells *in vitro* (Söderhäll *et al.,* 1986; Persson *et al.,* 1987) (Fig. 6) and the role of a 76-kDa protein in cellular communication events has been identified (Johansson and Söderhäll, 1985, 1989a; Kobayashi *et al.,* 1990). This 76-kDa protein is associated with the pPO system and is multifunctional since it acts as a cell adhesion factor for crayfish semigranular and granular cells (Johansson and Söderhäll, 1988), can degranulate both semigranular and granular cells by a regulated exocytosis (Johansson and Söderhäll, 1989b), and also functions to promote encapsulation (Kobayashi *et al.,* 1990). In insects, little is known about molecular events controlling cell–cell communication, although Rantamäki *et al.* (1991) partially purified a 90-kDa protein from *Blaberus craniifer* hemocytes that cross-reacted with a monospecific antiserum against the 76-kDa protein from crayfish blood cells. This 90-kDa protein, like the 76-kDa molecule in crayfish, also functions to enhance adhesion of the cockroach hemocytes to the substratum and to induce degranulation of these cells.

Other factors, apart from the 90-kDa protein, are likely to influence cell–cell communication and the behavior of insect hemocytes during immune reactions. Some of these substances have been described in detail by Lackie (1988) and include an encapsulation-promoting factor (EPF) (Ratner and Vinson, 1983; Davies *et al.,* 1988), chemokinins (Huxham and Lackie, 1988), and various wound factors (e.g., Cherbas, 1973). All of these substances are probably involved in the initiation of immunity by wounding but, apart from EPF, little is known about these molecules. EPF has been partially purified from the plasma of *Heliothis virescens* and is a small 3.5-kDa, heat-stable peptide possibly originating from the hemocytes. The function of EPF is to promote the adhesive behavior and encapsulation ability of plasmatocytes *in*

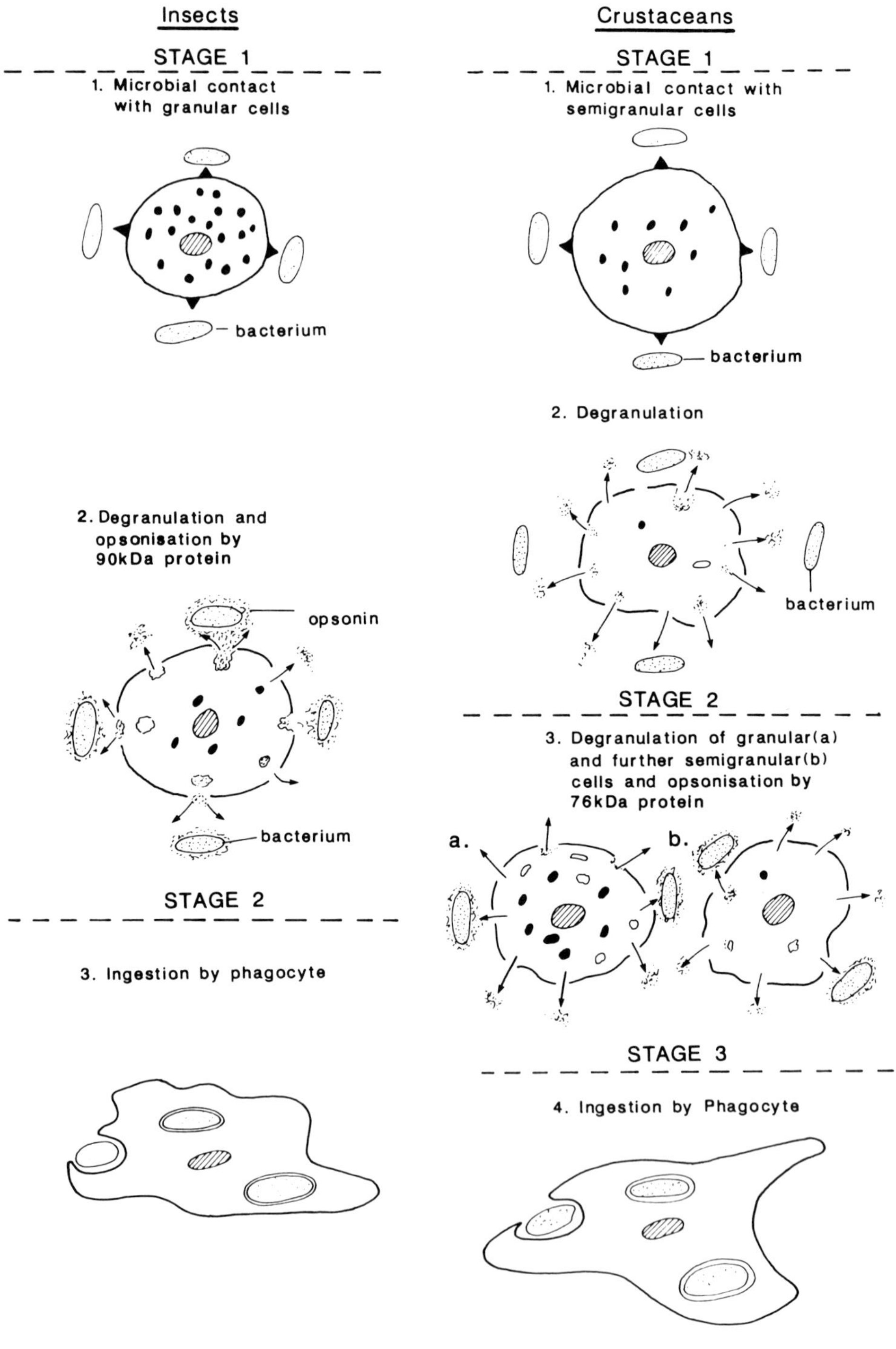

Insects
STAGE 1
1. Microbial contact with granular cells
bacterium
2. Degranulation and opsonisation by 90kDa protein
opsonin
bacterium
STAGE 2
3. Ingestion by phagocyte
Crustaceans
STAGE 1
1. Microbial contact with semigranular cells
bacterium
2. Degranulation
bacterium
STAGE 2
3. Degranulation of granular(a) and further semigranular(b) cells and opsonisation by 76kDa protein
a.
b.
STAGE 3
4. Ingestion by Phagocyte

vitro (Davies *et al.*, 1988). EPF may possibly be produced by activation of the pPO cascade, as may the chemokinins (Lackie, 1988).

D. Interrelationship of Cellular and Humoral Immunity

It is apparent from previous sections that the division of insect immunity into cellular and humoral components is a matter of convenience rather than of fact. Thus, plasma agglutinins are derived from the fat body and/or hemocytes (Takahashi *et al.*, 1985; Rowley *et al.*, 1986) and released into the plasma, where they undertake their immune regulatory role that results in the binding of foreign antigens and associated agglutinins onto the hemocyte surface (see Section III.B). Likewise, the pPO system is, at least partially, synthesized in the hemocytes and released by degranulation into the plasma. Various inhibitors and activators controlling the pPO system in arthropods have, however, been isolated from the plasma (Hergenhahn *et al.*, 1987, 1988; Söderhäll *et al.*, 1988) so that a complex interplay between plasma and cellular components controls the pPO cascade. Additionally, the antibacterial proteins have recently been shown in *Drosophila* and other species to be synthesized not only in the fat body but also in the hemocytes (Dickinson *et al.*, 1988; Matsuyama and Natori, 1988; Samakovlis *et al.*, 1990) for release into the hemolymph to form the humoral component of the immune response.

There is compelling evidence provided by hemocyte transfer experiments with *Galleria mellonella* larvae for a role for insect blood cells in induction of a humoral antibacterial response (de Verno *et al.*, 1983, 1984). In the first of these studies, hemocytes from *Galleria* previously immunized with *Pseudomonas aeruginosa* endotoxin (LPS) were injected into naive larvae before challenging these with live *P. aeruginosa* at various time intervals. Good pro-

Figure 6 Two hypothetical models for cell–cell cooperation in arthropod immunity. The insect model is based on observations with *Galleria mellonella* in which it is proposed that non-self recognition is carried out by the granular cells and ingestion by the plasmatocytes (phagocytes). The crustacean model differs from the insect scheme in having an amplification step at stage 2 in which adjacent granular and semigranular cells are induced to discharge their pPO system. Since most insects have many more granule-containing cells than *Galleria*, the crustacean scheme may well be present in such species. Circulating agglutinins are probably involved in enhancing contact and binding of the microbial invaders to the surfaces of the granular/semigranular cells at stage 1. Based on the work of Ratcliffe *et al.* (1984), Söderhäll and Smith (1986), Johansson and Söderhäll (1988), and colleagues. See text for additional references. Modified from Ratcliffe (1991) with permission of CRC Press, Inc.

tection was afforded by the transferred hemocytes provided that they were injected into the larvae within 4 hr of LPS inoculation of the donors. Hemocytes from non-LPS-immunized larvae provided no protection to recipients challenged with *P. aeruginosa*. De Verno *et al.* (1983) believe these results show that the hemocytes are active in the induction of humoral immunity early in the immune response and may be releasing factor(s) to stimulate the fat body cells to produce antibacterial proteins. Faye and Wyatt (1980) and Abu-Hakima and Faye (1981) also noticed an intimate association *in vitro* between hemocytes containing bacteria and the fat body cells from *Cecropia* pupae, and suggested that the phagocytic granular cells move to and disintegrate in the region of the fat body. In subsequent experiments, De Verno *et al.* (1984) using *Galleria* fat body *in vitro* also showed that maximal production of antibacterial activity was stimulated by the presence of LPS and/or hemocytes. They suggested that the LPS causes the release of hemocyte factor(s) resulting in enhanced production of antibacterial proteins by the fat body.

Similar experiments utilizing the fat body of *Manduca sexta* (Dunn *et al.*, 1985) and *Hyalophora cecropia* (Trenczek and Faye, 1988) produced additional support for a role for the hemocytes in inducing the synthesis of antibacterial proteins. In *Manduca,* Dunn *et al.* (1985) suggested that fragments of peptidoglycan liberated by digestion of the bacteria by plasma lysozyme or following phagocytosis by the hemocytes may serve as elicitors of protein production. This conclusion may differ slightly from that just described for *Galleria,* as the elicitor is derived from the bacteria and not from the hemocytes, although de Verno *et al.* (1984) did report that antibacterial activity could also be stimulated by LPS alone. Similarly, with *Hyalophora* fat body cultures, Trenczek and Faye (1988) reported that antibacterial protein production was stimulated by LPS and bacteria. They also showed, however, that hemocytes from injured or immunized pupae and not those from untreated insects had an elicitor effect.

Most of the studies described clearly indicate a role for the hemocytes in the induction of the humoral antibacterial proteins. The fact that the process of injuring the insects during dissection may elicit antibacterial protein synthesis shows, however, that signals other than those produced through hemocyte-processed bacterial substances must be involved (Trenczek and Faye, 1988; Samakovlis *et al.*, 1990). It may be of significance that peptidoglycan, LPS, and β-1,3-glucan not only induce antibacterial protein production (Dunn *et al.*, 1985; Samakovlis *et al.*, 1990) but also activate the pPO system. The latter too is enhanced by hemolymph lectins (H. Durrant, N. A. Ratcliffe, and C. C. Chen, unpublished results) so that an extremely complicated network is gradually emerging that may involve microbes, agglutinins, pPO, hemocytes, damaged tissues, and fat body cells. The situation is further complicated by the possible involvement of the endocrine system mediating immune reactivity events during the life of the insect (Shiraishi and Natori, 1989).

Finally, Faye (1990) has reported a possible connection between the recognition of bacteria and the activation of the immune protein genes in *Hyalophora*. This process involves the immune protein haemolin, previously called P4, which is present in normal hemolymph and binds to the bacterial surface *in vitro* to form a complex with another hemolymph protein with a mass of about 125 kDa (Sun *et al.*, 1990). This complex may function to deliver a recognition signal to the cells responsible for the synthesis of the immune proteins. The reason haemolin is believed to function in immune recognition and activation is, first, because it binds to the bacterial surface and, second, because it has four immunoglobulin domains and shares 38% homology with the immunoglobulin part of the insect cell recognition molecule, neuroglian (Faye, 1990).

E. Future Areas for Study

From this brief overview, it is apparent that, with the application of modern biochemical and molecular techniques, progress in our understanding of insect immunity has been accelerating. This progress will no doubt continue in the areas outlined in the foregoing. In addition, as knowledge has accrued of how invertebrates deal with invading parasites and pathogens, attention is beginning to focus on more applied aspects of the immune systems of invertebrates. Thus, the possible importance of the host defenses as determinants of infectivity in vector–parasite encounters and the use of changes in immune reactivity as monitors of environmental pollution are now being examined.

Regarding vector–parasitic interactions, evidence has been accumulating for the importance of the vector host defenses in determining the outcome of such associations (reviewed in Molyneux *et al.*, 1986; Ratcliffe and Rowley, 1987; Kaaya, 1989). For example, Collins *et al.* (1986) reported that in strains of *Anopheles gambiae* refractory and susceptible toward *Plasmodium cynomolgi*, there was a differential ability of the host mosquitoes to encapsulate the ookinetes in the midgut wall. The refractory strain encapsulated the parasites in a layer of melaninlike substance whereas in susceptible insects only a reduced reaction occurred (Paskewitz *et al.*, 1989). It was suggested that refractoriness is due to a greater ability of the host to recognize the living parasites. It is possible that the melaninlike substance resulted from enhanced levels of pPO in the refractory strain. Furthermore, in strains of the tsetse flies *Glossina palpalis palpalis* and *Glossina morsitans morsitans*, which are refractory and susceptible to *Tryanosoma brucei rhodesiense,* respectively, pPO activity in the hemolymph and midgut has also been shown to be far greater in the resistant flies. Even more significant were differences between the sexes of *G. m. morsitans* in which the females show 7% and males 27% salivary gland infections (I. Maudlin and S. C. Welburn, personal communication). The females had much higher levels than the males of pPO activation of the hemolymph when parasites were used as elicitors (Table 9) (Nigam and Ratcliffe, 1991; Y. Nigam, N. A. Ratcliffe, I. Maudlin, and S. C. Welburn, unpublished observations). The contribution, if any, of the hemocytes to these results has not yet been determined. Additional work is clearly required on the role of pPO immune recognition in these vector species.

The importance of lectins in the gut and hemolymph of tsetse flies in the transmission of trypanosomes has also been reported (Pereira *et al.,* 1981; Ibraham *et al.,* 1984; Molyneux *et al.,* 1986; Maudlin and Welburn, 1987, 1988a,b; Ingram and Molyneux, 1988; Welburn and Maudlin, 1989, 1990; Welburn *et al.,* 1989). These workers have detected agglutinins for trypanosomes in the gut and hemolymph of tsetse flies and believe that these molecules are important for the development of the parasite in the vector insect. Thus, a refractory line of *G. m. morsitans* killing 50% of trypanosomes in 170 hr had higher titers of midgut lectins than a susceptible line killing only 10% of parasites in this time (Welburn *et al.,* 1989). They concluded that the lectin prevented trypanosome establishment in the midgut perhaps by lysis of the parasites. The fact that blocking the lectin with its sugar ligand, D-glucosamine, significantly increased midgut infection rates of *Trypanosoma con-*

Table 9

Phenoloxidase Activity of Male and Female Flies of Susceptible G. *morsitans morsitans* 1/6 Strain

Treatment of hemolymph	Incubation time (hr)						
	0.25	0.5	1.0	2.0	4.0	6.0	8.0
Control[a] male	0.25[b,c]	0.34	0.79	1.10	1.16	1.26	1.33
Control[a] female	1.37	2.19	2.85	3.02	3.18	3.49	3.57
Procyclics[d] male	0.17	0.32	0.90	1.23	1.29	1.29	1.33
Procyclics[d] female	1.31	2.23	2.98	3.50	3.73	4.09	4.28

[a]Controls contained buffer, hemolymph, and L-dopa.
[b]Measured as units of phenoloxidase where 1U − 1A492 mg^{-1} protein.
[c]Means of three experiments but standard deviations not given here.
[d]Procyclics of *Trypanosoma brucei rhodesiense* in buffer incubated with hemolymph and L-dopa.

From Y. Nigam, N. A. Ratcliffe, I. Maudlin, and S. C. Welburn (unpublished results).

golense and *T. brucei rhodesiense* supports this idea (Maudlin and Welburn, 1987). Concomitantly, however, it has been reported that the midgut lectin and a hemolymph lectin in tsetse are involved in maturation of trypanosomes (Welburn and Maudlin, 1989, 1990). This apparent dual role for the lectins indicates the complexity of this vector–parasite association.

Additional evidence for a role of the vector immune system as a determinant of infectivity is provided by hemolymph transfer experiments with simuliids (blackflies) (Ham, 1986) and by numerous reports of the encapsulation of parasites in mosquitoes (reviewed in Ratcliffe, 1982; Kaaya, 1989). Ham (1986) described the presence of a parasite-induced or -derived factor in the hemolymph of simuliids infected with *Onchocerca lienalis* that confers protection when transferred to naive flies. Examples of encapsulation in mosquitoes include a report by Christensen and Forton (1986) that the initial stage in encapsulation of *Dirofilaria immitis* microfilariae injected into *Aedes aegypti* was the lysis of hemocytes at or near the surface of the parasite prior to deposition of a melaninlike pigment, and the observation of Chen and Laurence (1985) that the encapsulation of *Brugia pahangi* microfilariae in *Anopheles quadrimaculatus* involved both humoral and cellular events.

Once the details of the vector immune reactions are fully understood, it might be possible to manipulate them and increase the refractoriness of the host insects to parasitic invasion. This might involve the insertion of specific genes to increase the potency of the vector immune response to produce refractory strains of insects (Kaaya, 1989).

Finally, the fact that the immune systems of invertebrates are exquisitely sensitive and constantly sampling and responding to minute quantities of environmental molecules means that any contaminants present are likely to modify immunity in some way. If such modifications can be carefully documented then it may be possible to use the invertebrate immune systems as monitors of environmental pollution. For example, Miosky *et al.* (1989) in mollusks have demonstrated the value of monoclonal antibodies in monitoring the increased incidence of a leukemiclike condition in the soft-shell clam, *Mya arenaria,* collected from polluted waters of Maine. With insects, unfortunately, little is known of the action of xenobiotics on the immune system. Feir (1979) briefly reviewed information on insect immune responses to toxic substances such as insecticides, while Saxena and Saxena (1985) described cytopathological changes in the hemocytes of *Periplaneta americana* after exposure to the pesticide, malathion. More recently, Jones *et al.* (1989) also showed that the antibacterial activity of the hemolymph of the lepidopteran *Pieris brassicae* was significantly suppressed by exposure to 2,4,5-trichlorophenoxyacetic acid (2,4,5-T). Clearly, more work is required in order to determine the effects of environmental insults on a range of immune parameters.

Acknowledgments

I am grateful to the Nuffield Foundation, The British Council, and The Science and Engineering Research Council (Grants GR/F/17421, GR/G/60857, and GR/G/40224) for financial support.

References

Abu-Hakima, R., and Faye, I. (1981). An ultrastructural and autoradiographic study of the immune response in *Hyalophora cecropia* pupae. *Cell Tissue Res.* **217:**311–320.

Anderson, R. S., Holmes, B., and Good, R. A. (1973). *In vitro* bactericidal capacity of *Blaberus craniifer* hemocytes. *J. Invertebr. Pathol.* **22:**127–135.

Anggraeni, T., and Ratcliffe, N. A. (1991). Studies on cell–cell cooperation during phagocytosis by purified haemocyte populations of the wax moth, *Galleria mellonella. J. Insect Physiol.* **37:**453–460.

Ashida, M., and Yamazaki, H. I. (1990). Biochemistry of the phenoloxidase system in insects: With special reference to its activation. *In* "Molting and Metamorphosis" (E. Ohnishi and H. Ishizaki, eds.), pp. 239–265. Jpn. Sci. Soc. Press, Tokyo/Springer-Verlag, Berlin.

Ashida, M., Ochiai, M., and Niki, T. (1988). Immunolocalization of prophenoloxidase among hemocytes of the silkworm, *Bombyx mori. Tissue Cell* **20:**599–610.

Azambuja, P. D., Garcia, E. S., and Ratcliffe, N. A. (1991). Aspects of classification of hemiptera hemocytes from six triatomine species. *Mem. Inst. Oswaldo Cruz* **86:**1–10.

Beckage, N. E. (1990). Parasite effects on host development. *In* "New Directions in Biological Control: Alternatives for Suppressing Agriculture" (R. Baker and P. E. Dunn, eds.), pp. 497–515. Liss, New York.

Beeman, S. C., Wilson, M. E., Bulla, L. A., and Consigli, R. A. (1983). Structural characterisation of the hemocytes of *Plodia interpunctella. J. Morphol.* **175:**1–16.

Bohn, H. (1986). Hemolymph clotting in insects. *In* "Immunity in Invertebrates" (M. Brehélin, ed.), pp. 188–207. Springer-Verlag, Berlin.

Boman, H. G., and Hultmark, D. (1987). Cell-free immunity in insects. *Annu. Rev. Microbiol.* **41:**103–126.

Bradley, R. S., Stuart, G. S., Stiles, B., and Hapner, K. D. (1989). Grasshopper haemagglutinin: Immunochemical localization in haemocytes and investigation of opsonic properties. *J. Insect Physiol.* **35:**353–361.

Brehélin, M. (1990). Depression of immune reactions in insects. *Res. Immunol.* **141:**935–938.

Brehélin, M., and Zachary, D. (1986). Insect haemocytes: A new classification to rule out the controversy. *In* "Immunity in Invertebrates" (M. Brehélin, ed.), pp. 36–48. Springer-Verlag, Berlin.

Casteels, P., Ampe, C., Riviere, L., Damme, J. V., Elicone, C., Fleming, M., Jacobs, F., and Tempst, P. (1990). Isolation and characterization of abaecin, a major antibacterial response peptide in the honeybee (*Apis mellifera*). *Eur. J. Biochem.* **187:**381–386.

Chadwick, J. M., and Aston, W. P. (1978). An overview of insect immunity. *In* "Animal Models of Comparative and Developmental Aspects of Immunity and Disease" (M. E. Gershwin and E. L. Cooper, eds.), p. 1. Pergamon, New York.

Chain, B. M., and Anderson, R. S. (1983). Inflammation in insects: The release of plasmatocyte depletion factor following interaction between bacteria and haemocytes. *J. Insect Physiol.* **29:**1–4.

Chen, C. C., and Laurence, B. R. (1985). An ultrastructural study on the encapsulation of microfilariae of *Brugia pahangi* in the haemocoel of *Anopheles quadrimaculatus*. *Int. J. Parasitol.* **15:**421–428.

Chen, C. C., Ratcliffe, N. A., and Rowley, A. F. (1993). Detection, isolation and characterization of multiple lectins from the haemolymph of the cockroach, *Blaberus discoidalis*. *Biochem. J.* (in press).

Cherbas, L. (1973). The induction of an injury reaction in cultured haemocytes from saturniid pupae. *J. Insect Physiol.* **19:**2011–2023.

Christensen, B. M., and Forton, K. F. (1986). Hemocyte-mediated melanization of microfilariae in *Aedes aegypti*. *J. Parasitol.* **72:**220–225.

Collins, F. H., Sakai, R. K., Vernick, K. D., Paskewitz, S., Seeley, D. C., Miller, L. H., Collins, W. E., Campbell, C. C., and Gwadz, R. W. (1986). Genetic selection of a plasmodium-refractory strain of the malaria vector *Anopheles gambiae*. *Science* **234:**607–609.

Croft, S. L., East, J. S., and Molyneux, D. H. (1982). Anti-trypanosomal factor in the haemolymph of *Glossina*. *Acta Trop.* **39:**293.

Crossley, A. C. (1968). The fine structure and mechanism of breakdown of larval intersegmental muscles in the blowfly, *Calliphora erythrocephala*. *J. Insect Physiol.* **14:**1389–1407.

Davies, D. H., Hayes, T. K., and Vinson, S. B. (1988). Preliminary characterization and purification of *in vitro* encapsulation promoting factor: A peptide that mediates insect haemocyte adhesion. *Dev. Comp. Immunol.* **12:**241–254.

de Verno, P. J., Aston, W. P., and Chadwick, J. S. (1983). Transfer of immunity against *Pseudomonas aeruginosa* pp. 1–11 in *Galleria mellonella* larvae. *Dev. Comp. Immunol.* **7:**423–434.

de Verno, P. J., Chadwick, J. S., Aston, W. P., and Dunphy, G. B. (1984). The *in vitro* generation of an antibacterial activity from the fat body and hemolymph of non-immunised larvae of *Galleria mellonella*. *Dev. Comp. Immunol.* **8:**537–546.

Dickinson, L., Russell, V., and Dunn, P. E. (1988). A family of bacteria-regulated, cecropin D-like peptides from *Manduca sexta*. *J. Biol. Chem.* **263:**19424–19429.

Drif, L., and Brehélin, M. (1989). Agglutinin mediated immune recognition in *Locusta migratoria* (Insecta). *J. Insect Physiol.* **35:**729–736.

Dularay, B., and Lackie, A. M. (1985). Haemocytic encapsulation and the prophenoloxidase activation pathway in the locust *Schistocerca gregaria*. *Insect Biochem.* **15:**827–834.

Dunn, P. E., Dai, W., Kanost, M. R., and Geng, C. (1985). Soluble peptidoglycan fragments stimulate antibacterial protein synthesis by fat body from larvae of *Manduca sexta*. *Dev. Comp. Immunol.* **9:**559–568.

Dunphy, G. B., and Chadwick, J. S. (1989). Effects of selected carbohydrates and the contribution of the prophenoloxidase cascade system to the adhesion of strains of *Pseudomonas aeruginosa* and *Proteus mirabilis* to hemocytes of non immune larval *Galleria mellonella*. *Can. J. Microbiol.* **35:**524–527.

Faye, I. (1990). Acquired immunity in insects: The recognition of non-self and the subsequent onset of immune protein genes. *Res. Immunol.* **141:**927–931.

Faye, I., and Wyatt, G. R. (1980). The synthesis of antibacterial proteins in isolated fat body from cecropia silkmoth pupae. *Experientia* **36:**1325–1326.

Feir, D. (1979). Cellular and humoral responses to toxic substances. *In* "Insect Hemocytes: Development, Forms, Functions and Techniques" (A. P. Gupta, ed.), pp. 415–421. Cambridge Univ. Press, New York and London.

George, J. F., Howcroft, T. K., and Karp, R. D. (1987). Primary integumentry allograft reactivity in the American cockroach, *Periplanata americana*. *Transplantation* **43:**514–519.

Götz, P., and Boman, H. G. (1985). Insect immunity. *In* "Comprehensive Insect Physiology, Biochemistry and Pharmacology" (G. A. Kerkut and L. I. Gilbert, eds.), pp. 453–485. Pergamon, Oxford.

Green, E. I., and Carter, J. B. (1991). Increased hemocyte counts in *Tipula paludosa* (Diptera, Tipulidae) hemolymph after exposure of larvae to diethyl ether and acetic acid vapors. *J. Invertebr. Pathol.* **57:**437–438.

Gupta, A. P., ed. (1979). "Insect Hemocytes: Development, Forms, Functions and Techniques." Cambridge Univ. Press, New York and London.

Ham, P. J. (1986). Acquired resistance to *Onchocerca lienalis* infections in *Simulium ornatum* Meigen and *Simulium lineatum* Meigen following passive transfer of hemolymph from previously infected simuliids (Diptera, Simuliidae). *Parasitology* **92:**269–277.

Ham, P. J., Nolan, G., and Natori, S. (1991). Antibody blocking of microfilaricidal activity of mosquito and black fly immune haemolymph. *Br. Soc. Parasitol.*, Liverpool, England, Abstr., p. 53.

Hergenhahn, H.-G., Aspan, A., and Söderhäll, K. (1987). Purification and characterization of a high-Mr proteinase inhibitor of prophenoloxidase activation from crayfish plasma. *Biochem. J.* **248:**223–228.

Hergenhahn, H.-G., Häll, M., and Söderhäll, K. (1988). Purification and characterization of an α_2-macroglobulin-like proteinase inhibitor from plasma of the crayfish *Pacifastacus leniusculus*. *Biochem. J.* **255:**801–806.

Hinks, C. F., and Arnold, J. W. (1977). Haemopoiesis in Lepidoptera. II. The role of the haemopoietic organs. *Can J. Zool.* **55:**1740–1755.

Hoffmann, J. A., and Hoffmann, D. (1990). The inducible antibacterial peptides of dipteran insects. *Res. Immunol.* **141:**910–918.

Hoffmann, J. A., Zachary, D., Hoffmann, D., and Brehélin, M. (1979). Postembryonic development and differentiation: Hemopoietic tissues and their functions in some insects. *In* "Insect Hemocytes: Development, Forms, Functions and Techniques" (A. P. Gupta, ed.), pp. 29–82. Cambridge Univ. Press, New York and London.

Huxham, I. M., and Lackie, A. M. (1988). Behaviour *in vitro* of separated haemocytes from the locust, *Schistocerca gregaria*. *Cell Tissue Res.* **251:**677–684.

Ibrahim, E.A.R., Ingram, G. A., and Molyneux, D. H. (1984). Haemagglutinins and parasite agglutinins in haemolymph and gut of *Glossina*. *Tropenmed. Parasitol.* **35:**151–156.

Ingram, G. A., and Molyneux, D. H. (1988). Sugar specificities of anti-human ABO(H) blood group erythrocyte agglutinins (lectins) and haemolytic activity in the haemolymph and gut extracts of three *Glossina* species. *Insect. Biochem.* **18:**267–279.

Ingram, G. A., East, J., and Molyneux, D. H. (1984). Naturally occurring agglutinins against trypanosomatid flagellates in the haemolymph of insects. *Parasitology* **89:**435–451.

Johansson, M. W., and Söderhäll, K. (1985). Exocytosis of the prophenoloxidase activating system from crayfish haemocytes. *J. Comp. Physiol. B* **156:**175–181.

Johansson, M. W., and Söderhäll, K. (1988). Isolation and purification of a cell adhesion factor from crayfish blood cells. *J. Cell Biol.* **106:**1795–1803.

Johansson, M. W., and Söderhäll, K. (1989a). Cellular immunity in crustaceans and the proPO system. *Parasitol. Today* **5:**171.

Johansson, M. W., and Söderhäll, K. (1989b). A cell adhesion factor from crayfish hemocytes has degranulating activity towards crayfish granular cells. *Insect Biochem.* **19:**183–190.

Jones, J. C. (1962). Current concepts concerning insect hemocytes. *Am. Zool.* **2:**209–246.

Jones, J. C. (1970). Hemocytopoiesis in insects. *In* "Regulation of Hematopoiesis" (A. S. Gordon, ed.), pp. 7–65. Appleton, New York.

Jones, J. C., and Liu, D. P. (1968). A quantitative study of mitotic divisions of haemocytes of *Galleria mellonella* larvae. *J. Insect Physiol.* **14:**1055–1061.

Jones, R. D., Rowley, A. F., and Ratcliffe, N. A. (1989). Effect of exposure of *Pieris brassicae* larvae to 2,4,5-trichlorophenoxyacetic acid on the natural antibacterial activity of serum. *J. Invertebr. Pathol.* **54:**63–70.

Kaaya, G. P. (1989). A review of the progress made in recent years on research and understanding of immunity in insect vectors of human and animal diseases. *Insect Sci. Appl.* **10:**751–769.

Karp, R. D. (1990). Transplantation immunity in insects: Does allograft responsiveness exist? *Res. Immunol.* **141:**923–927.

Kawaguchi, N., Kamano, H., and Natori, S. (1991). Involvement of *Sarcophaga* lectin in the development of imaginal discs of *Sarcophaga peregrina* in an autocrine manner. *Dev. Biol.* **144:**86–93.

Knibiehler, B., Mirre, C., Cacchini, J.-P., and Le Parco, Y. (1987). Haemocytes accumulate collagen transcripts during *Drosophila melanogaster* metamorphosis. *Wilhelm Roux's Arch. Dev. Biol.* **196:**243–247.

Kobayashi, M., Johansson, M. W., and Söderhäll, K. (1990). The 76 kD cell adhesion factor from crayfish haemocytes promotes encapsulation *in vitro*. *Cell Tissue Res.* **260:**13–18.

Komano, H., Mizuno, D., and Natori, S. (1980). Purification of lectin induced in the hemolymph of *Sarcophaga peregrina* larvae on injury. *J. Biol. Chem.* **255:**2919.

Komano, H., Mizuno, D., and Natori, S. (1981). A possible mechanism of induction of insect lectin. *J. Biol. Chem.* **256:**7087–7089.

Kubo, T., Kawasaki, K., and Natori, S. (1990). Sucrose-binding lectin in regenerating cockroach (*Periplaneta americana*) legs: Its purification from adult hemolymph. *Insect Biochem.* **6:**585–591.

Lackie, A. M. (1988). Haemocyte behaviour. *Adv. Insect Physiol.* **21:**85–177.

Leonard, C., Ratcliffe, N. A., and Rowley, A. F. (1985a). The role of prophenoloxidase activation in non-self recognition and phagocytosis by insect blood cells. *J. Insect Physiol.* **31:**789–799.

Leonard, C., Söderhäll, K., and Ratcliffe, N. A. (1985b). Studies on prophenoloxidase and proteinase activity of *Blaberus craniifer* haemocytes. *Insect Biochem.* **15:**803–810.

Matsuyama, K., and Natori, S. (1988). Molecular cloning of cDNA for sapecin and unique expression of the sapecin gene during the development of *Sarcophaga peregrina*. *J. Biol. Chem.* **263:**17117–17121.

Maudlin, I., and Welburn, S. C. (1987). Lectin mediated establishment of midgut infections of *Trypanosoma congolense* and *Trypanosoma brucei* in *Glossina morsitans*. *Tropenmed. Parasitol.* **38:**167–170.

Maudlin, I., and Welburn, S. C. (1988a). The role of lectins and trypanosome genotype in the maturation of midgut infections in *Glossina morsitans*. *Tropenmed. Parasitol.* **39:**56–58.

Maudlin, I., and Welburn, S. C. (1988b). Tsetse immunity and the transmission of trypanosomiasis. *Parasitol. Today* **4:**109–111.

Mead, G. P., Ratcliffe, N. A., and Renwrantz, L. R. (1986). The separation of insect haemocyte types on Percoll gradients: Methodology and problems. *J. Insect Physiol.* **32:**167–177.

Minnick, M. F., Rupp, R. A., and Spence, K. D. (1986). A bacterial-induced lectin which triggers hemocyte coagulation in *Manduca sexta*. *Biochem. Biophys. Res. Commun.* **37:**729–735.

Miosky, D. L., Smolowitz, R. M., and Reinisch, C. L. (1989). Leukemia cell specific protein of the bivalve mollusc *Mya arenaria*. *J. Invertebr. Pathol.* **53:**32–40.

Molyneux, D. H., Takle, G., Ibrahim, E. A., and Ingram, G. A. (1986). Insect immunity in Trypanosomatidae. *In* "Immune Mechanisms in Invertebrate Vectors" (A. M. Lackie, ed.), pp. 117–144. Clarendon Press, Oxford.

Mullett, H. T., Ratcliffe, N. A., and Rowley, A. F. (1993). The generation and characterization of anti-insect blood cell monoclonal antibodies. *J. Cell Sci.* (in press).

Muta, T., Miyata, T., Misumi, Y., Tokunaga, F., Nakamura, T., Toh, Y., Ikehara, Y., and Iwanaga, S. (1991). An endotoxin-sensitive serine protease zymogen with a mosaic structure of complement-like, epidermal growth factor-like, and lectin-like domains. *J. Biol. Chem.* **266:**6554–6561.

Nappi, A. J., and Christensen, B. M. (1986). Hemocyte cell surface changes in *Aedes aegypti* in response to mirofilariae of *Dirofilaria immitis*. *J. Parasitol.* **72:**875–879.

Nappi, A. J., and Silvers, M. (1984). Cell surface changes associated with cellular immune reactions in *Drosophila*. *Science* **225:**1166–1168.

Nigam, Y., and Ratcliffe, N. A. (1991). Activation of the prophenoloxidase system in vectors of trypanosomes and its possible role in infectivity. *Br. Soc. Parasitol.* Liverpool, England, Abstr., p. 53.

Paskewitz, S. M., Brown, M. B., Collins, F. H., and Lea, A. O. (1989). Ultrastructural localization of phenoloxidase in the midgut of refractory *Anopheles gambiae* and association of the enzyme with encapsulated *Plasmodium cynomolgi*. *J. Parasitol.* **75:**594–600.

Pathak, J.P.N. (1983). Effect of endocrine glands on the unfixed total haemocyte counts of the bug, *Halys dentata*. *J. Insect Physiol.* **29:**91–94.

Pathak, J.P.N. (1986). Haemogram and its endocrine control in insects. *In* "Immunity in Invertebrates" (M. Brehélin, ed.), pp. 49–59. Springer-Verlag, Berlin.

Pendland, J. C., and Boucias, D. G. (1986). Characteristics of a galactose-binding hemagglutinin (lectin) from hemolymph of *Spodoptera exigua* larvae. *Dev. Comp. Immunol.* **10:**477–488.

Pendland, J. C., Heath, M. A., and Boucias, D. G. (1988). Function of a galactose-binding lectin from *Spodoptera exigua* larval haemolymph: Opsonization of blastopores from entomopathogenous hypomycetes. *J. Insect Physiol.* **34:**533–540.

Pereira, M.E.A., Andrade, A.F.B., and Ribeiro, J.M.C. (1981). Lectins of distinct specificity in *Rhodnius prolixus* react selectively with *Trypanosoma cruzi*. *Science* **211:**597–600.

Persson, M., Vey, A., and Söderhäll, K. (1987). Encapsulation of foreign particles *in vitro* by separated blood cells from crayfish, *Astacus leptodactylus*. *Cell Tissue Res.* **246:**409–415.

Price, C. D., and Ratcliffe, N. A. (1974). A reappraisal of insect haemocyte classification by the examination of blood from fifteen insect orders. *Z. Zellforsch. Mikrosk. Anat.* **147:**313–324.

Rantamäki, J., Durrant, H., Liang, Z., Ratcliffe, N. A., Duvič, B., and Söderhäll, K. (1991). Isolation of a 90 kD protein from haemocytes of *Blaberus craniifer* which has similar functional and immunological properties to the 76 kD protein from crayfish haemocytes. *J. Insect Physiol.* **37:**627–634.

Ratcliffe, N. A. (1982). Cellular defense reactions in insects. *In* "Immune Reactions to Parasites" (W. Frank, ed.), pp. 223–244. Fischer-Verlag, New York.

Ratcliffe, N. A. (1991). The prophenoloxidase system and its role in arthropod immunity. *In* "Phylogenesis of Immune Functions" (G. W. Warr and N. Cohen, eds.), pp. 46–65. CRC Press, Boca Raton, FL.

Ratcliffe, N. A., and Gagen, S. J. (1977). Studies on the *in vivo* cellular reactions of insects: An ultrastructural analysis of nodule formation in *Galleria mellonella*. *Tissue Cell* **9:**73–85.

Ratcliffe, N. A., and Rowley, A. F. (1974). *In vitro* phagocytosis of bacteria by insect blood cells. *Nature (London)* **225:**391–392.

Ratcliffe, N. A., and Rowley, A. F. (1979). Role of hemocytes in defense against biological agents. *In* "Insect Hemocytes: Development, Forms, Functions and Techniques" (A. P. Gupta, ed.), pp. 331–414. Cambridge Univ. Press, London and New York.

Ratcliffe, N. A., and Rowley, A. F. (1987). Insect responses to parasites and other pathogens. *In* "Immune Responses in Parasitic Infections: Immunology, Immunopathology, and Immunoprophylaxis" (E.J.L. Soulsby, ed.), Vol. 4, pp. 271–332. CRC Press, Boca Raton, FL.

Ratcliffe, N. A., Leonard, C., and Rowley, A. F. (1984). Prophenoloxidase activation: Nonself recognition and cell cooperation in insect immunity. *Science* **225:**557–559.

Ratcliffe, N. A., Rowley, A. F., Fitzgerald, S. W., and Rhodes, C. P. (1985). Invertebrate immunity: Basic concepts and recent advances. *Int. Rev. Cytol.* **97:**183.

Ratcliffe, N. A., Brookman, J. L., and Rowley, A. F. (1991). Activation of the prophenoloxidase cascade and initiation of nodule formation in locusts by bacterial lipopolysaccharides. *Dev. Comp. Immunol.* **15:**33–39.

Ratner, S., and Vinson, S. B. (1983). Encapsulation reactions *in vitro* by haemocytes of *Heliothis virescens. J. Insect Physiol.* **11:**855–863.

Richards, E. H., and Ratcliffe, N. A. (1990). Direct binding and lectin-mediated binding of erythrocytes to haemocytes of the insect, *Extatosoma tiaratum. Dev. Comp. Immunol.* **14:**269–281.

Richards, E. H., and Renwrantz, L. R. (1991). Two lectins on the surface of *Helix pomatia* haemocytes: A Ca^{2+}-dependent, GalNac-specific lectin and a Ca^{2+}-independent mannose 6-phosphate-specific lectin which recognises activated homologous opsonins. *J. Comp. Physiol. B* **161:**43–54.

Richards, E. H., Ratcliffe, N. A., and Renwrantz, L. (1989). The binding of lectins to carbohydrate moieties on haemocytes of insects, *Blaberus craniifer* (Dictyoptera) and *Extatosoma tiaratum* (Phasmida). *Cell Tissue Res.* **275:**445–454.

Rizki, T. M., and Rizki, R. M. (1982). Blood cell surface changes in *Drosophila* mutants with melanotic tumours. *Science* **220:**73–75.

Rowley, A. F., and Ratcliffe, N. A. (1979). An ultrastructural and cytochemical study of the interaction between latex particles and the haemocytes of the wax moth *Galleria mellonella in vitro. Cell Tissue Res.* **199:**129–137.

Rowley, A. F., and Ratcliffe, N. A. (1980). Insect erythrocyte agglutinins. *In vitro* opsonization experiments with *Clitumnus extradentatus* and *Periplaneta americana* haemocytes. *Immunology* **40:**483–492.

Rowley, A. F., and Ratcliffe, N. A. (1981). Insects. *In* "Invertebrate Blood Cells" (N. A. Ratcliffe and A. F. Rowley, eds.), Vol. 2, pp. 421–488. Academic Press, London.

Rowley, A. F., Ratcliffe, N. A., Leonard, C. M., Richards, E. H., and Renwrantz, L. (1986). Humoral recognition factors in insects, with particular reference to agglutinins and the prophenoloxidase system. *In* "Hemocytic and Humoral Immunity in Arthropods" (A. P. Gupta, ed.), pp. 381–406. Wiley, New York.

Rowley, A. F., Brookman, J. L., and Ratcliffe, N. A. (1990). Possible involvement of the prophenoloxidase system of the locust, *Locusta migratoria,* in antimicrobial activity. *J. Invertebr. Pathol.* **56:**31–38.

Samakovlis, C., Kimbrell, D. A., Kylsten, P., Engström, A., and Hultmark, D. (1990). The immune response in *Drosophila:* Pattern of cecropin expression and biological activity. *EMBO J.* **9:**2969–2976.

Saul, S. J., Bin, L., and Sugumaran, M. (1987). The majority of prophenoloxidase in the hemolymph of *Manduca sexta* is present in the plasma and not in the hemocytes. *Dev. Comp. Immunol.* **11:**479–485.

Saxena, S. C., and Saxena, P. N. (1985). Cytopathological and numerical changes in hemocytes of *Periplaneta americana* after treatment with malathion. *Bull. Environ. Contam. Toxicol.* **34:**842–849.

Schmidt, O., and Theopold, U. (1991). Immune defense and suppression in insects. *BioEssays* **13:**343–346.

Schmit, A. R., and Ratcliffe, N. A. (1977). The encapsulation of foreign tissue implants in *Galleria mellonella* larvae. *J. Insect Physiol.* **24:**511–521.

Schmit, A. R., Rowley, A. F., and Ratcliffe, N. A. (1977). The role of *Galleria mellonella* hemocytes in melanin formation. *J. Invertebr. Pathol.* **29:**232–234.

Shiraishi, A., and Natori, S. (1989). Humoral factor activating the *Sarcophaga* lectin gene in cultured fat body. *Insect Biochem.* **19:**261–267.

Smith, V. J., and Söderhäll, K. (1983). β-1,3-Glucan activation of crustacean hemocytes *in vitro* and *in vivo. Biol. Bull.* (*Woods Hole, Mass.*) **164:**299–314.

Söderhäll, K., and Smith, V. J. (1986). Prophenoloxidase-activating cascade as a recognition and defense system in arthropods. *In* "Hemocytic and Humoral Immunity in Arthropods" (A. P. Gupta, ed.), pp. 251–285. Wiley, New York.

Söderhäll, K., Häll, L., Unestam, T., and Nyhlen, L. (1979). Attachment of phenoloxidase to fungal cell walls in arthropod immunity. *J. Invertebr. Pathol.* **34:**285–294.

Söderhäll, K., Smith, V. J., and Johansson, M. W. (1986). Hemocyte lysate enhancement of fungal spore encapsulation by crayfish hemocytes. *Dev. Comp. Immunol.* **8:**23–30.

Söderhäll, K., Rögener, W., Söderhäll, I., Newton, R. P., and Ratcliffe, N. A. (1988). The properties and purification of a *Blaberus craniifer* plasma protein which enhances the activation of haemocyte prophenoloxidase by a β1,3-glucan. *Insect Biochem.* **18:**323–330.

Srivastava, S. C., and Richards, A. G. (1965). An autoradiographic study of the relation between haemocytes and connective tissue in the wax moth *Galleria mellonella. Biol. Bull.* (*Woods Hole, Mass.*) **128:**337–345.

Sun, S.-C., Lindström, I., Boman, H. G., Faye, I., and Schmidt, O. (1990). Hemolin: An insect immune protein belonging to the immunoglobin super family. *Science* **250:**1729–1732.

Takahashi, H., Komano, H., Kawaguchi, N., Kitamura, N., Nakanishi, S., and Natori, S. (1985). Cloning and sequencing of cDNA of *Sarcophaga peregrina* humoral lectin induced on injury of the body wall. *J. Biol. Chem.* **260:**12,228–12,233.

Takahashi, H., Komano, H., and Natori, S. (1986). Expression of the lectin gene in *Sarcophaga peregrina* during normal development and under conditions where the defense mechanism is activated. *J. Insect Physiol.* **32:**771–779.

Trenczek, T. (1986). Injury and immunity in insects. *Dev. Comp. Immunol.* **10:**627.

Trenczek, T., and Bennich, H. (1992). Characterization of haemocytes by monoclonal antibodies. *In* "Molecular Insect Science" (H. H. Hagedorn, J. G. Hildebrand, M. G. Kidwell, and J. H. Law, eds.), Plenum, New York.

Trenczek, T., and Faye, I. (1988). Synthesis of immune proteins in primary cultures of fat body from *Hyalophora cecropia. Insect Biochem.* **18:**299–312.

Vinson, S. B. (1988). Physiological studies of parasitoids reveal new approaches to the biological control of insect pests. *ISI Atlas Sci.: Anim. Plant Sci.*, pp. 25–31.

Walters, J. B., and Ratcliffe, N. A. (1981). A comparison of the immune response of the waxmoth, *Galleria mellonella* to pathogenic and nonpathogenic bacteria. *In* "Aspects of Developmental and Comparative Immunology I" (J. B. Solomon, ed.), pp. 147–152. Pergamon, Oxford.

Walters, J. B., and Ratcliffe, N. A. (1983). Studies on the *in vivo* cellular reactions of insects: Fate of pathogenic and nonpathogenic bacteria in *Galleria mellonella* nodules. *J. Insect Physiol.* **29:**417–424.

Welburn, S. C., and Maudlin, I. (1989). Lectin signalling of maturation of *T. congolense* infections in tsetse. *Med. Vet. Entomol.* **3:**141–145.

Welburn, S. C., and Maudlin, I. (1990). Haemolymph lectin and the maturation of trypanosome infections in tsetse. *Med. Vet. Entomol.* **4:**43–48.

Welburn, S. C., Maudlin, I., and Ellis, D. S. (1989). Rate of trypanosome killing by lectins in midguts of different species and strains of *Glossina. Med. Vet. Entomol.* **3:**77–82.

Zachary, D., and Hoffmann, D. (1984). Lyzozyme is stored in the granules of certain haemocyte types. *J. Insect Physiol.* **30:**405–411.

Chapter 13

The Response to Foreign Tissue Transplants in Insects

Richard D. Karp
Department of Biological Sciences
University of Cincinnati
Cincinnati, Ohio

I. Introduction: What Is the Significance of Studying Graft Rejection in Insects?

When one peruses the contemporary literature, it becomes quite evident that the use of insects as models for the study of evolution of immunity is increasing at a tremendous rate. The importance of insects in research is no mystery, since they are the dominant animal life-form on the planet, and therefore represent our major competitor in the ecosystem. They not only comprise a constant threat to our food supply, but also act as vectors for many major parasitic diseases that devastate humans around the world, particularly in third world countries. Thus, anything we can learn about how they function, and in particular defend themselves against the many insults of nature, might give us some insights as to how to develop methods for their control. From a purely basic scientific point of view, insects not only represent the major group of extant invertebrates, but also comprise one of the few invertebrate representatives that can be raised successfully in the laboratory in numbers sufficient to be used to design meaningful molecular and genetic studies to unravel significant biological questions. Thus, they are an obvious resource for developmental and comparative studies whose focus is to better understand the origins of adaptive immunity. One approach for such studies is the transplantation of foreign tissue. Grafting has become a very accepted way of studying cell-mediated immunity in invertebrates, and thus has been widely used in experimentation probing the limits of adaptive immunity in these animals. It was only natural that such techniques would be applied to insects in an effort to learn the extent of their defense capabilities. What follows in this chapter is the interesting history of the study of graft responsiveness in insects and the conclusions that can be drawn from the results.

II. Historical Perspective

Insects are known to have several major cellular defense mechanisms for dealing with foreign invaders. These include phagocytosis, nodule formation, and encapsulation. Nodule formation is a strategy used by many invertebrates to contain large numbers of bacteria by entrapping them with hemocytes, whereas encapsulation is the use of hemocytes to wall off objects that are too large to be phagocytosed, such as nematode parasites (as reviewed by Davies and Siva-Jothy, 1991). The latter, along with the transplantation of foreign tissue, has been used as a means of assaying for the ability of an insect to recognize non-self and respond to this challenge in a specific manner. The discovery of such responsiveness would be taken as evidence for the existence of a true immune response in these animals. Several laboratories have been interested in this question through the years. The following is a brief summary of some of the various approaches that have been used to test whether specific cell-mediated responsiveness exists in insects, and the conflicts that arose from the interpretation of the results generated by these studies.

A. The Encapsulation Response to Foreign Objects

The encapsulation response is a hemocyte-mediated activity that essentially walls off large particles from the rest of the hemocoel and its contents. Since it is commonly assumed that some type of recognition event initiates the process, some workers have utilized encapsulation as a means of determining if insects can recognize and react to foreign tissue. A classic example of the use of this technique is represented by the study of Carton (1976) in which insect eggs of varying foreignness were implanted into the hemocoels of *Pimpla instigator* larvae. The most impressive encapsulation reactions were induced by the eggs that were the most xenogeneic relative to the host. Allogeneic eggs also induced a significant response, but one that was not as intense as that for xenogeneic eggs. It has been argued that the response to allogeneic eggs was due to the presence of foreign secretions or viruslike particles coating the eggs from parasitized donors (Lackie, 1983b). However, the eggs used in the Carton study were dissected from *Pimpla* stock-cultured animals, and thus did not come from parasitized animals. Even if that were not the case, several studies have clearly shown that associated viruslike particles or polydnaviruses that are injected into the host along with the parasitoid eggs actually inhibit the encapsulation response rather than stimulate its induction (Salt, 1964, 1965; Edson *et al.*, 1980; Rizki and Rizki, 1984; See other chapters in this volume).

Although the use of encapsulation as a means of assaying for the recognition of foreignness would appear to be a very straightforward process, numerous reports indicate that there are many variables that exist that may cloud the interpretation of the data. Several studies have indicated that the ability to mount an encapsulation response can be greatly affected by species differences (Salt, 1963; Lackie, 1979; Lackie *et al.*, 1985; Tackle, 1988). Variations in responsiveness can be due to differences in the normal number of circulating hemocytes in a species (Lackie, 1979), the role of genetic heterogeneity within a species (Carton and Bouletreau, 1985), the degree of surface charge (Lackie, 1983a; Christensen *et al.*, 1987), and wettability of the implanted substances (Lackie, 1983a). Thus, the inability to generate an encapsulation response to foreign material may be due to innate species characteristics that have nothing to do with the presence or absence of adaptive immune mechanisms.

Scott (1971) reported on a study in which encapsulation was used to monitor the recognition of xenogeneic, allogeneic, and syngeneic nerve cord implants in the American cockroach, *Periplaneta americana*. Xenogeneic implants were heavily encapsulated, but allogeneic implants were encapsulated only if they had been exposed to proteolytic enzymes prior to implantation. One interpretation of these results could be that the alloantigens present on the nerve cord were cryptic and not available for recognition until exposed to the enzymatic treatment. If this is the case, then lack of responsiveness does not necessarily mean that the corresponding immune mechanism does not exist.

There is no question that the encapsulation reaction is an essential weapon in the defensive arsenal possessed by insects and serves as a highly effective means of dealing with invading organisms. However, based on the relatively long list of variables that can affect its performance, encapsulation leaves much to be desired as a reliable means of monitoring specific recognition of foreignness in studies designed to determine if adaptive cell-mediated immunity exists in insects.

B. The Response to the Transplantation of Foreign Tissue

A different approach that has been used with some regularity in the study of cell-mediated immunity in invertebrates is the response to foreign tissue grafts. The grafting studies in insects have not been without their share of controversy. For instance, although most workers agree that insects are able to recognize xenografts, there is some difference of opinion as to whether these animals can discriminate between allogeneic differences. There has also been a wide range of experimental approaches employed to detect recognition and responsiveness to foreign tissue, some of which have been questioned as to their effectiveness and accuracy. The data from these assays have been gathered either by macroscopic observation or by the use of histological techniques. This has added to the confusion since the latter is far more sensitive than the former, and thus one can arrive at very different conclusions depending on how the data are collected.

In the early 1980s, several laboratories published studies reporting on the transplantation of integument between insects. Thomas and Ratcliffe (1982) grafted various species of insects and followed their progress by gross observation of the transplants. Coloration, melanization, and disintegration were used as criteria for scoring rejection. Based on those parameters, xenografts appeared to be rapidly rejected and melanized, whereas allografts and autografts were not. However, when a few of the grafts were examined histologically, it was found that the allografts had stimulated cellular infiltration in the graft bed. Thus, gross observation of the allografts in this study may not have been sensitive enough to reveal an accurate picture of what was actually occurring in the host animal.

The transplantation studies of Jones and Bell (1982) relied solely on histological evaluation of the grafts. *Periplaneta americana* was used as the recipient of integumentary grafts from donors of varying phylogenetic distances from *Periplaneta*. Grafts were sampled after various periods of time and prepared histologically for examination. The grafts were scored by statistically analyzing the number of hemocytes infiltrating each graft site, and comparing this to the amount of infiltration under autograft controls. *Periplaneta americana* vigorously rejected xenografts from *Blaberus*, *Nauphoeta*, and *Leucophaea*. Of particular interest was the fact that the more sensitive histological procedure revealed a significant response to a closely related species, *Periplaneta brunnea*. Jones and Bell also compared the response to allografts as compared to autografts, but found only a slight difference that was not statistically significant.

Lackie (1983b) also employed *P. americana* in her grafting studies but was using a quite different approach for evaluating responsiveness. The experimental design was to graft foreign integument onto nymphs and follow acceptance or rejection by determining whether the donor tissue survived the next molt of the host. If the donor was not recognized as foreign, the unrejected epidermal layer underlying the cuticle would produce more cuticle of the donor type. This was determined by observing the color of the transplant following the molt: if the color was dark brown (the color of the *Blatta orientalis* donor cuticle), the graft was scored as accepted; if the color was reddish brown (the color of the *Periplaneta americana* host), the graft was scored as rejected. Although it was reported that the xenogeneic integument from *Blaberus craniifer* was rejected, cuticular transplants from the more closely related *Blatta orientalis* were not. It was concluded that the ability of the insect to recognize non-self was rather weak. However, these results were in direct conflict with those of Jones and Bell (1982), who convincingly demonstrated, using the more sensitive histological assay, that *P. americana* could significantly ($0.001 > P > 0.005$) reject integumentary grafts from *Bl. orientalis* donors. Other circumstances that might affect the interpretation of the results of the molting assay are: (1) As much as 90% of the molted cuticle may be resorbed following enzymatic digestion, and eventually be incorporated into the new cuticle (Hepburn, 1985). This could obviously have an effect on the color of the new cuticle regardless of the origin of the epidermal layer (2) There were no autograft controls performed to ensure that *Periplaneta* cuticle would not turn dark brown due to melanization caused by the trauma of the grafting procedure, thus giving false positive results.

Our initial studies on transplantation reactivity in *Periplaneta americana,* using the quantification of infiltrating hemocytes as a measure of responsiveness (similar to the protocol of Jones and Bell, 1982), revealed strong transplantation immunity against xenografts, but again there was no clear-cut evidence for the presence of allograft immunity (George *et al.*, 1984). It was our belief that the particular assay we were using was not sensitive enough to determine if allograft immunity existed in the roach, because it was based on quantifying the numbers of hemocytes in the graft bed with no indication of whether or not they were functioning as immunocytes. We needed a more direct approach for the observation of activity against foreign allografts. After reviewing the data from the previous study, we observed that all the xenografts were missing the subcuticular epidermal layer, which is largely responsible for the production of the exoskeleton (Fig. 1A). However, autograft controls, even though they might have some hemocytic infiltrate present, always showed intact epidermal layers underneath the cuticle (Fig. 1B).

A new grafting study was undertaken to test for the existence of allograft immunity using the new criterion of observing the fate of the subcuticular

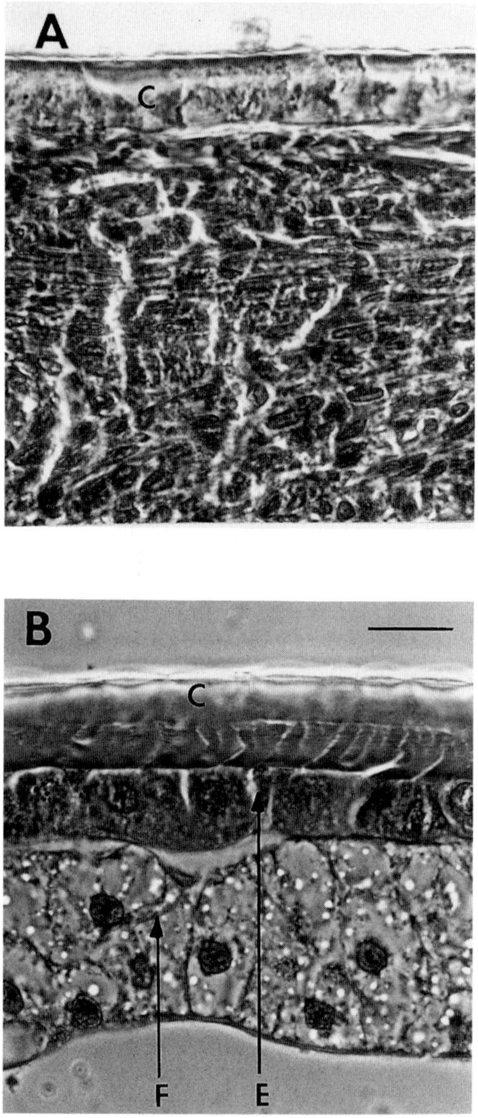

Figure 1 Scoring of graft rejection based on the fate of the subcuticular epidermal layer. (A) Five-day xenograft from *Eublaberus distanti* on *Periplaneta americana*. (B) Five-day autograft control from the same recipient as above. Note the absence of the subcuticular epidermal layer in the xenograft, whereas the epidermis is fully intact in the autograft control. C = cuticle; E = epidermal layer; F = fat body. Bar = 20 μm.

epidermal layer as a determination of whether or not the tissue had been rejected. The results of these studies (George *et al.*, 1987) clearly indicated that allografts were recognized as foreign, since 92% of the allogeneic epidermal layers had been destroyed by day 7 posttransplantation versus only 8% in autograft controls (Table 1). Although the allograft response can be considered to be acute, it is relatively slower than the response to xenografts (Table 1), which is the same pattern that is observed in mammals. A second study using implants seeded with allogeneic hemocytes labeled with [^{3}H]thymidine confirmed not only that allograft immunity existed in the roach, but that the response to the foreign tissue was cytotoxic in nature (Howcroft and Karp, 1987). Thus, the direct observation of the fate of living foreign tissue transplanted into the American cockroach indicated that this insect was fully capable of recognizing and reacting to allogeneic differences.

C. The Issue of Immunologic Memory

A long-established definition of true adaptive immunity is that such responses display both specificity and immunologic memory. We therefore were interested in determining if these important criteria were present in the roach allograft response. For this purpose animals received filter paper implants seeded with allogeneic donor hemocytes as a first-set graft. The filter paper technique was employed so that the same donor could be used again. Seven days later, the recipients received a second-set integumentary allograft from

Table 1

Rejection of Foreign Integumentary Grafts by *Periplaneta americana*

Days posttransplantation	Percentage of grafts rejected[a]			
	Xenografts	Autografts	Allografts	Autografts
1	94	10	38	15
3	100	16	60	9
5	100	6	80	9
7	100	11	92	8
10	—	—	95	5
20	—	—	93	7

[a]Scoring or rejection is based on the presence or absence of the subcuticular epidermal layer as determined by histological examination of the grafts. Values are an average of at least two trials with total sample sizes exceeding 30 animals (except for allograft days 10 and 20, which had sample sizes of 23 and 15, respectively). Data summarized from George *et al.* (1987).

the same donor as well as another allogeneic graft from an unrelated donor as a third-party control. Thus, we could test for the presence of both specificity and memory. Grafts were removed after various periods of time and scored histologically for the presence or absence of the subcuticular epidermal layer. The results (Hartman and Karp, 1989) indicated that the second-set allografts were rejected in an accelerated fashion, since peak reactivity occurred in 3 days, whereas third-party controls followed first-set rejection kinetics. These results indicated that not only was there at least short-term memory present, but that the response was highly specific since the recipient reacted to the second-set and third-party grafts in a differential manner. The recipient was thus able to distinguish between two different allogeneic donors. Thus, the allograft response in the American cockroach displays the hallmarks of true adaptive immunity.

III. Mediation and Regulation of the Roach Graft Response

We have successfully established that the American cockroach can recognize and reject xenografts and allografts. The next step in our studies is to find ways to determine how these responses are mediated and regulated. We have been developing some tools that may prove to be promising in this regard.

A. Inhibition of Graft Rejection by Gamma Irradiation

It is commonly known that the use of X irradiation will inhibit adaptive immunity in mammals due to the fact that these responses have a cellular basis. Since gross observation of graft beds in transplanted roaches revealed the presence of hemocyte infiltrates, we wondered if irradiation would have a similar effect on the roach response to foreign integument. Initial experiments revealed that when roaches were subjected to 5000 rads of gamma irradiation, the number of circulating hemocytes was reduced to one-third of normal. On the basis of this, we exposed recipients to 5000 rads of radiation and then grafted them with xenografts and autografts, either on the same day as the radiation treatment or 24 hr later. The grafts were then scored for rejection after various periods of time posttransplantation. The results indicated that animals receiving transplants 24 hr after irradiation, when hemocyte numbers were depleted, had the maximum response to xenografts delayed for 10 days as compared to untreated controls (Eppensteiner and Karp, 1989). There was a direct correlation between the recovery of hemocyte numbers and the recovery of the full capability to reject foreign tissue. This, taken with previous data indicating that labeled hemocytes are not killed when incubated with allogeneic cell-free hemolymph (Howcroft and Karp, 1987), indicated that graft rejection in the roach is cell-mediated in nature. In addition, these studies offer a new approach for determining which type of hemocyte mediates graft rejection in the roach. Roaches could be irradiated and, 24 hr later, reconstituted with different subpopulations of circulating hemocytes from both sensitized and nonsensitized animals to see which cell type will fully restore the response to foreign grafts. This would provide direct evidence for the existence of immunocytes in this insect.

B. The Effects of Eicosinoids on the Xenograft Response

Products of eicosinoid biosynthetic pathways, such as the prostaglandins, are well known to have powerful effects on mammalian immune responses (Strom *et al.*, 1977; Plaut, 1987). It was of interest to us to determine if eicosinoids had similar effector or regulatory effects on the roach immune response. The approach was to treat animals with dexamethasone (dex), which is a generalized inhibitor of eicosinoids since it interferes with arachadonic acid release, and observe what effect the inhibitor had on xenograft rejection in the roach. Animals were treated with two different doses of dex either 24 hr prior to grafting, simultaneously with grafting, or 12 hr postgrafting. Animals injected with 0.2 mg/ml of dex showed a severely reduced capability for rejecting xenografts, no matter when they received the dex (Karp *et al.*, 1993). Preinjected animals had only a 35% rate of rejection, simultaneous had only 17%, and postinjected had only 21%, whereas untreated controls showed normal rates of rejection. However, at a concentration of 0.02 mg/ml, dex showed a different pattern of results. When the low dose was given pregraft, dex actually enhanced graft rejection with a rate of 88%. When the low dose of dex was given simultaneously with grafting or 12 hr postgraft, the effects were similar to that of the high dose of dex, in that rejection was once more inhibited, with rates falling to 30 and 33%, respectively. By the use of direct cell counting, we know that dex is not directly cytotoxic to hemocytes and thus is not curtailing the graft response in this way. Rather, it would seem that dex is shutting down the graft response in some undetermined regulatory way. The use of more specific inhibitors in the future will allow us to identify whether it is the products of the cyclooxygenase or lipoxygenase pathways (or both) that are involved in this phenomenon. In addition, the differential effects of the low dose of dex might allow us to dissect the response into its various components (i.e., recognition versus effector function) so that we might eventually understand how insect immunocytes work.

IV. Discussion and Conclusions: Where Do We Go from Here?

The results from several different studies convincingly indicate that long-lived insects, such as cockroaches, have the ability to recognize and reject xenografts. Our extensive studies make it very clear that roaches also possess the ability to recognize allogeneic differences and respond to them via mechanisms that display the characteristics of true adaptive cell-mediated immunity (George *et al.*, 1987; Howcroft and Karp, 1987). The real significance of these findings goes beyond the observation that this animal has the ability to reject transplanted allogeneic tissue. More importantly, it implies that roaches have evolved a surveillance mechanism whereby they can recognize antigenic changes on their cells or tissues that have come about through mutation or some other type of unwanted modification. Since such modifications might constitute a threat to the well-being of the organism, a functioning surveillance system would recognize these changes and stimulate defensive countermeasures to remove them.

Our future studies will focus on the mechanisms involved in the cockroach graft response. We need to identify the cell type that actually mediates the response, as well as define the molecular components, such as receptors, that are responsible for recognition of foreignness. The use of radiation chimeras and passive transfer studies, as mentioned earlier, will hopefully prove useful in identifying the immunocyte that is responsible for graft rejection in the roach. The fact that such cells can specifically reject foreign grafts obviously means that a cell-surface receptor must be involved in distinguishing self from non-self. In this regard, recent studies in our lab indicate that xenografted roaches show an increased induction of a 102-kDa hemolymph protein 14 days posttransplantation (Harju and Karp, 1993). The 102-kDa protein band is the same one that is so closely associated with the adaptive humoral response in the roach that is induced by the injection of soluble proteins (Duwel-Eby *et al.*, 1991). This raises the possibility that the molecule associated with humoral immunity in the roach may also serve as a cell-surface receptor in cell-mediated responses. Follow-up studies, now in progress, will be able to make those distinctions.

With the prospects of so many important issues needing to be addressed, the future looks very bright for invertebrate transplantation immunology. We have excellent animal models at our disposal, such as insects, which can be cultured in the lab, thus avoiding the vicissitudes of studying animals from the wild. The technology that has been developed at the cell and molecular levels is now fully accessible for use in studies on invertebrates. Thus, we stand poised to ask some very sophisticated questions concerning the evolution of adaptive immunity, with the expectation of some very exciting and revealing results to come.

Acknowledgments

The work described in this chapter that has emanated from our laboratory was supported by NSF research grants PCM 8316140 and DCB 8702382, NIH research grant GM 39398, and the University of Cincinnati Foundation Fund for Comparative Immunology.

References

Carton, Y. (1976). Isogenic, allogenic and xenogenic transplants in insect species. *Transplantation* **21:**17–22.

Carton, Y., and Bouletreau, M. (1985). Encapsulation ability of *Drosophila melanogaster:* A genetic analysis. *Dev. Comp. Immunol.* **9:**211–219.

Christensen, B. M., Forton, K. F., Lafond, M. M., and Grieve, R. B. (1987). Surface changes on *Brugia pahangi* microfilariae and their association with immune evasion in *Aedes aegypti. J. Invertebr. Pathol.* **49:**14–18.

Davies, D. H., and Siva-Jothy, M. T. (1991). Encapsulation in insects: Polydnaviruses and encapsulation-promoting factors. *In* "Immunology of Insects and Other Arthropods" (A. P. Gupta, ed.), pp. 119–132. CRC Press, Boca Raton, FL.

Duwel-Eby, L. E., Faulhaber, L. M., and Karp, R. D. (1991). Adaptive humoral immunity in the American cockroach. *In* "Immunology of Insects and Other Arthropods" (A. P. Gupta, ed.), pp. 385–402. CRC Press, Boca Raton, FL.

Edson, K. M., Vinson, S. B., Stoltz, D. B., and Summer, M. D. (1980). Virus in a parasitoid wasp: Suppression of the cellular immune response in the parasitoid's host. *Science* **211:**582–583.

Eppensteiner, J. M., and Karp, R. D. (1989). The effect of gamma irradiation on the xenograft response in the American cockroach. *J. Insect Physiol.* **35:**81–86.

George, J. F., Karp, R. D., and Rheins, R. A. (1984). Primary integumentary xenograft reactivity in the American cockroach, *Periplaneta americana. Transplantation* **37:**478–484.

George, J. F., Howcroft, T. K., and Karp, R. D. (1987). Primary integumentary allograft reactivity in the American cockroach, *Periplaneta americana. Transplantation* **43:**514–519.

Harju, M. A., and Karp, R. D. (1993). *J. Insect Physiol.* (in press).

Hartman, R. S., and Karp, R. D. (1989). Short-term immunologic memory in the allograft response of the American cockroach, *Periplaneta americana. Transplantation* **47:**920–922.

Hepburn, H. R. (1985). Structure of the integument. *In* "Comprehensive Insect Physiology, Biochemistry and Pharmacology" (G. A. Kerkut and L. I. Gilbert, eds.), Vol. 3, pp. 1–58. Pergamon, New York.

Howcroft, T. K., and Karp, R. D. (1987). Demonstration of cell-mediated cytotoxicity to allogeneic and xenogeneic tissue in the American cockroach, *Periplaneta americana,* using a combination in vivo/in vitro assay. *Transplantation* **44:**129–135.

Jones, S. E., and Bell, W. J. (1982). Cell-mediated immune-type response of the American cockroach. *Dev. Comp. Immunol.* **6:**35–42.

Karp, R. D., Manbeck, M. L. Barge, J. L. (1993). (in preparation).

Lackie, A. M. (1979). Cellular recognition of foreignness in two insect species, the American cockroach and the desert locust. *Immunology* **36:**909–914.

Lackie, A. M. (1983a). Effect of substratum wettability and charge on adhesion in vitro and encapsulation in vivo by insect haemocytes. *J. Cell Sci.* **63:**181–190.

Lackie, A. M. (1983b). Immunological recognition of cuticular transplants in insects. *Dev. Comp. Immunol.* **7:**41–50.

Lackie, A. M., Tackle, G., and Tetley, L. (1985). Haemocyte encapsulation in the locust *Schistocerca gregaria* (Orthoptera) and in the cockroach *Periplaneta americana* (Dictyoptera). *Cell Tissue Res.* **240:**343–351.

Plaut, M. (1987). Lymphocyte hormone receptors. *Annu. Rev. Immunol.* **5:**621–669.

Rizki, R. M., and Rizki, T. M. (1984). Selective destruction of a host blood cell type by a parasitoid wasp. *Proc. Natl. Acad. Sci. U.S.A.* **81:**6154–6158.

Salt, G. (1963). Experimental studies in insect parasitism. XII. The reactions of six exopterygote insects to an alien parasite. *J. Insect Physiol.* **9:**647–669.

Salt, G. (1964). The ichneumonid parasite *Nemeritis canescens* (Gravenhorst) in relation to the wax moth *Galleria mellonella* (L.). *Trans. R. Entomol. Soc. London* **116:**1–14.

Salt, G. (1965). Experimental studies on insect parasitism. XIII. The haemocytic reaction of a caterpillar to eggs of its habitual parasite. *Proc. R. Soc. London, Ser. B* **162:**303–318.

Scott, M. T. (1971). Recognition of foreignness in invertebrates: Transplantation studies using the American cockroach (*Periplaneta americana*). *Transplantation* **11:**78–86.

Strom, T. B., Carpenter, C. B., Cragoe, E. J., Jr., Norris, S., Devlin, R., and Perper, R. J. (1977). Suppression of *in vivo* and *in vitro* alloimmunity by prostaglandins. *Transplant. Proc.* **9:**1075–1079.

Tackle, G. B. (1988). Studies on the cellular immune responses of insects toward the insect pathogen *Trypanosoma rangeli*. *J. Invertebr. Pathol.* **51:**64–72.

Thomas, I. G., and Ratcliffe, N. A. (1982). Integumental grafting and immunorecognition in insects. *Dev. Comp. Immunol.* **6:**643–654.

Chapter 14

Regulation of Insect Hemolymph Phenoloxidases

Manickam Sugumaran
Department of Biology
University of Massachusetts
at Boston
Harbor Campus
Boston, Massachusetts

Michael R. Kanost
Department of Biochemistry
Kansas State University
Manhattan, Kansas

Insects do not possess complicated immunoglobulins to tackle invading parasites and other microorganisms, but seem to have a wide spectrum of other defense mechanisms. During the defense reaction, the attacking foreign object is often found encapsulated and darkened. The dark pigment is invariably assumed to be melanin. Therefore, melanin and the enzymes that are responsible for its synthesis, namely, phenoloxidases, are considered as essential components of the insect's defense mechanism. Phenoloxidase usually occurs as an inactive proenzyme in the hemolymph and is activated during an immune response. Some authors have even invoked prophenoloxidase as part of the non-self recognition mechanisms of insects. Given such an important role in the insect's defense system, it is essential to understand the molecular mechanisms of phenoloxidase activation, the fate of activated enzyme, and the course of its reactions.

Upon recognition of foreignness, a specific, as yet unidentified factor in the hemolymph seems to be activated, which in turn activates an inactive serine protease. This serine protease activates another serine protease that serves as the direct activator of inactive prophenoloxidase by cleaving a peptide of M_r = 5000 from it. The activation of prophenoloxidase can also be achieved by cuticular protease(s) and by an alternate route involving phospholipids. The negative control seems to be exerted by serine protease inhibitors and other less characterized biochemicals. Activated phenoloxidase generates primarily quinones by acting on diphenolic substrates found in the hemolymph. Quinones, being highly cytotoxic, can directly kill the foreign organism or undergo polymerization to form part of the capsule. Newly discovered isomerases, namely, quinone isomerase and dopachrome tautomerase, seem to control the quinonoid reactivity by converting quinones to quinone methides and provide them either for detoxification or for other reactions. Reactions of quinones and quinone methides generate melanin and/or sclerotin that forms part of the capsule. The biochemistry of these processes is briefly discussed.

I. Introduction

Insects and other arthropods do not possess immunoglobulins to tackle invading parasites and other objects, yet they are able to defend themselves against foreign organisms by a variety of reactions, such as phagocytosis, encapsulation, nodule formation, agglutination, and synthesis of antibacterial proteins (Whitcomb *et al.*, 1974; Coombe *et al.*, 1984; Gotz and Boman, 1985; Ratcliffe *et al.*, 1985; Gotz, 1986; Christensen and Nappi, 1988; Gupta, 1988; Lackie, 1988; Boman *et al.*, 1991; Schmidt and Theopold, 1991). The common defense reaction against parasites and other foreign objects that are too large to be phagocytosed in the insect hemocoel is encapsulation by hemocytes. Cellular capsules that are formed by the hemocytes are often found darkened and hardened (Gupta, 1988). Similar observations during humoral encapsulation have also been made (Gotz, 1986). The chemical nature of the dark pigment within the capsule is yet to be established, but invariably researchers assume that this substance is melanin produced by the polymerization of phenoloxidase-generated quinones (Gupta, 1988). Therefore melanin and the enzyme that is responsible for its biosynthesis, that is, phenoloxidase, are considered as integral parts of the insect's immune system. Interestingly, phenoloxidases are present in both cuticle and hemolymph and are known to play a crucial role not only in melanization, but also in sclerotization of insect cuticle (Sugumaran, 1988, 1991a) and wound healing (Lai-Fook, 1966).

Numerous review articles on the role of phenoloxidases in insect immunity have appeared in the literature. The excellent monograph by Gupta (1988) sums up the literature until 1988, after which period the readers can refer to the articles by Sugumaran (1990) and Ashida and Yamazaki (1990). Because the subject matter of this book focuses on insects, pioneering studies carried out by Soderhall and his associates (Johansson and Soderhall, 1989) on the participation of phenoloxidase in crustacean immunity are not reviewed here.

II. Prophenoloxidase Activation

A. Prophenoloxidase

There are three kinds of phenoloxidases: monophenol monooxygenase, *o*-diphenoloxidase, and *p*-diphenoloxidase (laccase). Monophenol monooxygenases, such as tyrosinase, also possess the *o*-diphenoloxidase activity, but the *o*-diphenoloxidase does not necessarily possess monophenol monooxygenase activity. Finally, laccases, which show preference for *p*-diphenols, can also readily attack the *o*-diphenols. Therefore, they possess *o*-diphenoloxidase activity as well. While testing the phenoloxidase activity, if *o*-diphenols such as dopa or dopamine are used as substrates, it is inferred that the *o*-diphenoloxidase activity of all three enzymes will be detected. Such assays will not distinguish the three different phenoloxidase activities. Because of these complications, we will use only the common term phenoloxidase throughout this text.

Phenoloxidase usually occurs as an inactive proenzyme form in the hemolymph of most of the insects tested. Evidence for the presence of prophenoloxidases in the hemolymph of *Locusta migratoria* (Brehélin *et al.*, 1989), *Tenebrio molitor* (Heyneman, 1965), *Bombyx mori* (Ashida, 1971), *Antheraea pernyi* (Evans, 1967), *Manduca sexta* (Aso *et al.*, 1985), *Hyalophora cecropia* (Andersson *et al.*, 1989), *Galleria mellonella* (Pye, 1974), *Sarcophaga barbarta* (Hughes, 1976), *Calliphora vicina* (Thomson and Sin, 1970), *Musca domestica* (Tsukamoto *et al.*, 1986), *Drosophila melanogaster* (Seybold *et al.*, 1975), and *Aedes aegypti* (Ashida *et al.*, 1990) is well documented. Prophenoloxidases from *B. mori* (Ashida, 1971), *Calliphora* (Pau and Eagles, 1975), *M. domestica* (Tsukamoto *et al.*, 1986), and *T. molitor* (Heyneman, 1965) have been highly purified. In spite of its key role in insect immunity, circular sclerotization, and wound healing, it is rather surprising that the studies on this enzyme are so limited. However, understandably, the difficulties associated with the purification of this proenzyme have severely hampered progress in this line of research. Besides, problems arising from spontaneous activation of the enzyme, its absorption to glass and other surfaces including gels and other materials used for chromatography, rapid degradation, and eventual inactivation have deterred several investigators from undertaking a detailed study on this enzyme. Much of the detailed biochemistry of prophenoloxidase comes from the monumental work of Ashida and coworkers (Ashida and Yamazaki, 1990).

B. Prophenoloxidase Activating Enzyme

Ohnishi (1954) first demonstrated that in *Drosophila*, a protein factor could activate prophenoloxidase activity. Subsequent work by Schweiger and Karlson (1962) not only confirmed this observation but also provided evidence that the activation of the proenzyme is achieved by limited proteolysis. Later, similar results were reported in *B. mori* (Ashida and Dohke, 1980), *D. melanogaster* (Seybold *et al.*, 1975), *S. barbarta* (Hughes, 1976), *S. bullata* (Saul and Sugumaran, 1988), *M. sexta* (Aso *et al.*, 1985; Saul and Sugumaran, 1987), and *Blaberus craniifer* (Leonard *et al.*, 1985b). Working with the silkworm *B. mori,* Ashida and associates successfully purified the prophenoloxidase from the hemolymph and demonstrated that the activation of prophenoloxidase is achieved by proteolytic cleavage of the proenzyme (Ashida and Dohke, 1980). The activating protease from the larval cuticle of the same organism was isolated and characterized by Dohke (1973), and the presence of a similar enzyme in the hemolymph has been demonstrated by Ashida and Yoshida (1988). Availability of purified proenzyme and its activator from the same organism aided the establishment of the proteolytic activation mechanism (Ashida and Dohke, 1980). Activation by the cuticular protease severed a peptide of M_r = 5000 from the proenzyme, confirming unequivocally the participation of "limited proteolysis" as a key mechanisms of prophenoloxidase activation. In most other insects tested, this mode certainly seems to operate as evidenced by the inhibition of activation by serine protease inactivators such as diisopropyl phosphonofluoridate, *p*-nitrophenyl-*p*-guanidobenzoate, or phenylmethylsulfonyl fluoride. Ashida's group also demonstrated the occurrence of this reaction in the hemolymph (Ashida and Yoshida, 1988). Interestingly, the prophenoloxidase activating enzyme from the hemolymph itself seems to be present as a proenzyme, thereby indicating the presence of a cascade of enzyme activities associated with phenoloxidase (Yoshida and Ashida, 1986).

C. Prophenoloxidase Cascade

Although they did not define each activity specifically, Seybold *et al.* (1975) first reported the presence of a cascade of enzyme activities associated with phenoloxidase activation. Yoshida and Ashida (1986) observed the microbial activation of two serine proteases prior to prophenoloxidase activation in the silkworm hemolymph. One of the proteases was identified to be the prophenoloxidase activating enzyme, whereas the other protease hydrolyzing benzoylarginine ethyl ester was termed a benzoylarginine ethyl esterase. Interestingly, activity of benzoylarginine ethyl esterase appeared in the plasma prior to the appearance of prophenoloxidase activating enzyme (or phenoloxidase activity). Thus it seems that microbial products trigger eventually in the order benzoyl arginine ethyl esterase, prophenoloxidase activating enzyme, and phenoloxidase (Fig. 1). Therefore, it is fairly certain that prophenoloxidase activation occurs in a cascade of reactions. Evidence for the presence of a similar cascade in *M. sexta* (Saul and Sugumaran, 1987) and *S. bullata* (Saul and Sugumaran, 1988) has been provided by Sugumaran's

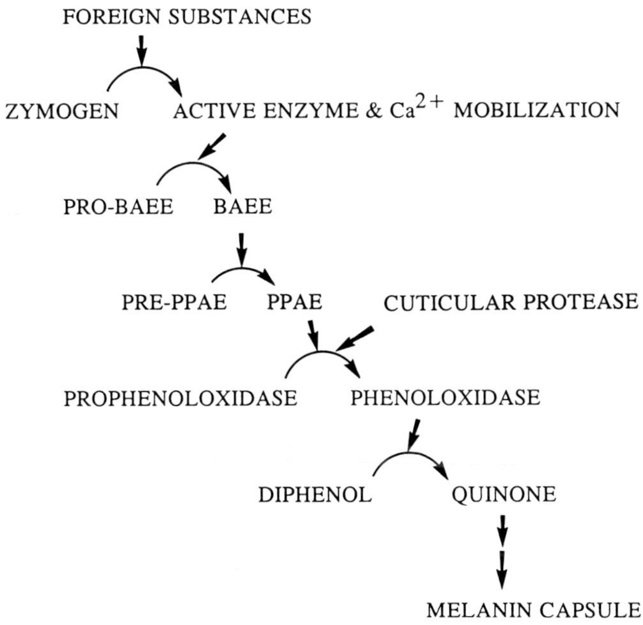

Figure 1 The prophenoloxidase cascade. LPS, peptidoglycan, β-1,3-glucans, and/or other molecules present on foreign substances trigger the prophenoloxidase system of the host by a cascade of reactions. The less-defined initial reaction leads to mobilization of calcium and the activation of enzyme activity that triggers the pro-benzoyl arginine ethyl esterase (PRO-BAEE) to active enzyme. This serine protease in turn activates preprophenoloxidase activating enzyme (PRE-PPAE) to the active form. The active PPAE proteolytically activates prophenoloxidase. Prophenoloxidase can also be activated by the cuticular protease directly. Active phenoloxidase catalyzes the conversion of diphenols to quinones and provides them for melanin capsule formation.

group. In both of these organisms, at least one serine protease seems to be activated prior to the activation of prophenoloxidase. Andersson *et al.* (1989) have shown the presence of two activator proteins in *H. cecropia.*

D. Triggering of the Cascade

At this point it is not clear how the first protease in the silkworm that seems to be present in the inactive proenzyme form is activated by microbial products. But there is no doubt that prophenoloxidase is triggered by different microbial products. For such triggering, divalent calcium seems to be essential (Leonard *et al.*, 1985a; Yoshida and Ashida, 1986; Saul and Sugumaran, 1987, 1988; Ashida and Yoshida, 1988; Brehélin *et al.*, 1989; Brookman *et al.*, 1989; Ashida and Yamazaki, 1990; Sugumaran, 1990). Therefore, it is probable that some calcium-dependent enzyme activity appears upon interaction of the foreign objects with the insect hemolymph components. Several microbial products, such as bacterial, fungal, and yeast cell-wall components, cause prophenoloxidase activation in the hemolymph (see Fig. 1).

It is well known among researchers working in the area of insect immunity that it is essential to maintain pyrogen-free conditions when handling insect blood cells. Yet, it is surprising to see several authors reporting the failure of lipopolysaccharide (LPS, a cell-wall component of gram-negative bacteria) to activate prophenoloxidase. In *B. mori* (Yoshida and Ashida, 1986), *G. mellonella* (Brookman *et al.*, 1989), *Schistocerca gregaria* (Dularay and Lackie, 1985), and *S. bullata* (Saul and Sugumaran, 1988), for example, LPS does not seem to activate the prophenoloxidase. However, in *M. sexta* (Saul and Sugumaran, 1987), rapid activation of prophenoloxidase by LPS has been reported.

One of us pointed out that LPS-mediated prophenoloxidase activation is transient in nature (Saul and Sugumaran, 1987) and one can easily miss the activity. Although it is not clear why this phenomenon occurs, it has been independently observed in another insect as well (Brehélin *et al.*, 1989). Since activated phenoloxidase is very sticky, it can adhere to foreign objects and other cell surfaces and generate cytotoxic quinonoid compounds to kill the intruder. Such toxic products of this enzyme activity are also dangerous for the host. Therefore, it is likely that the activated phenoloxidase is short-lived (Saul and Sugumaran, 1987), similar to some of the complement pathway components found in the blood of higher animals. Therefore, use of discontinuous assays can cause lack of detection of phenoloxidase activity. In addition, other factors can also cause misleading results as exemplified by the studies in locusts. Initially, Ratcliffe's group (Brookman *et al.*, 1989) reported a lack of LPS activation in locusts, whereas Brehélin *et al.* (1989) demonstrated that LPS caused prophenoloxidase activation in the same organism.

Since the latter workers directly collected the hemolymph without the use of any anticoagulants, whereas Brookman *et al.* (1989) used an anticoagulation buffer consisting of ethylene diamine tetra-acetate, citrate, and glucose, it is likely that some of the components of anticoagulant buffer interfered with the activation triggering mechanisms and inhibited the prophenoloxidase activation. Nevertheless, upon reinvestigation in the same organism, Ratcliffe's group did observe that LPS caused prophenoloxidase activation in whole blood (Ratcliffe *et al.*, 1991).

The LPS activation must be mediated by some sort of a recognition mechanism. Accordingly, an LPS binding protein from the hemolymph of the American cockroach has been isolated by Natori's group (Jomori *et al.*, 1990), but its relationship with the prophenoloxidase cascade has not yet been defined. Considerably more biochemical work has been done on the triggering of the cascade by gram-positive bacteria and yeast cellular components. Since a detailed account of these results is presented in the article by Ashida and Yamazaki (1990), only a short account is given here. The triggering substance from gram-positive bacteria seems to be peptidoglycan (Yoshida and Ashida, 1986; Brookman *et al.*, 1989). As little as a few nanograms per milliliter of peptidoglycan can activate the prophenoloxidase cascade very efficiently in the silkworm hemolymph. Ashida's group also established that the β-1,3-glucan receptors and the peptidoglycan receptors for triggering prophenoloxidase are present as separate entities in this organism (Yoshida *et al.*, 1986). The β-1,3-glucan binding protein from *B. craniifer* (Soderhall *et al.*, 1988) and *B. mori* (Ochiai and Ashida, 1988) has been purified and characterized, but the characterization of a peptidoglycan binding protein is yet to be reported. Also not clear at this point is whether LPS receptors are the same as those for peptidoglycan or β-1,3-glucan receptors or independent receptors. Obviously much work needs to be done in this area.

Based on the foregoing discussion, a tentative scheme containing all the components of the cascade can be devised as shown in Fig. 1. Upon recognition of foreign objects, certain components in the hemolymph seem to become activated, which in turn trigger in succession two serine proteases and prophenoloxidase. Activated phenoloxidase attacks its substrate and produces cytotoxic quinones to kill the intruder. Alternatively, the quinones can polymerize and form the capsule. The control mechanisms of this cascade are discussed in Section III.

E. Lipids as Modulators

Activation of insect prophenoloxidase has been extensively reviewed by Brunet (1980). Two types of activation processes have been characterized so far. One is the protease-mediated activation elaborated in earlier pages. The second is the spontaneous and sometimes unaccountable activation by physical techniques. As early as 1945, Bodine described the activation of latent phenoloxidase from the diapausing eggs of the grasshopper *Melanoplus differentialis* by surface-active reagents, heat, and heavy metals. Later Funatsu and Inaba (1962) outlined the activation of *Musca* prophenoloxidase by sodium dodecyl sulfate, whereas Heyneman and Vercauteren (1964) demonstrated the activation of latent phenoloxidase from *T. molitor* by sodium oleate. Among the fatty acids tested, oleate was the most potent activator of *Tenebrio* phenoloxidase (Heyneman and Vercauteren, 1968). Sodium dodecyl sulfate as well as oleate also caused the activation of *Lucilia cuprina* proenzyme (Hackmann and Goldberg, 1967), but subsequent work by numerous researchers led to the establishment of proteases as the true activators of prophenoloxidase. As a result, the early studies on activation mechanisms involving physical treatments were completely ignored for physiological purposes. However, very recently, while examining the properties of latent prophenoloxidase from the left colleterial gland of *Periplaneta americana,* Sugumaran and Nellaiappan (1990) found that this proenzyme is specifically activated by treatment with detergents such as sodium dodecyl sulfate below their critical micellar concentrations. They could not find any evidence for an endogenous protease activator in this system. A thorough search led to the discovery that low concentrations of phospholipids served as effective activators of not only cockroach prophenoloxidase, but also the prophenoloxidase from a variety of insects and crustaceans (M. Sugumaran and K. Nellaiappan, 1990, 1991; M. Sugumaran, unpublished results). Among the phospholipids tested, lysolecithin proved to be the most potent activator of lobster prophenoloxidase (Sugumaran and Nellaiappan, 1991). It appears that chemical detergents such as sodium dodecyl sulfate or biological detergents such as fatty acids and phospholipids at very low concentrations bind to the prophenoloxidase and alter its conformation so as to allow the enzyme to expose its active site for its substrates. Such accessibility is perhaps brought about by proteolytic enzymes by splitting off the "inhibitory peptide" region from the proenzyme. Thus, irrespective of what treatment it is, prophenoloxidase activation may be attained by conformational changes that lead to the exposure of the active site of the enzyme.

Physiologically, the lipid activation of prophenoloxidase has important consequences. During invasion, many researchers have demonstrated the presence of damaged hemocytes on or around the foreign objects. (Ratcliffe *et al.*, 1985). Cellular damage can cause the release of phospholipase A_2, which in turn liberates lysolecithin and free fatty acids. The released lysolecithin, which is one of the most potent biological detergents, can function as a direct activator of phenoloxidase, while the free fatty acids generated can be used for the synthesis of arachidonic acid and eicosanoids, which could trigger further immune reactions as demonstrated by others (Blomquist *et al.*, 1991; Stanley-Samuelson *et al.*, 1991).

F. Compartmentalization

Since prophenoloxidase constitutes an important component of the insect's defense mechanism, it is important to examine its cellular localization. Earlier studies by Evans (1967) in *G. mellonella* and Pye (1974) in *A. pernyi* indicated the presence of the prophenoloxidase in the plasma fraction of hemolymph. Subsequently, Ashida (1981), using cane sugar as a stabilizing agent, reported the presence of this enzyme in the plasma fraction of *B. mori*. However, Leonard *et al.* (1985b) questioned the suitability of this procedure and, using an anticoagulation buffer consisting of ethylene diamine tetraacetate, glucose, and citrate to separate the hemocytes from the plasma fraction, claimed that the prophenoloxidase is present in the hemocytes and not in the plasma fraction of *B. craniifer.* But, Iwama and Ashida (1986), during their studies on the biosynthesis of prophenoloxidase, demonstrated that oenocytoids synthesized this enzyme specifically and released it into medium. Hence, it is most likely that prophenoloxidase is synthesized by oenocytoids and secreted into the plasma. Accordingly, using immunolocalization, Ashida *et al.* (1988) conclusively demonstrated the presence of prophenoloxidase in the plasma under normal physiological conditions. In both *S. bullata* and *M. sexta,* one of us (Saul and Sugumaran, 1987, 1988; Saul *et al.*, 1987) demonstrated that the majority of prophenoloxidase is localized in the plasma and not in the cells. Though Ratcliffe's group ascertained the presence of the prophenoloxidase system in the cellular fraction of locusts (Brookman *et al.*, 1989), Brehélin *et al.*, (1989) demonstrated its presence in both the plasma and cellular fractions of the same organism. The difference in these results is no doubt due to the differences in experimental conditions employed by these workers. As stated earlier, Ratcliffe's group injected the ethylene diamine tetra-acetate, citrate, glucose buffer into the organisms before collecting the hemolymph, whereas Brehélin's group directly collected the hemolymph for fractionation studies. It appears that the anticoagulation buffer interferes with the cellular localization studies. This as well as discrepancies in LPS activation certainly questions the wisdom of using anticoagulant solutions in the study of the prophenoloxidase system.

G. Is the Prophenoloxidase System a Recognition Mechanism?

The speculation that prophenoloxidase systems could serve as a recognition mechanism of non-self matter in an attractive but unsubstantiated one (Soderhall, 1982; Ratcliffe *et al.*, 1984; Leonard *et al.*, 1985a). The association of phenoloxidase with defense reactions of insects is well established, however, there is hardly any evidence in the literature to attribute a role for this enzyme in non-self recognition. Although activation of the prophenoloxidase system in the wax moth *G. mellonella* by a microbial product enhanced the recognition of non-self material (Ratcliffe *et al.*, 1984; Leonard *et al.*, 1985a), this does not serve as proof for its role in non-self recognition. As stated earlier, none of the microbial products tested directly activates prophenoloxidase; rather they seem to bind to some components in plasma (or on the cell surface) and trigger the prophenoloxidase cascade. Therefore, if one accepts the premise that prophenoloxidase activation is achieved by a cascade of reactions in the insect's hemolymph, then naturally the role of prophenoloxidase in recognition of non-self matter becomes questionable. However, there is no doubt that prophenoloxidase activation is a consequence of non-self recognition. Accordingly, Rizki and Rizki (1990) have successfully demonstrated that *D. melanogaster* larvae lacking hemolymph phenoloxidase activity could readily encapsulate a foreign organism injected into its blood but could not melanize it. Therefore, the hypothesis that prophenoloxidase serves as a mechanism for recognition of foreignness should be reevaluated.

III. Control Mechanisms of the Cascade

A. Introduction

Co-occurrence of prophenoloxidase, its substrate(s), and activators in the hemolymph would cause undesired activation of prophenoloxidase and generation of cytotoxic quinones. Such reactions are deleterious to the organism and hence appropriate control mechanisms must exist in the organisms to avoid such unwanted reactions. Keeping the enzymes in their inactive proenzyme forms, time constraints on the activated components, protecting the substrates from reacting with their enzymes, secretion of inhibitors, and compartmentalization of different components are a few of the ways to achieve these goals. Available evidence indicates that all of these mechanisms are effectively used in insects.

Evidence for the presence of (a) proenzyme forms, (b) time constraints on phenoloxidase activity, and (c) compartmentalization has already been presented. Ample examples can be given for the protection of substrates. The catecholic substrates are stored in the hemolymph as phosphate, sulfate, or glycoside conjugates that do not serve as substrates for phenoloxidase (Kramer and Hopkins, 1987). Upon enzymatic deblocking of the protective groups, free catechols can be generated at desired places. Hydrolytic enzymes such as glycosides, phosphatases, and sulfatases are usually compartmentalized in cells. During invasion by a foreign organism, cellular contents can be discharged to ensure mixing of these enzymes with their substrates and the liberated free *o*-diphenols can serve as substrates for phenoloxidase. In addition, inhibitors seem to play a crucial role in controlling the cascade.

B. Protease Inhibitors

It is a common practice to add heat-inactivated hemolymph to insect hemocyte cultures to prevent darkening reactions (Wyatt, 1956). Based on the extreme thermal stability of serine protease inhibitors of Bovine Pancreatic Trypsin Inhibitor (BPTI) class, and the limited proteolytic mode of activation of prophenoloxidase, it is logical to expect the participation of thermostable serine protease inhibitors in controlling the activity of proteases involved in prophenoloxidase cascade. Therefore, studies were undertaken to isolate heat-stable protease inhibitors from the hemolymph of *M. sexta* and *S. bullata* (Sugumaran *et al.,* 1985; Saul and Sugumaran, 1986; Ramesh *et al.,* 1988). These attempts led to the isolation of three BPTI type serine protease inhibitors. Two of these were isolated from *M. sexta* and the other was isolated from *S. bullata*. Sugumaran *et al.* (1985) demonstrated for the first time drastic inhibition of protease-mediated prophenoloxidase activation by all three inhibitors. In *M. sexta,* one of these serine protease inhibitors could inhibit the cuticular protease-mediated activation of prophenoloxidase, confirming the proposed physiological role of this inhibitor (Saul and Sugumaran, 1986). Brehélin *et al.* (1991) have demonstrated the presence of a serine protease inhibitor that controls the prophenoloxidase activation in the hemolymph of *L. migratoria*.

Manduca inhibitor A is a homodimer with a molecular weight of 14,000, whereas *Manduca* inhibitor B is a monomer with a molecular weight of 8000. *Sarcophaga* inhibitor is much smaller and closer to the molecular weight of BPTI. The primary amino acid sequences (partial) of these inhibitors are shown in Fig. 2. The active site P1 residue for all three inhibitors is arginine as opposed to lysine in BPTI (Ramesh *et al.,* 1988; Sugumaran, 1990). Interestingly, the adjacent residue at the active site of *Manduca* inhibitors and BPTI is alanine, whereas lysine serves this purpose in the *Sarcophaga* inhibitor. Three chymotrypsin inhibitors of the BPTI class have been isolated from *B. mori* and sequenced (Sasaki, 1984, 1988). These inhibitors have phenylalanine as the P1 residue. A role for these inhibitors in regulation of prophenoloxidase activation has not been established.

Serine protease inhibitors from the serpin gene superfamily (Carrell and Boswell, 1986) have been identified and characterized in two lepidopteran insects, *M. sexta* and *B. mori*. Serpins are proteins of M_r = 40,000–60,000 that, in vertebrates, appear to function primarily as regulators of endogenous

```
------------------------------------------------------------------
Inhibitor       1                 10                  20
---------
Manduca A       A G L Y K P P N N I E S E N E V Y T G N
Manduca B                                         E D
Sarcophaga                                    D V K S
BPTI                                            R P D

               21                 30                  40
                                        *
Manduca A       I C F L P L E V G V C R A L F F R Y G Y
Manduca B       I C S L P P E V G P C R A G F L K # A Y
Sarcophaga      A C L Q P K E V G P C R K S D F V F F Y
BPTI            F C L E P P Y T G P C K A R I I R Y F Y

               41                 50                  60

Manduca A       D P A I K A C X E F M Y G G X Q - - - -
Manduca B       Y S E L N K C K L F @ Y G G C Q G N E N
Sarcophaga      N D A T K A C E E F L Y G G C R G N D N
BPTI            N A K A G L C Q T F V Y G G C R A K R N

               61                 70                  80

Manduca A       - - - - - - - - - - - - - - - - - - - -
Manduca B       N F E T L Q A C X Q A
Sarcophaga      R F N T K E E C E K L C L
BPTI            N F K S A E D C M R T C G G A
------------------------------------------------------------------
# : F or Y (position "38" of Manduca inhibitor B)
@ : T or G (position "51" of Manduca inhibitor B)
```

Figure 2 Alignment of the protein sequences of *Manduca* protease inhibitors, *Sarcophaga* protease inhibitor, and Bovine Pancreatic Trypsin Inhibitor (BPTI). The P1 residue of the reactive site region is indicated by an asterisk.

serine proteases. For example, antithrombin-III and C1-inhibitor from human plasma are involved in regulation of blood clotting and complement activation, respectively. Like prophenoloxidase activation, these are pathways in which a cascade of serine proteases leads to activation of an enzyme through cleavage of a proenzyme.

A trypsin inhibitor (M_r = 42,000) and a chymotrypsin inhibitor (M_r = 43,000) from the serpin superfamily have been isolated from the hemolymph of *B. mori* (Sasaki and Kobayashi, 1984; Eguchi and Shomoto, 1985). These proteins have a reactive site near their C-terminal ends, at which they form a complex with serine proteases. The complex is stable in sodium dodecyl sulfate but not at high pH (Sasaki, 1985; Sasaki *et al.*, 1987), properties in common with well-characterized human serpins. Four serpins have been isolated from hemolymph of *M. sexta* (Kanost, 1990a). Two of these proteins are chymotrypsin inhibitors, one is a trypsin inhibitor, and the other is an elastase inhibitor, all of approximately M_r = 47,000. Although these four proteins are the most abundant serpins present in *M. sexta* hemolymph, two-dimensional gel electrophoresis and immunoblotting identified approximately ten proteins of M_r = 45,000–48,000 that react with antibodies to the elastase inhibitor (Kanost, 1990b).

cDNA clones for *M. sexta* elastase inhibitor and the *B. mori* trypsin inhibitor have been isolated and sequenced (Kanost *et al.* ,1989; Takagi *et al.*, 1990). The deduced protein sequences are 56% identical with each other (Fig. 3) and each insect protein is 25–30% identical with vertebrate serpin sequences. In the reactive site near the C-terminus, which determines the specificity of the inhibitors, the two insect serpin sequences are quite different. This is consistent with the observation that in mammalian serpin genes, the region encoding the reactive site has diverged much more rapidly than the remainder of the gene, resulting in a family of inhibitors with a fairly constant framework but variable reactive sites resulting in diverse functions (Hill and Hastie, 1987). The *B. mori* trypsin inhibitor has a lysine at the P1 residue of the reactive site, which determines its specificity for trypsin, whereas the *M. sexta* elastase inhibitor has an alanine as the P1 residue, leading to specificity for elastase inhibition. According to a convention of naming serpins based on their P1 residue (Carrell and Boswell, 1986), these two proteins may be referred to as *B. mori* lysserpin and *M. sexta* alaserpin.

Serpins have not yet been isolated from nonlepidopteran insects. However, hemolymph from insects in seven different orders contain abundant serine protease inhibitor activity and proteins of M_r = 40,000–60,000 (M. Kanost, unpublished data). Preliminary experiments also suggest that serpins are present in hemolymph of *D. melanogaster* (Kanost, 1990b).

The hypothesis that serpins function as regulators of prophenoloxidase activation is based on mostly circumstantial evidence. Like blood clotting in vertebrates, prophenoloxidase activation in insects must be a rapid, local, and transient response that does not spread far from the original site. The fact that insect hemolymph contains serpins similar to those that regulate the serine proteases of activation cascades in vertebrates points toward the possibility of a comparable role in the prophenoloxidase cascade. In support of

```
Bombyx     1  MKTIICLFTIAIAAMAAVTNLSNVLKNGNDNFTARMFTEVVKNNPGKSIVLSAFSVLPPLAQLALASDGETHEELLKAIG   80
Manduca    1  MKIIMCIFGLAALAMAGETDLQKILRESNDQFTAQMFSEVVKANPGQNVVLSAFSVLPPLGQLALASVGESHDELLRALA   80

Bombyx    81  FPDDDAIRTEFASKSRDLRSIKGVELKMANKVYVHDGGKLDENFAVVSRDVFNSDVQNIDFSKNTVAAKSINDWVEENTN  160
Manduca   81  LPNDNVTKDVFADLNRGVRAVKGVDLKMASKIYVAKGLELNDDFAAVSRDVFGSEVQNVDFVKSVEAAGAINKWVEDQTN  160

Bombyx   161  NRIKDLVNPDSLSSATAAVLVNAIYFKGAWSSKFDERLTSDRDFYVSKDKTIKVPMMYKRGDYKYGESAVLNAQLIEIPY  240
Manduca  161  NRIKNLVDPDALDETTRSVLVNAIYFKGSWKDKFVKERTMDRDFHVSKDKTIKVPTMIGKKDVRYADVPELDAKMIEMSY  240

Bombyx   241  KGDQSSLIVVLPKDKDGITQLQEALKDPKTLETAQQSMYSTEVDLYLPKFKIETETNLKDVLSNMNVNKIFNND AQITR  319
Manduca  241  EGDQASMIIILPNQVDGITALEQKLKDPKALSRAEERLYNTEVEITLPKFKIETTTDLKEVLSNMNIKKLFTPGAARLEN  320

                                                      *
Bombyx   320  LLKG ESLSVSEAIQKAFIEINEEGAEAAAANAFTMTRSSKVYVRPPIVFNANKPFYYALQVDGVIMFNGIFIN  392
Manduca  321  LLKTKESLTVDAAIQKAFIEVNEEGAEAAAANAFGIVPASLILTPEVHI   DRPFYFELKIDGIPMFNGKVIEP  392
                                                    *
```

Figure 3 Alignment of the protein sequences of *B. mori* trypsin inhibitor and *M. sexta* elastase inhibitor. The P1 residue of the reactive site region is indicated by an asterisk.

this speculation, injection of antibodies against *M. sexta* alaserpin into *M. sexta* larvae results in increased phenoloxidase activity, suggesting that the antibodies disrupt regulation of the activation pathway (Kanost, 1990b). Further experiments with isolated components of the pathway will be required to confirm the role of serpins as regulators of prophenoloxidase activation.

C. Prophenoloxidase Inhibitor

Suppression of host phenoloxidase activity by invading parasites is an attractive mechanism for control of host defense reaction exhibited by way of prophenoloxidase activation (Salt, 1970).

Stoltz and Cook (1983) demonstrated the suppression of host phenoloxidase activity by hymenopteran parasitoids. The parasitoids introduce polydnavirus particles into the host that appear to inhibit phenoloxidase activity (Stoltz *et al.*, 1984; Beckage *et al.*, 1990). This result clearly indicates that the parasites are either attempting to evade or disrupt the host defense reaction for their own survival. Hardening of capsules by phenoloxidase reaction products can potentially limit the growth and development of the parasites and hence parasites might have adapted the inhibition of prophenoloxidase activation as a strategy to circumvent this problem. Thus, in a host–parasite interaction the inhibition of prophenoloxidase activity plays a crucial role. The inhibition, which could occur theoretically at different stages in the cascade, needs to be examined. The possible presence of a phenoloxidase inhibitor in such systems is likely, but further studies are essential to evaluate this contention.

D. Dopachrome Conversion Factor

Tyrosine and its hydroxylated product, dopa, serve as the precursors for eumelanin biosynthesis. Eumelanin pigments are widely distributed in the animal kingdom as skin and hair pigments. Because of their free radical nature, melanins protect animals from damaging solar radiation (Prota, 1988). Abnormalities in melanin biosynthesis are associated with vitiligo as well as melanoma cancer. Naturally, numerous workers have devoted much attention to study the mechanism of melanin biosynthesis. The initial steps for melanin biosynthesis were established by the classical work of Raper and Mason and hence it is called the Raper–Mason pathway (Mason, 1955). The key reactions of the Raper–Mason pathway are: (a) hydroxylation of tyrosine to dopa, (b) oxidation of dopa to dopaquinone, (c) intramolecular cyclization of dopaquinone and subsequent generation of dopachrome, (d) conversion of dopachrome to 5,6-dihydroxyindoles, and (e) oxidative polymerization of

5,6-dihydroxyindoles. Though the first two reactions are catalyzed by ubiquitously present tyrosinase (a phenoloxidase), the rest of the reactions are believed to proceed nonenzymatically, as chemical oxidation of dopa could generate melanin pigments without the need for any enzyme(s). However, studies done in the last decade paved the way to the identification of a second enzyme, namely, dopachrome conversion factor, as a new regulator of melanogenesis (Pawelek *et al.*, 1980; Leonard *et al.*, 1988; Aroca *et al.*, 1989, 1991; Aso *et al.*, 1989, 1990; Pawelek, 1990; Sugumaran and Semensi, 1991). Dopachrome conversion factor from mammalian sources catalyzes the generation of 5,6-dihydroxyindole-2-carboxylic acid (Aroca *et al.*, 1989, 1991; Pawelek, 1990), whereas the enzyme isolated from insects seems to produce 5,6-dihydroxyindole from the same substrate (Aso *et al.*, 1990; Sugumaran and Semensi, 1991). Recent investigations on the mechanism of this conversion reveal that the reaction proceeds through a quinone methide intermediate (Sugumaran and Semensi, 1991; Sugumaran, 1991b). Figure 4 shows the modified Raper–Mason pathway for the biosynthesis of melanin (Sugumaran, 1991b). Dopachrome in the absence of metal ions has a definite life time at neutral pH. Therefore, presence of dopachrome conversion factor in insects ensures the faster generation of melanin pigment by readily generating the melanin precursor, 5,6-dihydroxyindole.

E. Quinone Isomerase

Hemolymph of insect larvae contains numerous tyrosine and dopa derivatives, mainly in storage form for later use as cuticular tanning precursors (Kramer and Hopkins, 1987). In particular, derivatives of *N*-acetyldopamine and *N*-β-alanyldopamine are widely found in insects. Since phenoloxidases are nonspecific (Brunet, 1980), they will attack indiscriminately any of the catecholic compounds. Natural abundance of *N*-acetyldopamine and *N*-β-alanyldopamine in the hemolymph makes it likely that these compounds will be oxidized by phenoloxidase in preference to others. The quinones generated from these compounds do not polymerize instantaneously under physiological conditions, but undergo slow transformations. As a result, they diffuse slowly from the site of generation and can react with thiols and other nucleophiles in the hemolymph. Such reactions can have deleterious effects not only on the invading parasites but also on the host. To prevent damage to self tissue, insects must have a means to deactivate the quinonoid molecules. One such mechanism has been discovered (Saul and Sugumaran, 1989a,b, 1990). Quinone isomerase (Saul and Sugumaran, 1990), an enzyme exhibiting wide substrate specificity toward its substrates, 4-alkyl-*o*-benzoquinones, converts the long-lived quinones to short-lived 2-hydroxy-*p*-quinone methides. Quinone methides, being highly unstable, readily and rapidly undergo

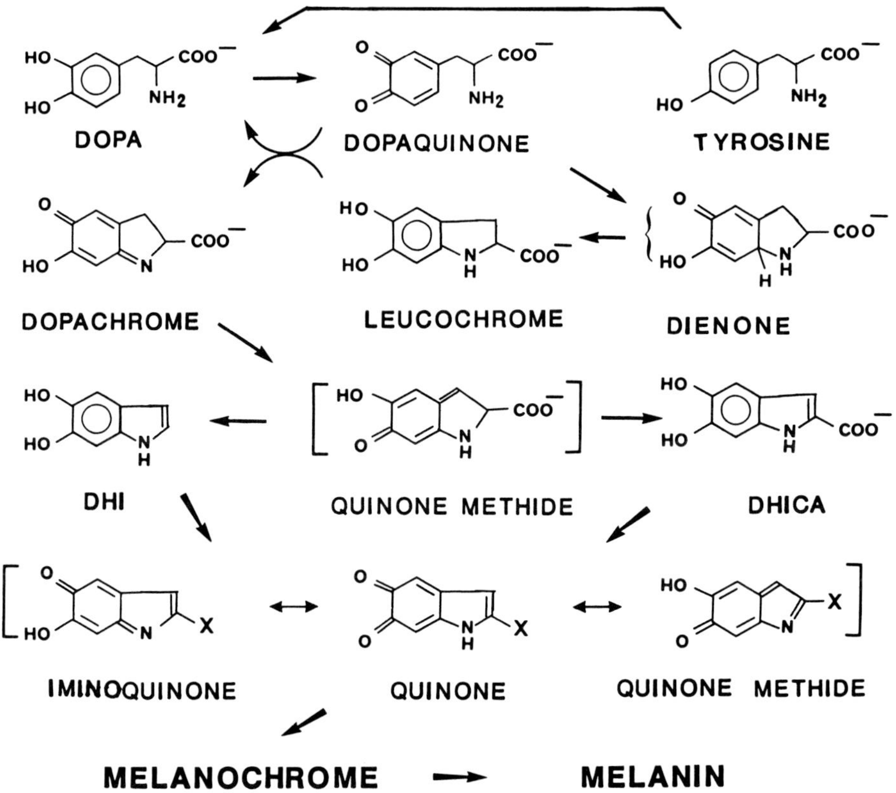

Figure 4 Modified Raper–Mason pathway of melanogenesis. Tyrosinase catalyzes the oxidation of tyrosine to dopa and dopa to dopaquinone. The latter undergoes rapid intramolecular cyclization to form leucochrome. Oxidation of leucochrome generates dopachrome. Dopachrome conversion factor tautomerizes this compound to the quinone methide, which can either lose carbon dioxide to form 5,6-dihydroxyindole (DHI) or simply isomerize to 5,6-dihydroxyindole-2-carboxylic acid (DHICA). Oxidation of these indoles to corresponding quinonoid species and their subsequent polymerization ensure the production of melanin via melanochrome.

Michael-1,6-addition reactions with surrounding nucleophiles with the regeneration of catechols. This reaction plays a crucial role in the sclerotization of insect cuticle (Sugumaran, 1988, 1991a). The role of quinone isomerase with respect to cuticular tanning is summarized in Fig. 5.

Sclerotization helps the soft-bodied insect to protect itself from the harsh environment by creating a formidable barrier. Functionally, this reaction can be used in insect immunity in two advantageous ways. When generated at the

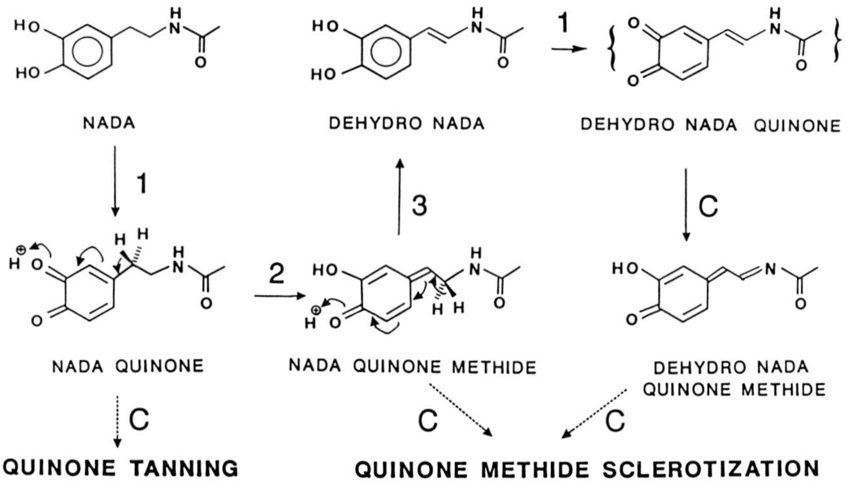

Figure 5 Mechanisms of cuticular tanning. Sclerotizing precursors such as *N*-acetyldopamine (NADA) are oxidized by phenoloxidases (**1**) to their corresponding quinones. These quinones participate in quinone tanning or serve as substrates for quinone isomerase (**2**). Quinone methides generated by the action of quinone isomerase participate in quinone methide sclerotization or are converted into dehydro dopamine derivatives by quinone methide tautomerase (**3**). Further oxidation of dehydro NADA generates quinone and quinone methide derivatives that participate in sclerotization. These reactions are used widely to harden insect cuticle and can also be used during encapsulation reactions. C = Chemical reactions.

site of a foreign object, the quinone methide could attach to the intruder's surface (Sugumaran, 1990). The catechol adducts thus formed could serve as a base for further reactions including deposition of melanin and sclerotin. Second, if the quinone methides are generated away from foreign objects, due to their inherent reactivity with water, they can yield side-chain hydroxylated catechols that are not as toxic as quinones or quinone methide. Therefore, the presence of quinone isomerase in insect hemolymph could serve to detoxify the quinones and to deposit the quinone methides on foreign objects (Fig. 6). Quinone isomerase might also be used efficiently during host–parasite interactions. If the host generates alkylquinones, the parasites can adapt the strategy of using the quinone isomerase to either detoxify these quinones or use them to make the capsule to protect themselves from further defensive reactions of the host.

To perform its task efficiently, it is also advantageous for the quinone isomerase to complex with phenoloxidase. Evidence for the existence of a quinone isomerase–phenoloxidase complex comes from molecular sieve chro-

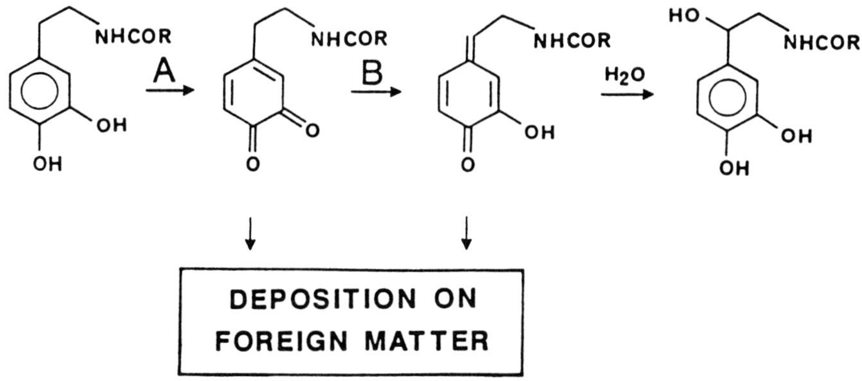

Figure 6 Use of quinone methide in insect immunity. 4-Alkylquinones generated by the action of activated phenoloxidase (**A**) on catecholamine derivatives such as *N*-acetyldopamine (R = CH_3) or *N*-β-alanyldopamine (R = $CH_2CH_2NH_2$) are converted by quinone isomerase (**B**) to quinone methides. Both quinones and quinone methides can be deposited on foreign objects to encapsulate them. Quinone isomerase in the hemolymph can also aid in destroying quinones, as the quinone methides formed react with water instantly to form nontoxic compounds.

matography (Saul and Sugumaran, 1989b) as well as inhibition studies. The quinone isomerase isolated from unactivated hemolymph of *S. bullata* elutes with an apparent M_r of 80,000, whereas the same enzyme from activated hemolymph co-chromatographs with phenoloxidase at a higher molecular weight ($M_r > 200{,}000$), indicating possible complex formation between these two proteins. This does not come as a surprise because activated phenoloxidase is known to be sticky and binds to glass, Sephadex gels, etc. But what is surprising is the ability of quinone isomerase to control the activity of phenoloxidase. Once complexed, the phenoloxidase activity is inhibited by as much as 50% by quinone isomerase (Fig. 7). Thus, quinone isomerase seems to serve as a negative control mechanism for phenoloxidase as well.

IV. Conclusion

Incorporation of these control mechanisms into the primary prophenoloxidase cascade leads to the overall scheme of reactions depicted in Fig. 8. According to this scheme, upon entry of foreign substances, recognition molecules (designated very broadly as zymogen) bind to them and trigger a variety of host-defense systems including calcium mobilization. The biochemical aspects of this stage are not yet clarified in detail, but some of the aspects have been

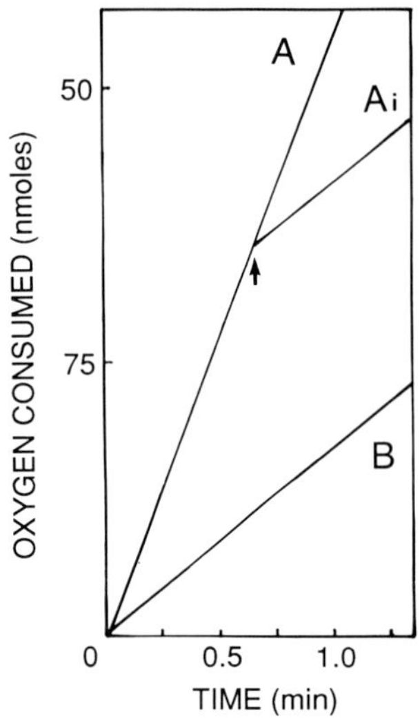

Figure 7 Inhibition of phenoloxidase by quinone isomerase. A reaction mixture containing 1 m*M* *N*-acetyldopamine and 10 μg of mushroom tyrosine in 1 ml of 0.1 *M* sodium phosphate buffer, pH 6.0, was incubated at 30°C and oxygen consumed during the reaction was continuously measured using a Clark type oxygen electrode (curve A). At the indicated time about 250 units of purified quinone isomerase from *S. bullata* was added (Ai). In curve B, quinone isomerase was present from the beginning of the reaction.

reviewed by Schmidt and Theopold (1991). What follows this process regarding prophenoloxidase activation is fairly well established. The activation of pro-benzoyl arginine ethyl ester, preprophenoloxidase activating enzyme, and prophenoloxidase occurs in sequence. The cellular damage and calcium mobilization can activate phospholipase A_2, which splits off arachidonic acid from the cellular phospholipids. Eicosanoids produced from arachidonic acid trigger further immune reaction, while lyophospholipids and phospholipids formed during cellular damage and phospholipase action directly activate prophenoloxidase. Protease inhibitors provide negative control by preventing prophenoloxidase activation, whereas quinone isomerase partially controls phenoloxidase activity. Positive control for melanin and/or sclerotin formation is provided by both quinone isomerase and dopachrome conversion factor. By unidentified mechanism(s) parasites and other predators seem to inhibit the prophenoloxidase cascade (not shown in Fig. 8).

This scheme represents only the control mechanisms know at this time and hence by no means can it be considered as complete. As additional research is completed, we are sure that this scheme might become more complex. At this point we also like to caution the readers regarding certain

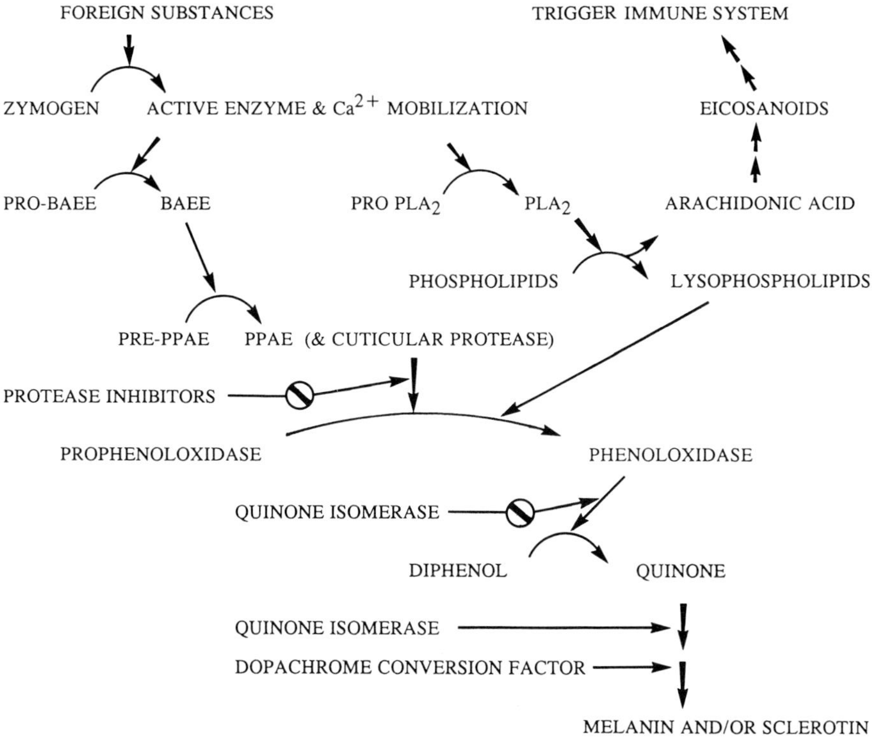

Figure 8 The prophenoloxidase cascade and its control mechanisms. Foreign substances activate prophenoloxidase cascade as outlined in Fig. 1. Protease inhibitors inhibit serine proteases, thus preventing them from activating the prophenoloxidase. Cellular damage and calcium mobilization can also trigger the inactive prophospholipase A_2 (PRO PLA_2) to active PLA_2. Lipids generated during cellular damage as well as PLA_2 action can directly bind to prophenoloxidase and activate it. Quinone isomerase can control the activity of active phenoloxidase negatively. In addition, this enzyme can play a positive role by providing the quinone methides for sclerotin formation. Dopachrome conversion factor, by facilitating the production of dihydroxyindoles, participates positively in melanin capsule formation. Phenoloxidase inhibitor could constitute another control mechanism (not shown in figure). The reactions of quinones and quinone methides lead to the production of melanin and/or sclerotin capsule.

complications. As pointed out in the Introduction, there are at least three different phenoloxidases known in various systems. These enzymes seem to serve three different functions in insects, namely, wound healing, cuticular sclerotization, and immune response. Because of the problems in assigning any particular role for one or more of the phenoloxidases, it is rather difficult to say precisely whether or not all the mechanisms outlined in Fig. 8 are applicable to all insects. It is possible that only part of the reactions operate in one or more insects. As for future directions, practically the whole area is wide open for a detailed study. Since phenoloxidase participates as a key component not only in insect immunity but also in sclerotizing machinery and wound-healing mechanisms, examining the detailed biochemistry of this vital enzyme system and its control mechanisms warrants considerable effort. Such studies may eventually have great commercial and life-saving potentials if novel vector control measures and insecticides can be made based on this work.

References

Andersson, K., Sun, S. C., Boman, H. G., and Steiner, H. (1989). Purification of the prophenoloxidase from *Hyalophora cecropia* and four proteins involved in its activation. *Insect Biochem.* **19:**629–637.

Aroca, P., Garcia-Borron, J. C., Solano, R., and Lozano, J. A. (1989). Regulation of mammalian melanogenesis. I. Partial purification and characterization of a dopachrome converting factor: Dopachrome tautomerase. *Biochim. Biophys. Acta* **1035:**266–275.

Aroca, P., Solano, F., Garcia-Borron, J. C., and Lozano, J. A. (1991). Specificity of dopachrome tautomerase and inhibition by carboxylated indoles. *Biochem. J.* **277:**393–397.

Ashida, M. (1971). Purification and characterization of prephenoloxidase from hemolymph of the silkworm *Bombyx mori. Arch. Biochem. Biophys.* **144:**749–762.

Ashida, M. (1981). A cane sugar factor suppressing activation of prophenoloxidase in haemolymph of the silkworm, *Bombyx mori. Insect Biochem.* **11:**57–65.

Ashida, M., and Dohke, K. (1980). Activation of prophenoloxidase by the activating enzyme of the silkworm, *Bombyx mori. Insect Biochem.* **10:**37–47.

Ashida, M., and Yamazaki, H. I. (1990). Biochemistry of the phenoloxidase system in insects with special reference to its activation. *In* "Molting and Metamorphosis" (E. Ohnishi and H. Ishizaki, eds.), pp. 239–265. Jpn. Sci. Soc. Press, Tokyo/Springer-Verlag, Berlin.

Ashida, M., and Yoshida, H. (1988). Limited proteolysis of prophenoloxidase during activation by microbial products in insect plasma and effect of phenoloxidase on electrophoretic mobility of plasma proteins. *Insect Biochem.* **18:**11–19.

Ashida, M., Ochiai, M., and Niki, T. (1988). Immunolocalization of prophenoloxidase among hemocytes of the silkworm, *Bombyx mori. Tissue Cell* **20:**599–610.

Ashida, M., Kinoshita, K., and Brey, P. T. (1990). Studies on prophenoloxidase activation in the mosquito, *Aedes aegypti* L. *Eur. J. Biochem.* **188:**507–515.

Aso, Y., Kramer, K. J., Hopkins, T. L., and Lookhart, G. L. (1985). Characterization of haemolymph protyrosinase and a cuticular activator from *Manduca sexta* (L). *Insect Biochem.* **15:**9–17.

Aso, Y., Imamura, Y., and Yamasaki, N. (1989). Further studies on dopa quinone imine conversion factor from cuticles of *Manduca sexta. Insect Biochem.* **19:**401–407.

Aso, Y., Nakashima, K., and Yamasaki, N. (1990). Changes in the activity of dopa quinone imine conversion factor during the development of *Bombyx mori. Insect Biochem.* **20:**685–690.

Beckage, N. E., Metcalf, J. S., Nesbit, D. J., Schleifer, K. W., Zetlan, S. R., and deBuron, I. (1990). Host hemolymph monophenoloxidase activity in parasitized *Manduca sexta* larvae and evidence for inhibition by wasp polydnavirus. *Insect Biochem.* **20:**285–295.

Blomquist, G. J., Borgeson, C. E., and Vundla, M. (1991). Polyunsaturated fatty acids and eicosanoids in insects. *Insect Biochem.* **21:**99–106.

Bodine, J. H. (1945). Tyrosine and phenols. Action of diversely activated tyrosinase on monohydric and o-dihydric phenols. *Proc. Soc. Exp. Biol. Med.* **58:**205–209.

Boman, H. G., Faye, I., Gudmundsson, G. H., Lee, J. Y., and Lidholm, D. A. (1991). Cell-free immunity in *Cecropia*. *Eur. J. Biochem.* **201:**23–31.

Brehélin, M., Drif, L., Buad, L., and Beomare, N. (1989). Insect hemolymph cooperation between humoral and cellular factors in *Locusta migratoria*. *Insect Biochem.* **19:**301–307.

Brehélin, M., Boigegrain, R. A., Drif, L., and Colleti-Previero, M. A. (1991). Purification of a protease inhibitor which controls prophenoloxidase activation in hemolymph of *Locusta migratoria* (Insecta). *Biochem. Biophys. Res. Commun.* **179:**841–846.

Brookman, J. L., Ratcliffe, N. A., and Rowley, A. F. (1989). Studies on the activation of the prophenoloxidase system of insects by bacterial cell wall components. *Insect Biochem.* **19:**47–57.

Brunet, P.C.J. (1980). The metabolism of the aromatic amino acids concerned in the crosslinking of insect cuticle. *Insect Biochem.* **10:**467–500.

Carrell, R. W., and Boswell, D. R. (1986). Serpins: The superfamily of plasma serine proteinase inhibitors. *In* "Proteinase Inhibitors" (A. J. Barrett and G. Salvesen, eds.), pp. 403–420. Elsevier, New York.

Christensen, B. M., and Nappi, A. J. (1988). Immune responses of arthropods. *ISI Atlas Sci.: Anim. Sci.*, pp. 15–19.

Coombe, D. R., Ey, P. L., and Jenkin, C. R. (1984). Self and nonself recognition in invertebrates. *Q. Rev. Biol.* **59:**231–255.

Dohke, K. (1973). Studies on prophenoloxidase-activating enzyme from cuticle of the silkworm *Bombyx mori*. *Arch. Biochem. Biophys.* **157:**203–209.

Dularay, B., and Lackie, A. M. (1985). Haemocytic encapsulation and the prophenoloxidase activation pathway in the locust *Schistocera gregaria* Forsk. *Insect Biochem.* **15:**827–834.

Eguchi, M., and Shomoto, K. (1985). Purification and properties of a chymotrypsin inhibitor from the silkworm haemolymph. *Comp. Biochem. Physiol. B* **81B:**301–307.

Evans, J.J.T. (1967). The activation of prophenoloxidase during melanization of the pupal blood of the Chinese oak silkmoth *Antheraea pernyi*. *J. Insect Physiol.* **13:**1699–1711.

Funatsu, M., and Inaba, T. (1962). Studies on tyrosinase in house fly. I. Protyrosinase in the pupae of the house fly and its activation. *Agric. Biol. Chem.* **26:**535–540.

Gotz, P. (1986). Mechanisms of encapsulation in dipteran hosts. *Symp. Zool. Soc. London* **56:**1–19.

Gotz, P., and Boman, H. G. (1985). Insect immunity. *In* "Comprehensive Insect Physiology, Biochemistry and Pharmacology" (G. A. Kerkut and L. I. Gilbert, eds.), Vol. 3, pp. 453–485. Pergamon, Oxford.

Gupta, A. P., ed. (1988). "Hemocytic and Humoral Immunity in Arthropods." Wiley, New York.

Hackmann, R. H., and Goldberg, M. (1967). The *o*-diphenoloxidase of fly larvae. *J. Insect Physiol.* **13:**531–544.

Heyneman, R. A. (1965). Final purification of a latent phenoloxidase with mono- and diphenoloxidase activity from *Tenebrio molitor*. *Biochem. Biophys. Res. Commun.* **21:**162–169.

Heyneman, R. A., and Vercauteren, R. E. (1964). Activation of the latent phenoloxidase of *Tenebrio molitor*. *Enzymologia* **28:**85–88.

Heyneman, R. A., and Vercauteren, R. E. (1968). Evidence for a lipid activator of prophenoloxidase in *Tenebrio molitor*. *J. Insect Physiol.* **14:**409–415.

Hill, R. E., and Hastie, N. D. (1987). Accelerated evolution in the reactive center regions of serine protease inhibitors. *Nature (London)* **326:**96–99.

Hughes, L. (1976). Haemolymphal activation of protyrosinase and the site of synthesis of haemolymph protyrosinase in larvae of the fleshfly, *Sarcophaga barbarta*. *J. Insect Physiol.* **22:**1005–1011.

Iwama, R., and Ashida, M. (1986). Biosynthesis of prophenoloxidase in hemocytes of larval hemolymph of the silkworm, *Bombyx mori*. *Insect Biochem.* **16:**547–555.

Johansson, M. W., and Soderhall, K. (1989). Cellular immunity in crustaceans and the prophenoloxidase system. *Parasitol. Today* **5:**171–176.

Jomori, T., Kubo, T., and Natori, S. (1990). Purification and characterization of lipopolysaccharide binding protein form hemolymph of American cockroach *Periplaneta americana*. *Eur. J. Biochem.* **190:**201–206.

Kanost, M. R. (1990a). Isolation and characterization of four serine proteinase inhibitors (serpins) from hemolymph of *Manduca sexta*. *Insect Biochem.* **20:**141–147.

Kanost, M. R. (1990b). Serine protease inhibitors from the serpin gene family in *Manduca sexta* and *Drosophila melanogaster*. *In* "Molecular Insect Science" (H. H. Hagedorn, J. G. Hilderbrand, M. G. Kidwell, and J. H. Law, eds.), pp. 139–146. Plenum, New York.

Kanost, M. R., Prasad, S. V., and Wells, M. A. (1989). Primary structure of a member of the serpin superfamily of proteinase inhibitors from an insect, *Manduca sexta*. *J. Biol. Chem.* **264:**965–972.

Kramer, K. J., and Hopkins, T. L. (1987). Tyrosine metabolism for insect cuticle tanning. *Arch. Insect Biochem. Physiol.* **6:**279–301.

Lackie, A. M. (1988). Haemocyte behaviour. *Adv. Insect Physiol.* **21:**85–178.

Lai-Fook, J. (1966). The repair of wounds in the integument of insects. *J. Insect Physiol.* **12:**195–226.

Leonard, C., Ratcliffe, N. A., and Rowley, A. F. (1985a). The role of prophenoloxidase activation in non-self recognition and phagocytosis by insect blood cells. *J. Insect Physiol.* **31:**789–799.

Leonard, C., Soderhall, K., and Ratcliffe, N. A. (1985b). Studies of prophenoloxidase and protease activity of *Blaberus craniifer* haemocytes. *Insect. Biochem.* **15:**803–810.

Leonard, L. J., Townsend, D. W., and King, R. A. (1988). Function of dopachrome oxidoreductase and metal ions in dopachrome conversion in the eumelanin pathway. *Biochemistry* **27:**6156–6161.

Mason, H. S. (1955). The phenolase complex. *Adv. Enzymol.* **16:**105–184.

Ochiai, M., and Ashida, M. (1988). Purification of a β-1,3-glucan recognition protein in the prophenoloxidase activating system from hemolymph of the silkworm, *Bombyx mori*. *J. Biol. Chem.* **263:**12056–12062.

Ohnishi, E. (1954). Activation of tyrosinase in *Drosophila virilis*. *Annot. Zool. Jpn.* **27:**88–93.

Pau, R. N., and Eagles, P.A.M. (1975). The isolation of an *o*-diphenoloxidase from third instar larvae of the blowfly *Calliphora erythrocephala*. *Biochem. J.* **149:**707–712.

Pawelek, J. M. (1990). Dopachrome conversion factor functions as an isomerase. *Biochem. Biophys. Res. Commun.* **166:**1328–1333.

Pawelek, J. M., Korner, A., Bergstrom, A., and Bologna, J. (1980). New regulators of melanin biosynthesis and the autodestruction of melanoma cells. *Nature* (*London*) **286:**617–619.

Prota, G. (1988). Progress in the chemistry of melanins and related metabolites. *Med. Res. Rev.* **8:**525–556.

Pye, A. E. (1974). Microbial activation of prophenoloxidase from immune insect larvae. *Nature* (*London*) **251:**610–613.

Ramesh, N., Sugumaran, M., and Mole, J. E. (1988). Purification and characterization of two trypsin inhibitors from the hemolymph of *Manduca sexta* larvae. *J. Biol. Chem.* **263:**11523–11527.

Ratcliffe, N. A., Leonard, C., and Rowley, A. F. (1984). Prophenoloxidase activation: Non-self recognition and cell cooperation in insect immunity. *Science* **226:**557–559.

Ratcliffe, N. A., Rowley, A. F., Fitzgerald, S. W., and Rhodes, C. P. (1985). Invertebrate immunity: Basic concepts and recent advances. *Int. Rev. Cytol.* **97:**183–325.

Ratcliffe, N. A., Brookman, J. L., and Rowley, A. F. (1991). Activation of the prophenoloxidase cascade and initiation of nodule formation in locusts by bacterial lipopolysaccharides. *Dev. Comp. Immunol.* **15:**33–39.

Rizki, R. M., and Rizki, T. M. (1990). Encapsulation of parasitoid eggs in phenoloxidase deficient mutants of *Drosophila melanogaster. J. Insect Physiol.* **36:**523–529.

Salt, G. (1970). "The Cellular Defense Reactions of Insects." Cambridge Univ. Press, Cambridge.

Sasaki, T. (1984). Amino acid sequence of a novel Kunitz-type chymotrypsin inhibitor from hemolymph of silkworm larvae, *Bombyx mori. FEBS Lett.* **168:**227–230.

Sasaki, T. (1985). The reactive site of silkworm hemolymph antichymotrypsin is located at the COOH-terminal region of the molecule. *Biochem. Biophys. Res. Commun.* **132:**320–326.

Sasaki, T. (1988). Amino acid sequences of two basic chymotrypsin inhibitors from silkworm larval hemolymph. *Hoppe-Seyler's Z. Physiol. Chem.* **369:**1235–1241.

Sasaki, T., and Kobayashi, K. (1984). Isolation of two novel proteinase inhibitors from hemolymph of silkworm larva, *Bombyx mori.* Comparison with human serum proteinase inhibitors. *J. Biochem.* (*Tokyo*) **95:**1009–1017.

Sasaki, T., Kobayashi, K., and Ozeki, T. (1987). Interaction of silkworm larval hemolymph antitrypsin and bovine trypsin. *J. Biochem.* (*Tokyo*) **102:**433–441.

Saul, S. J., and Sugumaran, M. (1986). Protease inhibitor controls prophenoloxidase activation in *Manduca sexta. FEBS Lett.* **208:**113–116.

Saul, S. J., and Sugumaran, M. (1987). Protease mediated prophenoloxidase activation in the hemolymph of the tobacco hornworm *Manduca sexta. Arch. Insect Biochem. Physiol.* **5:**1–11.

Saul, S. J., and Sugumaran, M. (1988). Prophenoloxidase activation in the hemolymph of *Sarcophaga bullata* larvae. *Arch. Insect Biochem. Physiol.* **7:**91–103.

Saul, S. J., and Sugumaran, M. (1989a). *o*-Quinone:quinone methide isomerase—A novel enzyme which prevents the destruction of self matter by phenoloxidase generated quinones during immune response in insects. *FEBS Lett.* **249:**155–158.

Saul, S. J., and Sugumaran, M. (1989b). Characterization of quinone tautomerase activity in the hemolymph of *Sarcophaga bullata. Arch. Insect Biochem. Physiol.* **12:**157–172.

Saul, S. J., and Sugumaran, M. (1990). 4-Alkyl-*o*-quinone/2-hydroxy-*p*-quinone methide isomerase from the larval hemolymph of *Sarcophaga bullata.* I. Purification and characterization of enzyme catalyzed reaction. *J. Biol. Chem.* **265:**16992–16999.

Saul, S. J., Bin, L., and Sugumaran, M. (1987). The majority of prophenoloxidase in the hemolymph of *Manduca sexta* is present in the plasma and not in the hemocytes. *Dev. Comp. Immunol.* **11:**479–485.

Schmidt, O., and Theopold, U. (1991). Immune defense and suppression in insects. *BioEssays* **13:**343–346.

Schweiger, A., and Karlson, P. (1962). Zum Tyrosinstoff wechsel der Insekten, X. Die Aktivierung der Praphenoloxydase und das Aktivator-enzym. *Hoppe-Seyler's Z. Physiol. Chem.* **329:**210–221.

Seybold, W. D., Meltzer, P. S., and Mitchell, H. K. (1975). Phenoloxidase activation in *Drosophila:* A cascade of reactions. *Biochem. Genet.* **13:**85–108.

Soderhall, K. (1982). Prophenoloxidase activating system and melanization. A recognition mechanism of arthropods? A review. *Dev. Comp. Immunol.* **6:**601–611.

Soderhall, K., Rogener, W., Soderhall, I., Newton, R. P., and Ratcliffe, N. A. (1988). The properties and purification of a *Blaberus craniifer* plasma protein which enhances the activation of haemocyte prophenoloxidase by a B-1,3-glucan. *Insect Biochem.* **18:**323–330.

Stanley-Samuelson, D. W., Jensen, E., Nickerson, K. W., Tiebel, K., Ogg, C. L., and Howard,

R. W. (1991). Insect immune response to bacterial infection is mediated by eicosanoids. *Proc. Natl. Acad. Sci. U.S.A.* **88:**1064–1068.

Stoltz, D. B., and Cook, D. I. (1983). Inhibition of host phenoloxidase activity by parasitoid hymenoptera. *Experientia* **39:**1022–1024.

Stoltz, D. B., Krell, P., Summers, M. D., and Vinson, S. B. (1984). Polydnaviridae—A proposed family of insect viruses with segmented, double-stranded, circular DNA genomes. *Intervirology* **21:**1–4.

Sugumaran, M. (1988). Molecular mechanisms for cuticular sclerotization. *Adv. Insect Physiol.* **21:**179–231.

Sugumaran, M. (1990). Prophenoloxidase activation and insect immunity. *UCLA Symp. Mol. Cell. Biol.* [*N. Ser.*] **121:**47–62.

Sugumaran, M. (1991a). Molecular mechanisms of sclerotization. *In* "The Physiology of Insect Epidermis" (K. Binnington and A. Retnakaran, eds.), pp. 141–168. CSIRO Publications, Victoria.

Sugumaran, M. (1991b). Molecular mechanisms for mammalian melanogenesis—Comparison with insect cuticular sclerotization. *FEBS Lett.* **293:**4–10.

Sugumaran, M., and Nellaiappan, K. (1990). On the latency and nature of phenoloxidase present in the left colleterial gland of the cockroach *Periplaneta americana*. *Arch. Insect. Biochem. Physiol.* **15:**165–181.

Sugumaran, M., and Nellaiappan, K. (1991). Lysolecithin—A potent activator of prophenoloxidase from the hemolymph of the lobster, *Homarus americanas*. *Biochem. Biophys. Res. Commun.* **176:**1371–1376.

Sugumaran, M., and Semensi, V. (1991). Quinone methides as intermediates in eumelanin biosynthesis. *J. Biol. Chem.* **266:**6073–6078.

Sugumaran, M., Saul, S. J., and Ramesh, N. (1985). Endogenous protease inhibitors prevent undesired activation of prophenolase in insect hemolymph. *Biochem. Biophys. Res. Commun.* **132:**1124–1129.

Takagi, H., Narumi, H., Nakamura, K., and Sasaki, T. (1990). Amino acid sequence of silkworm (*Bombyx mori*) hemolymph antitrypsin deduced from its cDNA nucleotide sequences: Confirmation of its homology with serpins. *J. Biochem.* (*Tokyo*) **108:**372–378.

Thomson, J. A., and Sin, Y. T. (1970). The control of prophenoloxidase activation in larval haemolymph of *Calliphora*. *J. Insect Physiol.* **16:**2063–2074.

Tsukamoto, T., Ishiguro, M., and Funatsu, M. (1986). Isolation of latent phenoloxidase from prepupae of the housefly, *Musca domestica*. *Insect Biochem.* **16:**573–581.

Whitcomb, R. F., Shapiro, M., and Granados, R. R. (1974). Insect defense mechanisms against microorganisms and parasitoids. *In* "The Physiology of Insecta" (M. Rockstein, ed.), pp. 447–484. Academic Press, New York.

Wyatt, S. S. (1956). Culture *in vitro* of tissue from the silkworm, *Bombyx mori*. *J. Gen. Physiol.* **39:**841–852.

Yoshida, H., and Ashida, M. (1986). Microbial activation of two serine enzymes and prophenoloxidase in the plasma fraction of hemolymph of the silkworm *Bombyx mori*. *Insect Biochem.* **16:**539–545.

Yoshida, H., Ochiai, M., and Ashida, M. (1986). B-1,3-Glucan receptor and peptidoglycan receptor are present as separate entities within insect prophenoloxidase activating system. *Biochem. Biophys. Res. Commun.* **141:**1177–1184.

Index to Volumes 1 and 2

Volume numbers are boldfaced, separated from the first page reference with a colon. Subsequent references to the same material are separated by commas.

ISBN 0-12-084441-9

90018

9 780120 844418